WAVELETS
Calderón–Zygmund and multilinear operators

Wavelets

Calderón–Zygmund and multilinear operators

Yves Meyer

Professor, Ceremade, Université Paris–Dauphine

Ronald Coifman

Yale University

Translated by David Salinger
University of Leeds

CAMBRIDGE
UNIVERSITY PRESS

CAMBRIDGE UNIVERSITY PRESS
Cambridge, New York, Melbourne, Madrid, Cape Town,
Singapore, São Paulo, Delhi, Tokyo, Mexico City

Cambridge University Press
The Edinburgh Building, Cambridge CB2 8RU, UK

Published in the United States of America by Cambridge University Press, New York

www.cambridge.org
Information on this title: www.cambridge.org/9780521794732

First published in English 1997

A catalogue record for this publication is available from the British Library

Library of Congress Cataloguing in Publication Data

Meyer, Yves.
[Ondelettes et opérateurs. Tome 2-3. English]
Wavelets: Calderón-Zygmund and multilinear operators / Yves Meyer
& Ron Coifman; translated by D. H. Salinger.
p. cm. - (Cambridge studies in advanced mathematics; 48)
Translation of vols. 2-3 of: Ondelettes et opérateurs.
Includes bibliographical references (p.   -   ) and index.
ISBN 0 521 42001 6 (hardback)
1. Wavelets (Mathematics) 2. Calderón-Zygmund operator.
I. Coifman, Ronald R. (Ronald Raphaël) II. Title. III. Series
QA403.3.M493513 1996
515'.2433--dc20 96-13536   CIP

ISBN 978-0-521-42001-3 Hardback
ISBN 978-0-521-79473-2 Paperback

Contents

"Ce à quoi l'un s'était failli, l'autre est arrivé et ce qui était inconnu à un siècle, le siècle suivant l'a éclairci, et les sciences et les arts ne se jettent pas en moule mais se forment et figurent en les maniant et polissant à plusieurs fois [...] Ce que ma force ne peut découvrir, je ne laisse pas de le sonder et essayer et, en retastant et pétrissant cette nouvelle matière, la remuant et l'eschaufant, j'ouvre à celui qui me suit quelque facilité et la lui rends plus souple et plus maniable. Autant en fera le second au tiers qui est cause que la difficulté ne me doit pas désespérer, ni aussi peu mon impuissance ..."

Montaigne, *Les Essais*, Livre II, Chapitre XII.

"Where someone failed, another has succeeded; what was unknown in one century, the next has discovered; science and the arts do not grind themselves into uniformity, but gain shape and regularity by carving and polishing repeatedly [...] What my own strength has not been able to uncover, I cease not from working at and trying out and, by reshaping and solidifying this new material, in moulding and heating it, I bequeath to him who follows some facility and make it the more supple and malleable for him. The second will do the same for the third, which is why difficulty does not make me despair, nor, any the more, my own weakness..."

Translator's note

This book is a translation of *Ondelettes et Opérateurs*, Volume II, *Opérateurs de Calderón-Zygmund*, by Yves Meyer, and Volume III, *Opérateurs multilinéaires*, by R.R. Coifman and Yves Meyer. The original numbering of the chapters and of the theorems has been retained, so that it is still possible to follow the forward references in Volume I. Chapter 12 is where Volume III of the French version started. The references to *Wavelets and Operators* (Cambridge University Press, 1992), are to the translation of Volume I, *Ondelettes*, by Yves Meyer.

David Salinger, Leeds, June 1996.

Preface to the English Edition

There has been great progress during the few years which separate the first edition of this work from the translation.

1. In the area of **multilinear operators**, Pierre Louis Lions made the following conjecture. We consider two (arbitrary) vector fields $E(x) = \bigl(E_1(x), \ldots, E_n(x)\bigr)$ and $B(x) = \bigl(B_1(x), \ldots, B_n(x)\bigr)$, satisfying the following conditions:

$$E_j(x) \in L^p(\mathbb{R}^n) \qquad \text{and} \qquad B_j(x) \in L^q(\mathbb{R}^n) \,,$$

where $1 \leq j \leq n$, $1 < p < \infty$, $1/p + 1/q = 1$, and

$$\operatorname{div} E(x) = 0 \qquad \text{and} \qquad \operatorname{curl} B(x) = 0 \,,$$

in the sense of distributions.

Then the scalar product $E \cdot B(x) = E_1(x)B_1(x) + \cdots + E_n(x)B_n(x)$, of the vector fields must belong to the Hardy space $\mathbb{H}^1(\mathbb{R}^n)$.

The reader may refer to [E8], where there are several proofs of this result and to [E11], where an application to partial differential equations is to be found.

2. For the **Cauchy Integral** on Lipschitz curves, Mark S. Melnikov and Joan Verdera have found a new, extraordinary proof. This proof is astonishing because it makes no use of the BMO space or of Carleson measures. Instead, an essential rôle is played by a very simple and surprising lemma about the geometry of the complex plane [E18].

3. For more about **potential theory** on Lipschitz domains and Verchota's thesis, the reader may turn to C. Kenig's excellent recent work [E15].

4. Can wavelets play a part in the study or the understanding of the
   Navier-Stokes equations? We still do not know the answer to
   this question. P. Federbush [E13] has undertaken the ambitious programme of solving the Navier-Stokes equations by a method of Faedo-
   Galerkin, using wavelet bases in which the wavelets have zero divergence. These bases were constructed by G. Battle and P. Federbush
   and, independently, by P.G. Lemarié-Rieusset [E17].

   Federbush's approach is analysed and discussed in M. Cannone's elegant text [E5].

5. Paradifferential operators are flourishing as a tool in the study and
   solution of non-linear partial differential equations. The reader may
   consult [E6] and [E21].

6. P. Auscher and Ph. Tchamitchian continue the study of problems
   raised by Kato's conjecture. These authors have obtained remarkable
   results in this direction without, however, entirely solving the original
   problem [E2].

7. In the even more applied areas of **Numerical Analysis** and **Statis-
   tics**, essential results are rooted in the analysis of Calderón-Zygmund
   operators in appropriate wavelet bases [E3], [E10].

Yves Meyer, Paris, May 1996.

Introduction

Confronted with an orthonormal basis e_j, $j \in J$, of the Hilbert space $H = L^2(\mathbb{R}^n)$, it is impossible to resist the temptation to study those operators $T : H \to H$ which are diagonal, or almost diagonal, with respect to that particular basis. We can then hope that those operators are familiar from other contexts, which would give the whole situation that coherence beloved of scientists. Unfortunately, until now, the attempt has been in vain. Diagonal operators corresponding to the usual orthonormal bases are generically pathological, and therefore not of interest. For example, in the trigonometric system, the diagonal operators are those satisfying $T(e^{ikx}) = m_k e^{ikx}$, where m_k is a bounded sequence; they are pathological unless, for example, Marcinkiewicz's condition $|m_{k+1} - m_k| \leq C/|k|$ is satisfied. In this case, T is related to a pseudo-differential operator and constitutes a first example of what we shall call **Calderón-Zygmund operators**.

Orthonormal wavelet bases provide the first and—as far as we know—the unique example of an orthonormal basis with interesting diagonal, or almost-diagonal, operators. Those operators are already known, in another context, as Calderón-Zygmund operators. This remarkable fact explains the success of wavelet series compared to other orthonormal series. A change of wavelet coefficients $a_k \to m_k a_k$, where m_k is a bounded sequence, does not have uncontrollable consequences for the sum $f(x) = \sum a_k w_k(x)$ of the wavelet series. The sum $f(x)$ is transformed into $g(x) = T[f](x)$ and the remarkable properties of the Calderón-Zygmund operators let us give a precise description of what happens to $f(x)$. Wavelet decompositions are thus robust, and their robustness comes from the relationship between wavelets and operators which we have just described.

But Calderón-Zygmund operators had been studied well before orthonormal wavelet bases came to light: they are the subject of an independent theory which we shall expound completely and autonomously in Chapters 7 to 11. So the reader may attempt this volume straight away, retaining from *Wavelets and Operators* just the existence of orthonormal wavelet bases. Calderón-Zygmund operators have a special relationship with wavelets and with classical pseudo-differential operators, of which they are a remarkable generalization. In fact, the Calderón-Zygmund operators are found "beyond the pseudo-differential operators".

Multilinear analysis is one of the routes into the non-linear problems studied in Chapters 12 to 16. This route is only possible for those non-linear problems with a holomorphic structure, enabling them to be decomposed into a series of multilinear terms of increasing complexity. This approach needs some care, because the holomorphic structure can only be established after the event, that is, once we have shown that the series of multilinear terms converges. A. Calderón was the pioneer of this method of attack, and some of the examples we give here are part of his programme. Others are due to T. Kato: the discovery, by A. McIntosh, of the relationship between Calderón's programme and Kato's has been a source of progress in recent years.

The multilinear operators encountered in the problems referred to above turn out to be Calderón-Zygmund operators, whose continuity is established using the earlier chapters of this volume. Wavelets make a final appearance—as eigenfunctions of certain realizations of para-products—in our last chapter, which is devoted to J.M. Bony's theory of paradifferential operators.

For the greater convenience of the reader, the Introduction to *Wavelets and Operators*, which also serves as a general introduction to the present volume, is reproduced in the next few pages.

Introduction to *Wavelets and Operators*

For many years, the sine, cosine and imaginary exponential functions have been the basic functions of analysis. The sequence $(2\pi)^{-1/2}e^{ikx}$, $k = 0, \pm 1, \pm 2, \ldots$ forms an orthonormal basis of the standard space $L^2[0, 2\pi]$; Fourier series are the linear combinations $\sum a_k e^{ikx}$. Their study has been, and remains, an unquenchable source of problems and discoveries in mathematical analysis. The problems arise from the absence of a good dictionary for translating the properties of a function into those of its Fourier coefficients. Here is an example of the kind of difficulty that occurs. J.P. Kahane, Y. Katznelson and K. de Leeuw have shown ([150]) that, to get a continuous function $g(x)$ from an arbitrary square-summable function $f(x)$, it is sufficient to increase—or leave unchanged—the moduli of the Fourier coefficients of $f(x)$ and to adjust their phases judiciously. It is thus impossible to predict the properties (size, regularity) of a function solely from knowledge of the order of magnitude of its Fourier coefficients. Indeed it is still difficult if we know the Fourier coefficients explicitly, and many problems are still open.

At the beginning of the 1980s, many scientists were already using "wavelets" as an alternative to traditional Fourier analysis. This alternative gave grounds for hoping for simpler numerical analysis and more robust synthesis of certain transitory phenomena. The "wavelets" of J.S. Liénard or of X. Rodet ([167], [206]) were used for numerical treatment of acoustic signals (words or music) and those of J. Morlet ([124]) for stocking and interpreting seismic signals gathered in the course of oil prospecting expeditions. Among mathematicians, research was just as active: to mention only the most striking, R.R. Coifman and G. Weiss ([75]) invented the "atoms" and "molecules" which were to form the basic building blocks of various function spaces, the rules of assembly being clearly defined and easy to use. Certain of these atomic decompositions could, moreover, be obtained by making a discrete version of a well-known identity, due to A. Calderón, in which "wavelets" were implicitly involved. That identity was later rediscovered by Morlet and his collaborators .... Lastly, L. Carleson used functions very similar to "wavelets" in order to construct an unconditional basis of the H^1 space of E.M. Stein and G. Weiss.

These separate investigations had such a "family resemblance" that it seemed necessary to gather them together into a coherent theory, mathematically well-founded and, at the same time, universally applicable. The **orthonormal wavelet bases**, whose construction is given in the present volume, are a replacement for the empirical "wavelets" of Liénard, Morlet and Rodet.

The same orthonormal wavelet bases give direct access to the "atomic decompositions" of Coifman and Weiss, which are thus—for the first time—related to constructions of **unconditional bases** of the standard spaces of functions and distributions. The wavelet bases are universally applicable: "everything that comes to hand", whether function or distribution, is the sum of a wavelet series and, contrary to what happens with Fourier series, the coefficients of the wavelet series translate the properties of the function or distribution simply, precisely and faithfully.

So we have a new tool at our command, an instrument that lets us perform, without thinking, the delicate constructions that could not formerly be achieved without recourse to lacunary, or random, Fourier series. The exceptional properties of the sums of these special series become the everyday properties of generic sums of wavelet series.

The algorithms for analysis by, and synthesis of, orthogonal wavelet series will, doubtless, play an important rôle in many different branches of science and technology. Mathematicians, physicists, and engineers who want to know everything about wavelets now have the present volume (*Wavelets and Operators*) of this work at their disposal.

Wavelets: Calderón-Zygmund and Multilinear Operators is addressed more specifically to an audience of mathematicians. It deals with the operators associated with wavelets. G. Weiss has shown that the study of the operators acting on a space of functions or distributions can become very simple when the elements of the space admit "atomic decompositions". He writes "many problems in analysis have natural formulations as questions of continuity of linear operators defined on spaces of functions or distributions. Such questions can often be answered by rather straightforward techniques if they can first be reduced to the study of the operator on an appropriate class of simple functions which, in some convenient sense, generate the entire space." When these "simple elements" were the functions e^{ikx} of the trigonometric system, the bounded operators T on L^2, which were diagonal with respect to the trigonometric system, did not have any other interesting property (with the exception of translation-invariance, which follows immediately from the definition). It was then necessary to impose quite precise conditions on the eigenvalues of T in order to extend such an operator to other function spaces: the first results in this direction were obtained by J. Marcinkiewicz.

However, the bounded operators which can be diagonalized exactly or approximately, with respect to the wavelet basis, form an algebra **A** of bounded operators on L^2 and the well-known **Calderón-Zygmund real-variable methods** enable the operators of **A** to be extended to other spaces of functions or distributions. The algebra **A**, which extends the pseudo-differential operators in a natural sense, is strictly contained in the set **C** of operators whose study has been recommended by Calderón. Work on these operators should enable us to solve several outstanding classical problems in complex analysis and partial differential equations.

Here is a slightly more precise description of the set **C**, the delicate construction of which we have called "Calderón's programme". After having invented, together with A. Zygmund, what was to become the classical pseudo-differential calculus, Calderón intended to extend the field of application systematically, by weakening, as far as possible, the regularity hypotheses necessary for the algorithms to work.

The fundamental—and unexpected—discovery made by Calderón was the existence of a limit to the search for minimal hypotheses of regularity. There is a "natural boundary" which cannot be transgressed, and the extension of operators to this boundary is precisely the analytic extension of holomorphic functions on certain Banach spaces, as we shall show in Chapter 13.

Chapters 7–9 are devoted to the study and then the construction of the set **C** of operators of Calderón's programme. We call them the **Calderón-Zygmund operators**, although they are very different from the "historical operators" studied by Calderón and Zygmund in the 1950s and 60s.

Just like these "historical operators", those we consider can be defined by singular integrals, in a new sense which we clarify in Chapter 7. To go beyond the context of convolution operators, it becomes indispensable to have a criterion for L^2 continuity, without which the theory collapses like a house of cards. One such criterion is the well-known $T(1)$ theorem of G. David and J.L. Journé, which we shall prove in Chapter 8. The $T(1)$ Theorem replaces the Fourier transform, whose use remains restricted to convolution operators.

Unfortunately, the $T(1)$ theorem, although giving a necessary and sufficient condition, is not directly applicable to the most interesting operators of the set **C** of Calderón's programme. We do not know why that is. The operators in question have, however, a very special non-linear structure, which, when correctly exploited, allows us to pass from the "local" results given by David and Journé's theorem to the "global" theorems necessary for the functioning of Calderón's programme.

In Chapters 12–16, and also in Chapter 9, we have given the most beautiful of the applications of Calderón's programme. First comes the celebrated pseudo-differential calculus, initiated by Calderón, which, at present, has interesting and important applications to non-linear partial differential equations.

Then we pass to complex analysis and the Hardy spaces associated with Lipschitz domains of the complex plane. The object of Chapter 12 is the study of the Cauchy operator on rectifiable curves. We then examine the problem, posed by T. Kato, of determining the domain of the square roots of second order pseudo-differential operators, in the accretive case.

After that, we give an account of the results of B. Dahlberg, D. Jerison, C. Kenig and G. Verchota relative to the Dirichlet and Neumann problems in Lipschitz open sets.

We end with a brief presentation of J.M. Bony's paradifferential operators, which serve to analyse the regularity of non-linear partial differential operators.

Wavelets reappear, in a surprising way, as the eigenfunctions of certain paradifferential operators. Correctly handled, they remain present in the study of Hardy spaces and the Cauchy operator on a Lipschitz curve: the operator is "almost diagonal" with respect to a wavelet basis specially designed for complex analysis on that curve (Chapter 11). The construction of wavelet bases is thus sufficiently supple to be adaptable to differing geometric situations: we also obtain "non-orthogonal wavelet bases". At present, there is no universal basis which can simultaneously be used in the analysis of all the operators of Calderón's class **C**.

J.O. Strömberg was the first to construct an orthonormal basis of $L^2(\mathbb{R})$, of the form $2^{j/2}\psi(2^j x - k)$, $j, k \in \mathbb{Z}$, where, for each $m \in \mathbb{N}$, the function $\psi(x)$ was of class C^m and decreased exponentially at infinity.

Subsequent work on orthogonal wavelets has not followed the path discovered by Strömberg. That work is, essentially, due to I. Daubechies, P.G. Lemarié, S. Mallat, and the author. It is given here, with care and with complete proofs.

As far as operators are concerned, the well-known results of Calderón, Zygmund, and Cotlar will be described in Chapter 7, in the new context of the set **C**.

The other names the reader of this work will often encounter are J.M. Bony, G. David, P. Jones, J.L. Journé, C. Kenig, T. Murai, and S. Semmes.

The division into two volumes will allow this work to be read in several ways. As we have already suggested, the reader may wish to go no further than *Wavelets and Operators*, which surveys our present knowledge about wavelets. But the first part of *Wavelets: Calderón-Zygmund and Multilinear Operators* may also be read directly, assuming only the results quoted in the introductions of the first six chapters. Finally, the reader can go straight to Chapters 12, 13, 14, 15, or 16, because each of them forms a coherent account of a subject of independent interest (complex analysis, holomorphic functionals on Banach spaces, Kato theory, elliptic partial differential equations in Lipschitz domains, and, lastly, non-linear partial differential equations). The thread linking these different themes is, quite clearly, the use of wavelets in Calderón's programme in operator theory.

These books have been written at a level appropriate for first-year postgraduates, and we have tested them in France, and in the U.S.A., on various audiences of mathematicians and engineers. To read these books, it is therefore not necessary to have studied the remarkable treatises by E. Stein and G. Weiss ([221)], E. Stein ([217]), or J. Garcia-Cuerva and J.L. Rubio de Francia ([115]), not to mention the fundamental text and reference by Zygmund ([239]).

R. Coifman helped me to recognize the importance of Calderón's programme. Since the summer of 1974, our scientific collaboration has been devoted to its realization, and these books have been one of our projects. If our own work no longer appears in its original form here, it is because our zeal and enthusiasm have communicated themselves to younger research workers, who have found more elegant solutions to the problems we had been determined to resolve.

7

The new Calderón-Zygmund operators

1 Introduction

The Calderón-Zygmund operators we discuss in this chapter differ significantly from those which Calderón and Zygmund considered more than thirty years ago. We shall try to describe and explain how the definition of these operators has evolved.

The theory of Calderón-Zygmund operators began in the 1950s when Calderón and Zygmund systematically studied convolution operators appearing in elliptic partial differential equations.

The best-known example is given by the Riesz transforms

$$R_j = -i \frac{\partial}{\partial x_j} (-\Delta)^{-1/2}$$

for $1 \leq j \leq n$, where

$$\Delta = \frac{\partial^2}{\partial x_1^2} + \cdots + \frac{\partial^2}{\partial x_n^2}.$$

Riesz transforms arise when we study the Neumann problem in the upper half-plane. More precisely, let $\Omega \subset \mathbb{R}^{n+1}$ be the open set defined by $t > 0$ and $x \in \mathbb{R}^n$. We consider harmonic functions in the Sobolev space $H^2(\Omega)$, that is, those functions $u(x,t)$ on Ω such that $\Delta u + \partial^2 u/\partial t^2 = 0$. Then we can define the boundary values of $u(x,t)$ and the gradient of $u(x,t)$ on $t = 0$ (the boundary $\partial\Omega$ of Ω). In this context, the Riesz transforms R_j, $1 \leq j \leq n$, enable us to pass from the boundary values of the normal derivative $(-\partial u/\partial t)|_{t=0}$ to the boundary values of the tangential derivatives $(\partial u/\partial x_j)|_{t=0}$.

The continuity of the Riesz operators on $L^2(\mathbb{R}^n)$ follows immediately. We can either use Green's formula or conjugate each R_j by the Fourier transform $\mathcal{F}$. In the second case we get $\mathcal{F}R_j f(\xi) = \xi_j |\xi|^{-1} \hat{f}(\xi)$; it is enough to observe that $|\xi_j| \leq |\xi|$ in order to conclude that $\|R_j(f)\|_2 \leq \|f\|_2$. More precisely, we have $\sum_1^n \|R_j(f)\|_2^2 = \|f\|_2^2$.

On the other hand, it is not at all obvious that the R_j, $1 \leq j \leq n$, are continuous on $L^p(\mathbb{R}^n)$, $1 < p < \infty$. That result is achieved by the real variable methods of Calderón and Zygmund, which we now describe.

Calderón and Zygmund considered the more general problem of operators on $L^2(\mathbb{R}^n)$ defined by the following algorithm. We start with a function $\Omega(x)$ in $C(\mathbb{R}^n \setminus \{0\})$ which is homogeneous of degree 0 and which satisfies the condition $\int_{S^{n-1}} \Omega(x) \, d\sigma(x) = 0$, where $d\sigma$ is the rotation-invariant probability measure on the unit sphere S^{n-1}.

The first generation Calderón-Zygmund operators are convolution operators. We take the distribution $S = \mathrm{PV}\,\Omega(x)|x|^{-n}$ defined by

$$\langle S, \phi \rangle = \lim_{\varepsilon \downarrow 0} \int_{|x| \geq \varepsilon} \frac{\Omega(x)}{|x|^n} \phi(x) \, dx, \qquad \phi \in \mathcal{D}(\mathbb{R}^n).$$

Then the Calderón-Zygmund operator T is given by $T(f) = S \star f$, for each $f \in \mathcal{D}(\mathbb{R}^n)$. In other words,

$$(1.1) \qquad (T(f))(x) = \lim_{\varepsilon \downarrow 0} \int_{|y| \geq \varepsilon} f(x - y) \frac{\Omega(y)}{|y|^n} \, dy.$$

The limit exists if f is continuous, of Hölder exponent $\gamma > 0$, and square-summable.

Then $\mathcal{F}S = m(\xi)$, where $m(\xi) \in C^\infty(\mathbb{R}^n \setminus \{0\})$, where $m(\lambda \xi) = m(\xi)$, for $\lambda > 0$, and where the mean of $m(\xi)$ on the unit sphere is zero. For the Riesz transforms, $m(\xi) = \xi_j / |\xi|$.

This theory gave a unified treatment of earlier work of J. Marcinkiewicz (1938) and G. Giraud (1936). Calderón and Zygmund showed that the Marcinkiewicz multipliers associated with elliptic partial differential equations were the Fourier coefficients of the (periodified) distributions $\mathrm{PV}\,\Omega(x)|x|^{-n}$ used by Giraud. An essential tool in that unification was the notion of the Fourier transform of a tempered distribution (introduced by L. Schwartz between 1943 and 1945).

By these means, Calderón and Zygmund rediscovered, in a very natural way, the rules for composing two Calderón-Zygmund operators, T_1 and T_2: it is enough to employ the usual product of the corresponding multipliers $m_1(\xi)$ and $m_2(\xi)$. However, the symbol $m_3(\xi) = m_1(\xi)m_2(\xi)$ does not always have zero integral on the unit sphere, so it is necessary to consider the algebra of operators $cI + T$, where c is a constant and T is defined by (1.1).

After doing this, Calderón and Zygmund tried to extend the continuity properties of Calderón-Zygmund operators to $L^p(\mathbb{R}^n)$, $1 < p < \infty$. The complex variable method, introduced by Littlewood and Paley, and used by Marcinkiewicz to prove his multiplier theorem, did not work in the n-dimensional case. It was this obstacle which led Calderón and Zygmund to invent the real variable method. The essential ingredient is the Calderón-Zygmund decomposition which, for any parameter $\lambda > 0$, lets us write any function f in $L^1(\mathbb{R}^n)$ as a sum $g + h$, where $g \in L^2(\mathbb{R}^n)$ satisfies $\|g\|_2 \leq C\sqrt{\lambda}$ and where h is the sum of a series of oscillatory terms, each with support in a certain cube Q_j, such that the sum of the measures of the Q_j does not exceed C/λ. These oscillatory terms foreshadow the "atoms" of Coifman and Weiss, as well as wavelets.

The L^2 estimate for an operator T, together with quite a weak hypothesis on the kernel $K(x, y)$ of T ($\int_{|x-y| \geq 2|y'-y|} |K(x, y') - K(x, y)|\, dx \leq C$ is enough) gives the L^1-weak-L^1 estimate which we shall describe explicitly in this chapter. This, together with the L^2 estimate, allows us to derive the L^p estimates, for $1 < p \leq 2$, by interpolation. We get the L^q estimates, for $2 \leq q < \infty$, by duality, provided that we impose corresponding conditions where the roles of x and y in $K(x, y)$ are interchanged.

A second preoccupation of Calderón and Zygmund was to give a pointwise meaning to (1.1), when $f(x)$ was an arbitrary function in $L^2(\mathbb{R}^n)$, by proving the existence, for almost every $x \in \mathbb{R}^n$, of

$$\lim_{\varepsilon \downarrow 0} \int_{|y| \geq \varepsilon} f(x - y) \frac{\Omega(y)}{|y|^n}\, dy\,.$$

This problem leads to the *maximal operator T^**, defined by

$$(1.2) \qquad T^*f(x) = \sup_{\varepsilon > 0} \left| \int_{|y| \geq \varepsilon} f(x - y) \frac{\Omega(y)}{|y|^n}\, dy \right|,$$

and to the proof of the *maximal inequality*

$$(1.3) \qquad \|T^*f\|_2 \leq C\|f\|_2, \qquad f \in L^2(\mathbb{R}^n)\,.$$

The convolution operators which we have introduced correspond to elliptic problems with constant coefficients. To deal with partial differential equations with variable, but extremely regular, coefficients, Calderón and Zygmund invented what we shall call "second generation" Calderón-Zygmund operators.

Second generation Calderón-Zygmund operators are not convolution operators, but are given by slightly more elaborate distribution-kernels. More precisely, for a test function f (square-summable and sufficiently regular),

$$(1.4) \quad Tf(x) = \text{PV} \int K(x,y)f(y)\,dy = \lim_{\epsilon \downarrow 0} \int_{|x-y|\geq\epsilon} K(x,y)f(y)\,dy\,,$$

where $K(x,y)$ satisfies the following properties:–

$$(1.5) \qquad\qquad K(x,y) = L(x, x-y)\,;$$

$$(1.6) \qquad\qquad L(x,\lambda u) = \lambda^{-n} L(x, u)\,,$$

for every $x \in \mathbb{R}^n$, every $u \in \mathbb{R}^n \setminus \{0\}$ and every $\lambda > 0$;

$$(1.7) \qquad |\partial_u^\alpha \partial_x^\beta L(x,u)| \leq C\,, \quad \text{when } |u| = 1,\ |\alpha| \leq N \text{ and } |\beta| \leq m;$$

$$(1.8) \qquad \int_{S^{n-1}} L(x,u)\,d\sigma(u) = 0 \qquad \text{identically in } x \in \mathbb{R}^n.$$

By separation of variables (involving expansion in spherical harmonics with respect to the variable $u \in S^{n-1}$), when N was sufficiently large compared to n, Calderón and Zygmund reduced the operators of the second generation to absolutely convergent series $\sum_0^\infty M_j T_j$, where the T_j were first generation operators and the M_j were operators of pointwise multiplication by bounded functions.

The integer m is irrelevant, and may even be zero, as long as we restrict ourselves to studying the L^2 or L^p continuity of second generation operators. On the other hand, m plays an essential part as soon as we seek to compose the operators in question, in order to obtain an algebra with a precise symbolic calculus. To get this, we need to work modulo the ideal $\mathcal{I}$ of regularizing operators of order 1, that is, the operators S which are bounded on $L^2(\mathbb{R}^n)$ and are such that $(\partial/\partial x_j)S$ and $S\partial/\partial x_j$ are also bounded, for $1 \leq j \leq n$. Calderón and Zygmund proved that, modulo $\mathcal{I}$, the second generation Calderón-Zygmund operators formed a commutative algebra, as long as the integers m and N were large enough. Here again, it was necessary to include the operators of pointwise multiplication by those bounded functions $b(x)$ which had all their derivatives bounded.

Several years later (1965), Calderón took another look at the problem of the symbolic calculus when he sought conditions of minimal regularity, with respect to x, on kernels $L_1(x,u)$ and $L_2(x,u)$, corresponding to operators T_1 and T_2, which would imply the existence of a symbolic calculus modulo the regularizing operators of order 1. In applications to partial differential equations, the regularity with respect to x is given by that of the coefficients $c_\alpha(x)$ of the differential operators $\sum c_\alpha(x)\partial^\alpha$ considered, whereas the kernels themselves are infinitely differentiable with respect to the variable $u \in \mathbb{R}^n \setminus \{0\}$.

The symbolic calculus problem that we have just raised reduces to the study of the commutators $[A, T]$, where A is the operator of pointwise multiplication by a function $a(x)$, whose regularity is measured by the integer m of (1.7), and where $T = (\partial/\partial x_1)T_1 + \cdots + (\partial/\partial x_n)T_n$, the T_j being first generation Calderón-Zygmund operators.

In dimension 1, T_1 is replaced by the Hilbert transform H and T is replaced by the "Calderón operator" $\Lambda = DH$, where $D = -i(d/dx)$.

In 1965, Calderón showed that the commutator $[A, \Lambda]$ was bounded on $L^2(\mathbb{R})$ if and only if the function $a(x)$ was Lipschitz, that is, if there was a constant C such that $|a(x) - a(y)| \leq C|x - y|$, for all $x, y \in \mathbb{R}$. It is easy to see that this condition is necessary. The reverse implication is deep, and Calderón's proof relies on the characterization, established by Calderón for this purpose, of the complex $\mathbb{H}^1$ space by the integrability of Lusin's area function. This characterization opened the way to the study of the operators on $\mathbb{H}^1$ and of the dual of the real version of $\mathbb{H}^1$ ([109]).

To pass to dimension n, Calderón used the method of rotations (invented a few years earlier by Calderón and Zygmund) and showed that the set of operators defined by (1.4) to (1.8), with $m = 1$ and N large enough, becomes a commutative Banach algebra, once it is reduced modulo the regularizing operators of order 1.

This remarkable theorem will be proved in Chapter 9. It leads to a substantial improvement of most of the results obtained by applying the classical pseudo-differential calculus (where the symbols $\sigma(x, \xi)$ are infinitely differentiable with respect to x).

The transition from the L^2 continuity to the L^p continuity ($1 < p < \infty$) of the commutator $[A, \Lambda]$ is achieved by repeating the proof used for the case of the Riesz transforms.

From 1966, Calderón proposed studying commutators of higher order $T_k = [A, [A, \ldots, [A, D^k H] \ldots]]$ and proving that the T_k were continuous on $L^2(\mathbb{R})$ whenever A was an operator of pointwise multiplication by a Lipschitz function $a(x)$. The generating series $\sum_0^\infty \xi^k T_k$ is the operator defined by the Cauchy integral PV $\int_\Gamma (z - w)^{-1} f(w) \, dw$ (the reader will recall that PV means "principal value"), where Γ is the graph of a Lipschitz function and where f lies in $L^2(\Gamma, ds)$. Calderón posed the problem of the continuity of this latter operator on the Hilbert space $L^2(\Gamma, ds)$.

A further operator belonging to "Calderón's programme" is related to the classical method of using a double layer potential to solve the Dirichlet or the Neumann problem in a Lipschitz domain. The operator involved is then given in local coordinates by

$$Tf(x) = \frac{1}{\omega_n} \operatorname{PV} \int_{\mathbb{R}^n} K(x,y)f(y)\,dy\,,$$

where

$$K(x,y) = \frac{a(x) - a(y) - (x-y)\cdot\nabla a(y)}{(|x-y|^2 + (a(x)-a(y))^2)^{(n+1)/2}}\,,$$

and where $f \in L^2(\mathbb{R}^n)$. The Lipschitz domain is defined (locally) by $t > a(x)$, where $x \in \mathbb{R}^n$, $t \in \mathbb{R}$, and the function $a(x)$ is Lipschitz.

If $n = 1$, this kernel is precisely the real part of the Cauchy kernel and in dimension n it can be studied by the method of rotations, once we know that the Cauchy kernel is bounded on $L^2(\mathbb{R})$.

In 1977, Calderón proved the continuity of the operator defined by the Cauchy integral for every Lipschitz graph $y = a(x)$, where $\|a'\|_\infty < \varepsilon$; the method used did not give the value of the mysterious constant ε.

The third generation Calderón-Zygmund operators generalize the examples we have just described. In 1976, Calderón asked the following question. Let $K(x,y)$ be a function of two real variables, defined for $x \neq y$, which satisfies the estimates $|K(x,y)| \le C_0|x-y|^{-1}$ and $|(\partial/\partial x)K(x,y)| \le C_1(x-y)^{-2}$ and the condition $K(y,x) = -K(x,y)$. Is it true that the operator $T : \mathcal{D} \to \mathcal{D}'$ whose kernel is $\operatorname{PV} K(x,y)$ can be extended to a continuous operator on $L^2(\mathbb{R})$? Such a result would have given "for free" the continuity of the Cauchy kernel on all Lipschitz curves, etc. It is unfortunately not the case: J.L. Journé discovered in 1982 that it was necessary to assume that $T(1)$ belonged to BMO, this latter condition being necessary and sufficient for the L^2 continuity of the operator T. To define $T(1)$, we let g_ε denote the Gaussian $e^{-\varepsilon x^2}$ and we prove that there are renormalization constants $c(\varepsilon)$, such that $T(g_\varepsilon) - c(\varepsilon)$ converges, in the sense of distributions, to what we shall call $T(1)$. This distribution is, therefore, only defined modulo the constant functions, which is, after all, the case for elements of BMO.

The "$T(1)$ theorem" was finally proved by G. David and Journé in 1983. It is a remarkable result because it reduces all the previous results of Calderón, and the authors of this book, to simple integrations by parts. In particular, it immediately gives the continuity of the Cauchy kernel on Lipschitz graphs of small slope.

The extension to general Lipschitz curves is obtained by new real methods discovered by David and then simplified by T. Murai.

To define the "third generation Calderón-Zygmund operators", we replace $\mathbb{R}$ by $\mathbb{R}^n$ and drop the antisymmetry hypothesis $K(y,x) = -K(x,y)$. It is then no longer possible to define the operator $T : \mathcal{D} \to \mathcal{D}'$

just by the function $K(x, y)$ (which is defined on $\mathbb{R}^n \times \mathbb{R}^n \setminus \Delta$, where Δ is the diagonal).

We proceed in the reverse direction and start with a linear (and continuous) operator $T : \mathcal{D} \to \mathcal{D}'$. We say that this operator corresponds to a singular integral if the distribution-kernel of T, restricted to $\mathbb{R}^n \times \mathbb{R}^n \setminus \Delta$, is a function $K(x, y)$ satisfying the following hypotheses:

$$(1.9) \qquad |K(x, y)| \leq C_0 |x - y|^{-n},$$

$$(1.10) \qquad |\frac{\partial K(x, y)}{\partial x_j}| \leq C_1 |x - y|^{-n-1}, \qquad 1 \leq j \leq n,$$

$$(1.11) \qquad |\frac{\partial K(x, y)}{\partial y_j}| \leq C_1 |x - y|^{-n-1}, \qquad 1 \leq j \leq n.$$

In particular, if f is a test function and if x does not lie in the (compact) support of f, then

$$(1.12) \qquad Tf(x) = \int K(x, y) f(y) \, dy.$$

The fundamental problem (which will be solved in Chapter 8) is to find a necessary and sufficient condition for the continuity of T on the space $L^2(\mathbb{R}^n)$. We then say that T is a Calderón-Zygmund operator. We shall prove the existence of a bounded measurable function $m(x)$ and of a sequence ε_j of measurable functions on $\mathbb{R}^n$, taking positive values $\varepsilon_j(x) > 0$, such that, for every function $f(x)$ in $L^2(\mathbb{R}^n)$, we have

$$(1.13) \qquad Tf(x) = m(x) f(x) + \lim_{j \to \infty} \int_{|x-y| \geq \varepsilon_j(x)} K(x, y) f(y) \, dy,$$

for almost all $x \in \mathbb{R}^n$.

In the case of a convolution operator, $\varepsilon_j(x)$ does not depend on x; if ε_j can also be taken to be an arbitrary sequence tending to 0, then we are back to the classical idea of the principal value of a singular integral. The representation (1.13) explains the sense in which the operator T "corresponds to a singular integral".

Besides the examples we have already mentioned (the Calderón commutators, the Cauchy integral on Lipschitz curves and the double-layer potential), third generation Calderón-Zygmund operators arise in a completely different context. To show that a wavelet basis ψ_λ, $\lambda \in \Lambda$, constructed by the algorithms of Chapter 3 of *Wavelets and Operators*, is an unconditional basis of a classical space of functions or distributions B, we need to show that the operators $T : L^2(\mathbb{R}^n) \to L^2(\mathbb{R}^n)$ which are diagonal in our wavelet basis are also continuous on the space B. Now such operators are automatically Calderón-Zygmund operators, whose continuity on B will be established by the real variable methods developed in this chapter and in Chapter 10.

2 Definition of Calderón-Zygmund operators corresponding to singular integrals

As we stated in the introduction, we do not want to define a Calderón-Zygmund operator as the principal value of a singular integral.

More precisely, we also want to use kernels $K(x,y)$ for which the limit

$$\lim_{\varepsilon \downarrow 0} \int_{|x-y|\geq\varepsilon} K(x,y)f(y)\,dy$$

might not exist in the usual sense, even when $f(y)$ is a test function or $K(x,y)$ is antisymmetric. Now, in the antisymmetric case, that limit exists in the sense of distributions. This leads us to base the theory on the bilinear form $J(f,g) = \langle T(f), g \rangle = \langle S, g \otimes f \rangle$, where f and g are test functions, T is a linear operator from $\mathcal{D}$ into $\mathcal{D}'$ and S is the distribution-kernel of T. Then K is derived from S by the condition that K is the restriction of S to the complement of the diagonal.

Here is an example to clarify these distinctions. Let $\theta(x)$ be an odd function in $\mathcal{D}(\mathbb{R})$ such that $\int_{-\infty}^{\infty} \theta(x)\sin x\,dx = 1$. Further, let the support of θ be the union of the intervals $[-4/3, -2/3]$ and $[2/3, 4/3]$. Then we define the kernel $K(x,y)$ by

$$K(x,y) = \sum_{0}^{\infty} 2^k \theta(2^k(x-y))e^{i2^k(x+y)}.$$

We immediately get $K(y,x) = -K(x,y)$, $|K(x,y)| \leq C_0|x-y|^{-1}$ and $|\partial k/\partial x| \leq C_1|x-y|^{-2}$. If f is in $\mathcal{D}(\mathbb{R})$ and equals 1 on $[-10, 10]$, the existence of the limit $\lim_{\varepsilon\downarrow 0}\int_{|x-y|\geq\varepsilon} K(x,y)f(y)\,dy$, for $-1 \leq x \leq 1$, would imply that the series $\sum_0^\infty e^{i2^k x}$ converged. This series diverges everywhere (pointwise), but converges in the sense of distributions.

This leads us to consider

$$J_\varepsilon(f,g) = \iint_{|x-y|\geq\varepsilon} K(x,y)f(y)g(x)\,dy\,dx,$$

where $K(x,y)$ is an antisymmetric kernel with $|K(x,y)| \leq C_0|x-y|^{-1}$ and where f and g are functions in $\mathcal{D}(\mathbb{R})$. Now

$$J_\varepsilon(f,g) = \frac{1}{2}\iint_{|x-y|\geq\varepsilon} K(x,y)(f(y)g(x) - f(x)g(y))\,dy\,dx$$

and we can define $J(f,g)$ by the absolutely converging integral

$$J(f,g) = \frac{1}{2}\iint K(x,y)(f(y)g(x) - f(x)g(y))\,dy\,dx.$$

Then $J(f,g) = \lim_{\varepsilon\downarrow 0} J_\varepsilon(f,g)$, which enables us to complete the construction of $T : \mathcal{D} \to \mathcal{D}'$ by the formula $\langle T(f), g \rangle = J(f,g)$.

The case of non-antisymmetric kernels is more subtle: we can no longer use a kernel to define the operator. Reversing rôles, we start with an operator $T : \mathcal{D} \to \mathcal{D}'$. Let $S \in \mathcal{D}'(\mathbb{R}^n \times \mathbb{R}^n)$ denote the kernel of T. The existence of S is guaranteed by Schwartz's kernel theorem and S is related to T by the identity $\langle T(f), g \rangle = \langle S, g \otimes f \rangle$. The left-hand side involves the duality between $\mathcal{D}(\mathbb{R}^n)$ and $\mathcal{D}'(\mathbb{R}^n)$, whereas the duality on the right-hand side is that between $\mathcal{D}(\mathbb{R}^n \times \mathbb{R}^n)$ and $\mathcal{D}'(\mathbb{R}^n \times \mathbb{R}^n)$.

We let $\Omega = \{(x,y) \in \mathbb{R}^n : x \neq y\}$ and we consider the restriction $K(x,y)$ of S to Ω.

Definition 1. *Let $T : \mathcal{D}(\mathbb{R}^n) \to \mathcal{D}'(\mathbb{R}^n)$ be a continuous linear operator. We say that T is a Calderón-Zygmund operator if there are constants C_0, C_1, C_2, and an exponent $\gamma \in (0,1]$ such that the following four conditions are satisfied:*

(2.1) *$K(x,y)$ is a locally integrable function on Ω and satisfies*
$$|K(x,y)| \leq C_0 |x-y|^{-n};$$

(2.2) *if $(x,y) \in \Omega$ and $|x'-x| \leq |x-y|/2$, then*
$$|K(x',y) - K(x,y)| \leq C_1 |x'-x|^\gamma |x-y|^{-n-\gamma};$$

(2.3) *if $(x,y) \in \Omega$ and $|y'-y| \leq |x-y|/2$, then*
$$|K(x,y') - K(x,y)| \leq C_1 |y'-y|^\gamma |x-y|^{-n-\gamma};$$

(2.4) *T extends to a continuous linear operator on $L^2(\mathbb{R}^n)$ with*
$$\|T\| \leq C_2.$$

If conditions (2.2) and (2.3) are satisfied for an exponent $\gamma > 0$, they are satisfied, a fortiori, for every exponent $\gamma' \in (0,\gamma)$.

If $\gamma = 1$, conditions (2.2) and (2.3) can be written more simply as
$$\left|\frac{\partial K}{\partial x_j}\right| + \left|\frac{\partial K}{\partial y_j}\right| \leq C_1 |x-y|^{-n-1}, \qquad 1 \leq j \leq n.$$

Let us consider the "second generation" Calderón-Zygmund operators of the introduction. Then the kernel $K(x,y)$ is $L(x,x-y)$, where $L(x,u)$ satisfies (1.5), (1.6) and (1.7). Such a kernel never satisfies the estimate (2.2), unless $L(x,u)$ is independent of x, in which case the operator is a convolution operator, in other words, a "first generation" Calderón-Zygmund operator.

To justify our remark, we simply observe that the kernels $K(x,y)$ and $\lambda^n K(\lambda x \lambda y)$ satisfy (2.2) with the same constant: C_1. Using the homogeneity of $L(x,u)$ in u, we find that, if (2.2) holds, then, for every $\lambda \geq 1$ and for every u of length 1, we have
$$|L(\lambda x, u) - L(\lambda x', u)| \leq C_1 |x'-x|^\gamma,$$
when $|x'-x| \leq 1/2$. Letting λ tend to infinity concludes the argument. Thus condition (2.2) fails when $|x-y|$ is large.

A subset $\mathcal{B} \subset \mathcal{L}(L^2(\mathbb{R}^n), L^2(\mathbb{R}^n))$ is called a *bounded* set of Calderón-Zygmund operators if the operators $T \in \mathcal{B}$ satisfy (2.1), (2.2), (2.3) and (2.4) with the same exponent γ and the same constants C_0, C_1 and C_2.

Let $\mathcal{G}$ denote the group of unitary isomorphisms of $L^2(\mathbb{R}^n)$ of the form $Uf(x) = \delta^{-n/2} f(\delta^{-1}(x - x_0))$ for $x_0 \in \mathbb{R}^n$ and $\delta > 0$. Then, if T is a Calderón-Zygmund operator, the collection UTU^{-1}, $U \in \mathcal{G}$, is a bounded set of Calderón-Zygmund operators. The kernel of UTU^{-1} is $\delta^{-n} K(\delta^{-1}(x - x_0))$ and we note again that conditions (2.1) to (2.4) are invariant under change of scale.

We shall use the following "weak compactness theorem".

Proposition 1. *Let T_j, $j \in \mathbb{N}$, be a bounded sequence of Calderón-Zygmund operators. Then there exist a Calderón-Zygmund operator T and a subsequence $T_{j(m)}$ such that*

$$(2.5) \qquad \langle T_{j(m)}(f), g \rangle \to \langle T(f), g \rangle,$$

for $f \in L^2(\mathbb{R}^n)$ and $g \in L^2(\mathbb{R}^n)$.

We shall write $T_{j(m)} \rightharpoonup T$ in the above situation.

Since the operator norms of the $T_j : L^2(\mathbb{R}^n) \to L^2(\mathbb{R}^n)$ form a bounded sequence, there is a subsequence $T_{j(m)}$ which converges weakly to a linear operator $T : L^2(\mathbb{R}^n) \to L^2(\mathbb{R}^n)$.

It remains to prove that T is a Calderón-Zygmund operator. We let S_m and S denote the kernels of $T_{j(m)}$ and T. Then S is the limit of the S_m in the sense of convergence of distributions in $\mathcal{D}'(\mathbb{R}^n \times \mathbb{R}^n)$. The restrictions of the S_m to each compact subset of Ω are continuous functions satisfying Hölder conditions uniformly in m. Ascoli's theorem gives the uniform convergence of S_m to S on those compact subsets. Conditions (2.1), (2.2) and (2.3) on S are obtained by passing to the limit in the inequalities.

Conversely, every Calderón-Zygmund operator T can be written, as in (2.5), as the limit of a bounded sequence of Calderón-Zygmund operators T_m, whose kernels S_m are not only distributions, but also infinitely differentiable, bounded functions. More precisely, we have

Proposition 2. *Let $T : L^2(\mathbb{R}^n) \to L^2(\mathbb{R}^n)$ be a Calderón-Zygmund operator with kernel $K(x, y)$. Then there is a sequence $K_m(x, y)$ in $L^\infty(\mathbb{R}^n \times \mathbb{R}^n)$ such that $\partial K_m / \partial x_j$ and $\partial K_m / \partial y_j$ are also in $L^\infty(\mathbb{R}^n \times \mathbb{R}^n)$ and the following properties hold:*

(2.6) *the operators T_m defined by $T_m(x) = \int K_m(x, y) f(y) \, dy$ form a bounded sequence of Calderón-Zygmund operators;*

(2.7) *for every function $f \in L^2(\mathbb{R}^n)$,*

$$\lim_{m \to \infty} \| T(f) - T_m(f) \|_2 = 0.$$

To see this, we let $\phi \in \mathcal{D}(\mathbb{R}^n)$ denote a radial function of mean 1, we put $\phi_m(x) = m^n \phi(mx)$, and we write R_m for the operation of convolution by ϕ_m. Consider $T_m = R_m T R_m$. Let $S \in \mathcal{D}'(\mathbb{R}^n \times \mathbb{R}^n)$ denote the kernel of T and let S_m denote that of T_m. Then

$$(2.8) \quad S_m(x, y) = \int\!\!\int \phi_m(x-u) S(u, v) \phi_m(v-y)\, du\, dv = \langle T\tau_y \phi_m, \tau_x \phi_m \rangle,$$

where τ_x denotes translation by x.

We can suppose that the support of ϕ is contained in $|x| \leq 1$: we shall just deal with the case where $\gamma = 1$ in (2.2) and (2.3).

If $|x - y| \geq 4/m$, the supports of $\tau_y \phi_m$ and $\tau_x \phi_m$ are disjoint, so that

$$S_m(x, y) = \int\!\!\int \phi_m(x - u) K(u, v) \phi_m(v - y)\, du\, dv$$

$$= \int\!\!\int K\left(x - \frac{u}{m}, y - \frac{v}{m}\right) \phi(u)\phi(v)\, du\, dv.$$

Properties (2.1), (2.2), and (2.3) are thus satisfied by $S_m(x, y)$, with uniform constants, as long as $|x - y| \geq 4/m$.

For $|x - y| < 4/m$, we use the continuity of T on $L^2(\mathbb{R}^n)$ to get $|S_m(x, y)| \leq C m^n$. Further,

$$\frac{\partial}{\partial x_j} S_m(x, y) = \int\!\!\int \frac{\partial \phi_m}{\partial x_j}(x - u) S(u, v) \phi_m(v - y)\, du\, dv$$

$$= -\left\langle T\tau_y \phi_m, \tau_x \frac{\partial}{\partial x_j} \phi_m \right\rangle,$$

which gives

$$\left| \frac{\partial}{\partial x_j} S_m(x, y) \right| \leq C m^{n+1}.$$

These estimates give (2.1) and (2.2) for $|x - y| < 4/m$. (2.3) is done similarly. The convergence of $T_m(f)$ to $T(f)$ in $L^2(\mathbb{R}^n)$ follows immediately and concludes the proof of Proposition 2.

One last general property merits attention.

Let T be a Calderón-Zygmund operator. Suppose that the kernel K of T is zero. Then T is the operator of pointwise multiplication by a function $m(x) \in L^\infty(\mathbb{R}^n)$.

To prove this, we approximate as above, but in a more rudimentary fashion. Let $\mathcal{Q}_j$ denote the collection of dyadic cubes $2^{-j}k + 2^{-j}[0, 1)^n$, $k \in \mathbb{Z}^n$, and let $V_j \subset L^2(\mathbb{R}^n)$ be the subspace of functions constant on each cube $Q \in \mathcal{Q}_j$. Lastly, let $E_j : L^2(\mathbb{R}^n) \to V_j$ be the corresponding conditional-expectation operator. If T is bounded on $L^2(\mathbb{R}^n)$ and local (that is, the support of $T(f)$ is included in the support of f, for $f \in L^2(\mathbb{R}^n)$) then the same holds for $E_j T E_j : L^2(\mathbb{R}^n) \to V_j$. As a consequence, $E_j T$, restricted to V_j, is the operator of pointwise multiplication by a function $m_j(x)$, constant on each cube $Q \in \mathcal{Q}_j$. Moreover, we have

$$\|m_j\|_\infty = \|E_j T E_j\| \leq \|T\|.$$

The conditions of compatibility between $E_j T$ and $E_{j+1} T$ immediately give $m_j = E_j(m_{j+1})$. So the sequence of functions $m_j(x)$ is a uniformly-bounded martingale which converges to $m(x) \in L^\infty(\mathbb{R}^n)$. A simple passage to the limit shows that T is pointwise multiplication by $m(x)$.

To conclude: we could have proved Proposition 2 by replacing the regularization operators R_m by the operators E_j corresponding to a multiresolution approximation of regularity $r \geq 1$. The approximation of T by a bounded sequence of Calderón-Zygmund operators T_j would then have been given by $E_j T E_j$.

By Proposition 2, the theory of Calderón-Zygmund operators can be written using only absolutely converging integrals (in the first place) and then finishing by passing to the limit. This point of view is taken in [76] which is, to our knowledge, the first major reference to the general theory which we are presenting here.

The second approximation technique for Calderón-Zygmund operators is based on the truncation of kernels. For this we use a function $\phi \in \mathcal{D}(\mathbb{R}^n)$ which is radial, equal to 1, when $|x| \leq 1$, and to 0, when $|x| \geq 2$. We then replace $K(x,y)$ by $K_\delta(x,y) = K(x,y)(1 - \phi(\delta^{-1}(x-y)))$. It follows that $K_\delta(x,y) = 0$ when $|x-y| \leq \delta$ and the singularity of the kernel has disappeared. We let T_δ be the operator defined by the kernel K_δ. The relationship between T and T_δ is described by the following proposition.

Proposition 3. *The operators T_δ form a bounded set of Calderón-Zygmund operators. There exist a sequence $\delta_j \to 0$ and a function $m(x) \in L^\infty(\mathbb{R}^n)$ such that, for every pair (f,g) of functions of $L^2(\mathbb{R}^n)$, we have*

$$(2.9) \qquad \langle Tf, g \rangle = \lim_{j \to \infty} \langle T_{\delta_j} f, g \rangle + \int m(x) f(x) g(x)\, dx.$$

We start by verifying the first assertion. We have

$$\phi(u) = (2\pi)^{-n} \int e^{iu \cdot \xi} \hat{\phi}(\xi)\, d\xi$$

and

$$K_\delta(x,y) = \left(1 - \phi\left(\frac{x-y}{\delta}\right)\right) S(x,y),$$

where $S(x,y)$ is the distribution-kernel of T. We now separate the variables by representing $\phi(\delta^{-1}(x-y))$ as a Fourier transform:

$$\phi\left(\frac{x-y}{\delta}\right) = \frac{1}{(2\pi)^n} \int e^{ix \cdot \xi \delta^{-1}} e^{-iy \cdot \xi \delta^{-1}} \hat{\phi}(\xi)\, d\xi.$$

This gives

$$K_\delta(x, y) = S(x, y) - R_\delta(x, y),$$

where

$$R_\delta(x, y) = \frac{1}{(2\pi)^n} \int e^{ix \cdot \xi \delta^{-1}} S(x, y) e^{-iy \cdot \xi \delta^{-1}} \hat\phi(\xi) \, d\xi.$$

Let $\mathcal{R}_\delta$ denote the operator whose kernel is the distribution $R_\delta(x, y)$ and let $M_{(\xi)}$ be the operator of pointwise multiplication by $e^{ix \cdot \xi}$. We have

$$\mathcal{R}_\delta = \frac{1}{(2\pi)^n} \int M_{(\delta^{-1}\xi)} T M^*_{(\delta^{-1}\xi)} \hat\phi(\xi) \, d\xi$$

and the operator norm of $\mathcal{R}_\delta : L^2(\mathbb{R}^n) \to L^2(\mathbb{R}^n)$ is bounded above by $(2\pi)^{-n} \|T\| \|\hat\phi\|_1$.

The estimates (2.1), (2.2) and (2.3) are easy to verify as long as the cases $|x - y| \le \delta$, $\delta \le |x - y| \le 2\delta$ and $|x - y| \ge 2\delta$ are dealt with separately.

To prove the second assertion, we start by taking a subsequence δ_j such that $\langle T_{\delta_j} f, g \rangle \to \langle Lf, g \rangle$, for a certain Calderón-Zygmund operator L. If f and g are test functions of disjoint support, then $\langle T_{\delta_j} f, g \rangle = \langle Tf, g \rangle$, for large enough j. It follows that $\langle Tf, g \rangle = \langle Lf, g \rangle$. The kernel corresponding to $T - L$ vanishes and (2.9) is verified.

We shall see later that the approximation technique described by Proposition 2 is significantly better than that which we have just used. For example, if T is bounded on the Hardy space $H^1(\mathbb{R}^n)$, or its dual BMO($\mathbb{R}^n$), the same holds for the operators T_m of Proposition 2, but not for the approximations of Proposition 3.

3 Calderón-Zygmund operators and L^p spaces

The purpose of this section is to study the continuity of Calderón-Zygmund operators on $L^p(\mathbb{R}^n)$, when $1 < p < \infty$, and to describe what happens in the limiting cases $p = 1$ and $p = \infty$.

Let $K(x, y)$ be a function which is locally integrable on $\mathbb{R}^n \times \mathbb{R}^n$. Suppose that the operator $Tf(x) = \int K(x, y) f(y) \, dy$ is continuous on $L^\infty(\mathbb{R}^n)$. Then it necessarily follows that

$$(3.1) \qquad \operatorname*{ess\,sup}_{x \in \mathbb{R}^n} \int |K(x, y)| \, dy < \infty$$

and $\|T\|_{\infty,\infty}$ is given by the left-hand side of (3.1).

Similarly, if we assume that the operator T is continuous on $L^1(\mathbb{R}^n)$, then $\operatorname{ess\,sup}_{x \in \mathbb{R}^n} \int |K(x, y)| \, dx$ is finite and equals $\|T\|_{1,1}$.

Calderón-Zygmund operators do not satisfy these conditions because, in general, (2.1) allows the kernel to have too large a singularity.

A Calderón-Zygmund operator is thus not continuous as an operator from L^1 to L^1 itself. However, it can be extended as a continuous linear operator from $L^1(\mathbb{R}^n)$ to weak L^1, which space we shall now define.

We shall throughout use $|E|$ to denote the Lebesgue measure of the measurable set E.

Definition 2. *A measurable function $f(x)$ on $\mathbb{R}^n$ belongs to weak L^1 if*

$$\|f\|_w = \sup_{\lambda > 0} \lambda |\{x : |f(x)| > \lambda\}|$$

is finite.

The essential point is that $\|f + g\|_w \leq 2\|f\|_w + 2\|g\|_w$, even though $\|f\|_w$ is not a norm. The verification of the inequality is straightforward, indeed, if $|f(x) + g(x)| > \lambda$, then either $|f(x)| > \lambda/2$ or $|g(x)| > \lambda/2$. This gives the following relationship between sets:

$$\{x : |f(x) + g(x)| > \lambda\} \subset \{x : |f(x)| > \lambda/2\} \cup \{x : |g(x)| > \lambda/2\}$$

finally leading to

$$|\{x : |f(x) + g(x)| > \lambda\}| \leq \frac{2}{\lambda}\|f\|_w + \frac{2}{\lambda}\|g\|_w \,.$$

This implies that weak L^1 is a complete metric vector space. But weak L^1 is not a Banach space. To see this, we look at dimension 1 and observe that $|x - x_0|^{-1}$ belongs to weak L^1 with norm equal to 2. But the sequence of convex combinations $2^{-j} \sum_{0 \leq k < 2^j} |x - 2^{-j}k|^{-1}$ of such functions does not form a bounded sequence in weak L^1, because their value is greater than cj on the interval $[0, 1]$, where $c > 0$ is a constant.

To show that weak L^1 is complete, it is enough to verify that, if $\|f_k\|_w \leq 2^{-k}$, then $\sum_0^\infty f_k = f$ belongs to weak L^1. To see this, we repeat the argument that led to (3.1). If $|f(x)| > \lambda$, there is an integer k such that $|f_k(x)| > \alpha\lambda 2^{-k/2}$, where $\alpha = 1 - 1/\sqrt{2}$. But

$$|\{x : |f_k(x)| > \alpha\lambda 2^{-k/2}\}| \leq (\alpha\lambda)^{-1} 2^{k/2}\|f_k\|_w \leq (\alpha\lambda)^{-1} 2^{-k/2},$$

which gives what we want.

Let us proceed to the relationship between Calderón-Zygmund operators, L^1, and weak L^1.

Theorem 1. *Let T be a Calderón-Zygmund operator such that (2.1), (2.3) and (2.4) are satisfied with $C_0 = C_1 = C_2 = 1$ ((2.2) is irrelevant). Then there is a constant $C = C(n, \gamma)$ such that, for every function $f \in L^1 \cap L^2$ and every $\lambda > 0$,*

$$(3.2) \qquad |\{x \in \mathbb{R}^n : |Tf(x)| > \lambda\}| \leq \frac{C}{\lambda}\|f\|_1 \,.$$

If weak L^1 were a Banach space, Theorem 1 would be obvious with (2.1) as the only hypothesis, since the integral $\int K(x,y)f(y)\,dy$ could then be considered as a superposition of functions $x \to K(x,y)$ belonging to weak L^1 uniformly in y.

To prove Theorem 1, Calderón and Zygmund invented a remarkable (non-linear) decomposition of the functions $f \in L^1(\mathbb{R}^n)$. This is now called the "Calderón-Zygmund decomposition". Theorem 2 describes its properties.

Theorem 2. *Let $f(x)$ be a function in $L^1(\mathbb{R}^n)$ and let λ be a postive real number, called the threshold of the Calderón-Zygmund decomposition of f. Then f can be written as the sum of a function g, belonging to $L^1 \cap L^2$, and of a series of highly oscillating and localized terms b_j, $j \in J$.*

More precisely, there exist disjoint cubes Q_j, whose union is denoted by Ω, such that the support of each function b_j is contained in the corresponding cube Q_j and such that

(3.3) $$f = g + \sum_{j \in J} b_j\,, \qquad \text{throughout } \mathbb{R}^n,$$

(3.4) $$|g(x)| = |f(x)| \le \lambda\,, \qquad \text{for almost all } x \notin \Omega,$$

(3.5) $$|g(x)| \le 2^n\lambda\,, \qquad \text{for all } x \in \Omega,$$

(3.6) $$|\Omega| \le \lambda^{-1}\|f\|_1\,,$$

(3.7) $$\|g\|_2 \le 2^n\lambda^{1/2}\|f\|_1^{1/2}\,,$$

(3.8) $$\int_{Q_j} |b_j(x)|\,dx \le 2^{n+1}\lambda|Q_j|\,,$$

(3.9) $$\int_{Q_j} b_j(x)\,dx = 0\,.$$

Before proving this result, let us give an amusing example of it. We let $f(x) \in L^1(\mathbb{R}^n)$ denote the characteristic function of the interval $[0,1]$ and we want to carry out the Calderón-Zygmund decomposition of f with threshold $2^{-(m+1)}$, $m \in \mathbb{N}$. Then Ω is the interval $[0,2^m)$ with $g(x) = 2^{-m}$ on this interval and 0 elsewhere. The difference $b(x) = f(x) - g(x)$ is clearly a function whose integral is zero, with support in the unique "cube" of the decomposition, that is, the interval $[0,2^m)$, and satisfies property (3.8).

We now come to the proof of Theorem 2. We start with the collection $\mathcal{Q}$ of dyadic cubes $Q \subset \mathbb{R}^n$ and define $\mathcal{C} \subset \mathcal{Q}$ by the condition $Q \in \mathcal{C}$ if and only if $|Q|^{-1} \int_Q |f(x)|\,dx > \lambda$.

We first consider the case when $\mathcal{C}$ is empty. Then, for each dyadic cube Q, $|Q|^{-1} \int_Q |f(x)|\,dx \le \lambda$, which implies that $\|f\|_\infty \le \lambda$, as we observed in Section 2. In this case, the Calderón-Zygmund decomposition is $f = g$: there are no oscillatory terms b_j, $j \in J$.

If $\mathcal{C}$ is non-empty, the volume $|Q|$ of each cube $Q \in \mathcal{C}$ satisfies $|Q| < \lambda^{-1}\|f\|_1$ and it follows that the union Ω of the cubes $Q \in \mathcal{C}$ is precisely the disjoint union of the maximal cubes, which latter we shall denote by Q_j, $j \in J$.

For each dyadic cube $Q \in \mathcal{Q}$, we let $\tilde{Q}$ denote the "parent" of Q, that is, the dyadic cube containing Q and twice the dimension of Q.

Since the Q_j are maximal, their parents $\tilde{Q}_j$ do not lie in $\mathcal{C}$ and so $|\tilde{Q}_j|^{-1} \int_{\tilde{Q}_j} |f(x)|\, dx \leq \lambda$. This implies that $\int_{\tilde{Q}_j} |f(x)|\, dx \leq 2^n \lambda |Q_j|$ and hence that $|m_{Q_j} f| \leq 2^n \lambda$, where m_{Q_j} is the mean of f on Q_j.

We define $b_j(x)$ by $b_j(x) = f(x) - m_{Q_j} f$ for $x \in Q_j$ and $b_j(x) = 0$ otherwise. Then $g(x) = f(x)$, for $x \notin \Omega$, whereas $g(x) = m_{Q_j} f$, for $x \in Q_j$. These choices give (3.3), (3.8), and (3.9).

We now verify (3.4): that $|f(x)| \leq \lambda$ if $x \notin \Omega$. Now, if x is not in Ω, then $|Q|^{-1} \int_Q |f(y)|\, dy \leq \lambda$ for each dyadic cube containing x. Let E_j denote the conditional-expectation operator with respect to the σ-field generated by the dyadic cubes of side 2^{-j} (in other words, E_j is the operation of taking the mean on each of the generating cubes). Then $|E_j(f)(x)| \leq \lambda$ on $\mathbb{R}^n \setminus \Omega$. But $E_j(f)$ converges almost everywhere to f, as j tends to infinity, and it follows that $|f(x)| \leq \lambda$ almost everywhere on $\mathbb{R}^n \setminus \Omega$.

Property (3.5) has already been established. Let us proceed to (3.6). We have

$$|\Omega| = \sum_{j \in J} |Q_j| \leq \lambda^{-1} \int_{Q_j} |f(y)|\, dy \leq \lambda^{-1}\|f\|_1 .$$

As far as (3.7) is concerned, we have

$$\int_{\mathbb{R}^n} |g|^2 \, dx = \int_\Omega |g|^2 \, dx + \int_{\mathbb{R}^n \setminus \Omega} |g|^2 \, dx$$

$$\leq \sum_{j \in J} |m_{Q_j} f|^2 |Q_j| + \lambda \int_{\mathbb{R}^n \setminus \Omega} |f(x)|\, dx$$

$$\leq \sum_{j \in J} 4^n \lambda^2 |Q_j| + \lambda \int_{\mathbb{R}^n \setminus \Omega} |f(x)|\, dx$$

$$\leq 4^n \lambda \sum_{j \in J} \int_{Q_j} |f(x)|\, dx + \lambda \int_{\mathbb{R}^n \setminus \Omega} |f(x)|\, dx$$

$$= 4^n \lambda \int_\Omega |f(x)|\, dx + \lambda \int_{\mathbb{R}^n \setminus \Omega} |f(x)|\, dx \leq 4^n \lambda \|f\|_1 .$$

We have proved Theorem 2 and may now return to Theorem 1.

We should observe that conditions (2.1) and (2.3) are not necessary conditions. While keeping (2.4), we may replace (2.1), and (2.3) by the single condition

$$(3.10) \qquad \int_{\{x:|x-y'|\geq 2|y-y'|\}} |K(x,y') - K(x,y)|\, dx \leq C_1\,,$$

and give the proof under these hypotheses. Clearly, we may suppose that $C_1 = C_2 = 1$ and that $\|f\|_1 = 1$.

We now carry out a Calderón-Zygmund decomposition of f of threshold λ. This gives the cubes Q_j, $j \in J$, as described by Theorem 2. Let $Q_j^\star$ denote the "doubled" cube: the centre of $Q_j^\star$ is that of Q_j and the diameter of $Q_j^\star$ is twice that of Q_j. Then $\Omega^\star$ is the union of the $Q_j^\star$ and we get

$$|\Omega^\star| \leq \sum_{j\in J} |Q_j^\star| \leq 2^n \sum_{j\in J} |Q_j| \leq 2^n \lambda^{-1}\,.$$

Let E be the set of x that are not in $\Omega^\star$ and for which $|Tf(x)| > \lambda$. If we show that $|E| \leq C\lambda^{-1}$, the proof will be concluded, because the set of all x such that $|Tf(x)| > \lambda$ is contained in $E \cup \Omega^\star$ and we know that $|\Omega^\star| \leq 2^n \lambda^{-1}$.

In (3.3), we set $b(x) = \sum_{j\in J} b_j(x)$ and then let E_1 and E_2 denote the subsets of E where, respectively, $|Tg(x)| > \lambda/2$ and $|Tb(x)| > \lambda/2$. Then $E \subset E_1 \cup E_2$, which leads us to find upper bounds for the measures of E_1 and E_2.

For $|E_1|$, we simply use the continuity of T on $L^2(\mathbb{R}^n)$. We have $\|Tg\|_2 \leq \|g\|_2 \leq 2^n \lambda^{1/2}$. The Bienaymé-Chebyshev inequality gives

$$\frac{\lambda^2}{4}|E_1| \leq \|Tg\|_2^2 \leq 4^n \lambda$$

and thus $|E_1| \leq 4^{n+1}\lambda^{-1}$.

We proceed to E_2. Now $b = \sum_{j\in J} b_j$ with the sum converging in $L^2(\mathbb{R}^n)$ (as well as pointwise) if $f \in L^1 \cap L^2$, as we have assumed. Hence $Tb(x) = \sum_{j\in J} Tb_j(x)$. For $x \notin \Omega^\star$, we can use the representation of T by its kernel. This gives

$$Tb_j(x) = \int_{Q_j} K(x,y)b_j(y)\, dy$$

$$= \int_{Q_j} (K(x,y) - K(x,y_j))b_j(y)\, dy\,,$$

where y_j is the centre of Q_j. We then estimate the integral

$$I = \int_{\mathbb{R}^n \setminus \Omega^\star} |Tb(x)|\, dx$$

$$\leq \sum_{j\in J} \int_{\mathbb{R}^n \setminus Q_j^\star} \left\{ \int_{Q_j} |K(x,y) - K(x,y_j)||b_j(y)|\, dy \right\} dx\,.$$

We apply Fubini's theorem to change the order of integration in each of the double integrals. This freezes the variables y and y_j and, on applying (3.10), gives the bound $C_1 \sum_{j \in J} \|b_j\|_1$. This, in turn, can be estimated by using (3.6) and (3.8). So we get $\int_{E_2} |Tb(x)| \, dx \leq 2^{n+1} C_1$ and it is enough, once more, to use the Bienaymé-Chebyshev inequality to conclude.

Theorem 1 is thus proved.

We continue to follow Calderón and Zygmund, who show that T is continuous as an operator on L^p, $1 < p \leq 2$, by observing that T is continuous as an operator from L^1 to weak L^1 and from L^2 to L^2 and then by applying Marcinkiewicz' interpolation theorem ([217], p. 21, Theorem 5).

For $2 \leq p < \infty$, we observe that the adjoint T^* of T is again a Calderón-Zygmund operator: the corresponding kernel $K^*(x, y)$ is $\bar{K}(y, x)$. Letting $q \in (1, 2]$ denote the conjugate exponent of $p \in [2, \infty)$, the continuity of T^* on $L^q(\mathbb{R}^n)$ implies that of T on $L^p(\mathbb{R}^n)$.

We now return to the case $p = 1$, but replace L^1 by the space H^1 of Stein and Weiss.

To show that every Calderón-Zygmund operator extends to a continuous linear operator from H^1 to L^1, we use the atomic decomposition of the functions in H^1. We let B be a ball in $\mathbb{R}^n$, of volume $|B|$, and let $a(x)$ be an atom with support in B, that is, $\|a\|_2 \leq |B|^{-1/2}$, $\int_B a(x) \, dx = 0$ and $a(x) = 0$ outside B.

We intend to show that $\|T(a)\|_1 \leq C$, where C is a constant depending only on the constants C_0, C_1 and C_2 of Definition 1 and on the dimension n. Once again, the hypotheses of Definition 1 are stronger than necessary and may be replaced by (3.10).

We let B^* denote the "doubled" ball of B and write

$$\|T(a)\|_1 = \int_{B^*} |T(a)| \, dx + \int_{\mathbb{R}^n \setminus B^*} |T(a)| \, dx = I + J.$$

We estimate I by using the Cauchy-Schwarz inequality to give

$$I \leq |B^*|^{1/2} \|T(a)\|_2 \leq 2^{n/2} |B|^{1/2} \|T\| \|a\|_2 \leq 2^{n/2} \|T\|.$$

As far as J is concerned, we again apply the representation of T by the kernel K, and, writing x_0 for the centre of B and r for its radius, we get

$$J \leq \int_{|x - x_0| \geq 2r} \left\{ \int_{|y - x_0| \leq r} |K(x, y) - K(x, x_0)| |a(y)| \, dy \right\} dx.$$

Condition (3.10) enables us to integrate first with respect to x and then with respect to y, using the obvious upper bound $\|a\|_1 \leq 1$.

Having shown that $\|T(a)\|_1 \leq C$, for each atom $a(x)$ of H^1, it might appear that the proof of the continuity of $T : H^1 \to L^1$ is over. That is by no means the case. The difficulty lies in defining T on H^1, given that the atomic decomposition is not unique. To this end, we use the following lemma.

Lemma 1. *Let $a_j(x)$ be atoms of H^1, let λ_j be coefficients such that $\sum_0^\infty |\lambda_j| < \infty$ and let $\sum_0^\infty \lambda_j a_j = 0$. Then the series $\sum_0^\infty \lambda_j T(a_j)$ converges to 0 in $L^1(\mathbb{R}^n)$ norm.*

(There would be nothing to prove if the Calderón-Zygmund operators were bounded on $L^1(\mathbb{R}^n)$.)

To prove the lemma, we use the approximations T_m given by Proposition 2. The kernels K_m of T_m belong to $L^\infty(\mathbb{R}^n \times \mathbb{R}^n)$ and it follows that $\sum_{j=0}^\infty \lambda_j T_m(a_j) = 0$.

Then we observe that the T_m form a bounded sequence of Calderón-Zygmund operators. Hence $\|T_m(a_j)\|_1 \leq C$, by the argument above.

It is easy to check that, for each j, $T_m(a_j)$ converges in L^1 norm to $T(a_j)$. Finally, we get $\sum_0^\infty \lambda_j T(a_j) = 0$, because $\sum_0^\infty \lambda_j (T(a_j) - T_m(a_j))$ tends to 0 in L^1 norm, as m tends to infinity.

This proves that Calderón-Zygmund operators are (H^1, L^1) continuous.

As a corollary, we can define the extension of Calderón-Zygmund operators to L^∞ by using the (L^1, L^∞) and (H^1, BMO) dualities. More precisely, if $b(x) \in L^\infty(\mathbb{R}^n)$, we define $T(b)$ as a continuous linear form on H^1 by

$$(3.11) \qquad \langle T(b), u \rangle = \langle b, T^*(u) \rangle, \qquad u \in H^1 .$$

The right-hand term makes sense because the transpose T^* of T is again a Calderón-Zygmund operator and, thus, $T^*(u) \in L^1$. So $T(b)$ is in $\text{BMO}(\mathbb{R}^n)$.

It is worthwhile to make the connection between this definition and the action of T on $L^2(\mathbb{R}^n)$. To do that, we observe that, if $b \in L^\infty \cap L^2$ and if u is an atom, then (3.11) becomes an identity. Further, if $b \in L^\infty(\mathbb{R}^n)$, let us define the functions $b_j(x)$ by $b_j(x) = b(x)$, when $|x| \leq j$, and $b_j(x) = 0$, otherwise. Then $T(b_j)$ is defined by the action of T on $L^2(\mathbb{R}^n)$. If u is an atom of H^1, then

$$\langle T(b_j), u \rangle = \langle b_j, T^*(u) \rangle \to \langle b, T^*(u) \rangle .$$

Indeed, $\|b_j\|_\infty \leq \|b\|_\infty$, $b_j(x) \to b(x)$ almost everywhere, and $T^*(u) \in L^1(\mathbb{R}^n)$, so that we can apply Lebesgue's dominated convergence theorem.

The functions $T(b_j)$ thus form a bounded sequence in $\text{BMO}(\mathbb{R}^n)$ and this sequence converges to $T(b)$ in the $\sigma(\text{BMO}, H^1)$ topology. The following lemma describes this convergence precisely.

Lemma 2. *If T is a Calderón-Zygmund operator and if the functions b_j are formed by truncating the function $b \in L^\infty(\mathbb{R}^n)$, then there is a sequence $c(j)$ of constants such that $T(b_j) - c(j)$ converges, uniformly on compacta, to a function in $\mathrm{BMO}(\mathbb{R}^n)$ which is a representative of $T(b)$ modulo the constant functions.*

Indeed, we put $c(j) = \int_{1 \leq |y| \leq j} K(0,y) b(y) \, dy$ and we need to show that $T(b_j) - c(j)$ converges uniformly on the compact ball $|x| \leq R$. To this end, we split b_j into $g + h_j$, where $g(x) = b(x)$, when $|x| \leq 2R$, and $g(x) = 0$, otherwise. Taking $j > 2R$, we get, for $|x| \leq R$,

$$T(b_j)(x) = T(g)(x) + T(h_j)(x)$$

$$= T(g)(x) + \int_{2R \leq |y| \leq j} K(x,y) b(y) \, dy$$

$$= T(g)(x) + \int_{2R \leq |y| \leq j} (K(x,y) - K(0,y)) b(y) \, dy + c(j) - C(R),$$

where $C(R) = \int_{1 \leq |y| \leq 2R} K(0,y) b(y) \, dy$. But

$$|K(x,y) - K(0,y)| = O(|y|^{-n-\gamma}),$$

as $|y| \to \infty$, so the integral converges uniformly on $|x| \leq R$.

This new definition of $T(b)$ is compatible with the definition obtained by duality. Indeed, if u is an atom of H^1, u is necessarily of compact support and zero mean. Thus $\langle T(b_j) - c(j), u \rangle = \langle T(b_j), u \rangle = \langle b_j, T^*(u) \rangle \to \langle b, T^*(u) \rangle$. If f is the limit (uniformly on compacta) of the functions $T(b_j) - c(j)$, we get $\langle f, u \rangle = \langle b, T^*(u) \rangle = \langle T(b), u \rangle$ and f is a representative of $T(b)$.

The explicit definition of $T(b)$ given by Lemma 2 leads naturally to a direct proof that $T(b)$ belongs to BMO, when b is in $L^\infty(\mathbb{R}^n)$ and the kernel $K(x,y)$ corresponding to T satisfies the condition

$$(3.12) \qquad \int_{|x-y| \geq 2|x'-x|} |K(x',y) - K(x,y)| \, dy \leq C_1.$$

To show that $T(b)$ belongs to BMO, we let B denote an arbitrary ball, with centre x_0 and radius R, and let B^* denote the ball with the same centre and of radius $2R$. We then write $b = b_1 + b_2$, where b_1 is the product of b by the characteristic function of B^*. Then what we have to do is to define $T(b_2)(x)$, if $x \in B$, by the absolutely convergent integral

$$(3.13) \qquad T(b_2)(x) = \int_{|x_0-y| \geq 2R} (K(x,y) - K(x_0,y)) b(y) \, dy.$$

This definition of $T(b_2)(x)$ leads to a definition of $T(b)(x)$, because $T(b_1)$ poses no difficulty, indeed, $b_1 \in L^2(\mathbb{R}^n)$. Further, this definition of $T(b)(x)$ is compatible (in the sense of only differing by a constant, depending on x_0 and R) with the pointwise definition of Lemma 2.

Once we have taken these precautions, the proof that $T(b)$ belongs to BMO is quite simple. By (3.12), we have

$$\|T(b_2)\|_\infty \le \|b\|_\infty \int_{|y-x_0|\ge 2|x-x_0|} |K(x,y) - K(x_0,y)|\, dy \le C\|b\|_\infty$$

and, further,

$$\|T(b_1)\|_2 \le \|T\|\|b_1\|_2 \le 2^{n/2}\|b\|_\infty |B|^{1/2}\|T\|.$$

We thus get

$$\left(\int_B |T(b)(x)|^2\, dx\right)^{1/2} \le C\|b\|_\infty |B|^{1/2} + 2^{n/2}\|b\|_\infty |B|^{1/2}\|T\|$$

$$\le C'|B|^{1/2}\|b\|_\infty,$$

which proves that $T(b)$ belongs to BMO.

Let the function f and the sequence of functions f_j be in BMO. As BMO is the dual of H^1, we shall write $f_j \rightharpoonup f$ (in the $\sigma(\text{BMO}, H^1)$ topology) if $\langle f_j, h\rangle \to \langle f, h\rangle$ (as $j \to \infty$) for every $h \in H^1$.

With this notation, the following simple lemma will often be useful.

Lemma 3. *Let T_j, $j \in \mathbb{N}$, be a sequence of Calderón-Zygmund operators such that $T_j \rightharpoonup T$ in the sense of Proposition 1. Then, for each function $f \in L^\infty(\mathbb{R}^n)$, $T_j(f) \rightharpoonup T(f)$ in BMO with the $\sigma(\text{BMO}, H^1)$ topology.*

To prove this, we consider the transpose operators ${}^tT_j$ and ${}^tT$ of T_j and T, respectively. We need to show that $\langle f, {}^tT_j(g)\rangle$ converges to $\langle f, {}^tT(g)\rangle$, for every $g \in H^1$ and $f \in L^\infty$.

We can clearly reduce to the case where g is an atom with support in a ball B, centre x_0 and radius R. We let $\tilde{B}$ denote the ball, centre x_0 and radius $2R$, and we split f into $f_1 + f_2$, where $f_1(x) = f(x)$, if $b \in \tilde{B}$, and $f_1(x) = 0$, otherwise.

Since the operators $T_j : L^2(\mathbb{R}^n) \to L^2(\mathbb{R}^n)$ converge weakly to T, we get $\langle f_1, {}^tT_j(g)\rangle = \langle T_j(f_1), g\rangle \to \langle T(f_1), g\rangle$,

Now, the ${}^tT_j(g)$ satisfy the inequality $|{}^tT_j(g)(x)| \le C|x|^{-n-\gamma}$, uniformly in j, when $x \notin \tilde{B}$.

Further, Ascoli's theorem enables us to suppose that there is a subsequence j_m of j such that $K_{j_m}(x,y)$ converges uniformly to $K(x,y)$ when x and y belong to mutually disjoint compact subsets of $\mathbb{R}^n$. As a result, ${}^tT_{j_m}(g)(x)$ converges to ${}^tT(g)(x)$, when $x \notin \tilde{B}$.

Thus, Lebesgue's dominated convergence theorem gives

$$\lim_{m\to\infty} \langle f_2, {}^tT_{j_m}(g)\rangle = \langle f_2, {}^tT(g)\rangle.$$

But the limit does not depend on the choice of subsequence, so, finally,

$$\lim_{j\to\infty} \langle f_2, {}^tT_j(g)\rangle = \langle f_2, {}^tT(g)\rangle.$$

4 The conditions $T(1) = 0$ and ${}^tT(1) = 0$ for a Calderón-Zygmund operator

A natural problem is to characterize the Calderón-Zygmund operators which can be extended as continuous linear operators from $H^1(\mathbb{R}^n)$ to $H^1(\mathbb{R}^n)$ (not just from $H^1(\mathbb{R}^n)$ to $L^1(\mathbb{R}^n)$). Another is to characterize the Calderón-Zygmund operators which can be extended as continuous linear operators from BMO to BMO (not just from L^∞ to BMO).

Definition 3. *We say that a Calderón-Zygmund operator T satisfies the condition ${}^tT(1) = 0$ if $\int_{\mathbb{R}^n} T(a)\,dx = 0$, for each atom $a \in H^1$.*

This condition makes sense because $T(a)$ belongs to $L^1(\mathbb{R}^n)$. The notation ${}^tT(1) = 0$ arises from the following formal manipulation. We write $\int_{\mathbb{R}^n} T(a)\,dx = \langle T(a), 1 \rangle = \langle a, {}^tT(1) \rangle$. To get 0, this suggests writing ${}^tT(1) = 0$, but one might just as well have written ${}^tT(1) = 4$, or any other constant. Here ${}^tT$ denotes the transpose of T, but it is easy to see that the adjoint would do just as well.

Theorem 3. *A Calderón-Zygmund operator T extends to a continuous linear operator on $H^1(\mathbb{R}^n)$ if and only if ${}^tT(1) = 0$.*

For this statement, condition (3.10) will not be enough and we shall need to use (2.3).

The condition ${}^tT(1) = 0$ is necessary. Indeed, for each atom $a(x)$, $T(a)(x)$ must belong to $H^1(\mathbb{R}^n)$ and therefore must have zero mean.

To show that the condition ${}^tT(1) = 0$ is sufficient, we use the atomic decomposition of $H^1(\mathbb{R}^n)$. Following one of the authors and Weiss [75], it is enough to show that, up to multiplication by a constant, T turns the atoms $a(x)$ of H^1 into molecules. More precisely, if B is the ball corresponding to the atom $a(x)$, if x_0 is the centre and d the radius of B, we show that $m(x) = Ta(x)$ is a molecule of centre x_0 and width d, satisfying condition (5.1) of *Wavelets and Operators*, Chapter 5, for $n < s < n + 2\gamma$.

Here is how we show the condition is satisfied. We let $\tilde{B}$ denote the ball with centre x_0 and radius $2d$. If $x \notin \tilde{B}$ we get $T(a)(x) = \int K(x,y)a(y)\,dy = \int (K(x,y) - K(x,x_0))a(y)\,dy$, because $a(x)$ has zero mean. Using (2.3), it follows that $|T(a)(x)| \leq Cd^\gamma |x - x_0|^{-n-\gamma}$ and we thus get

$$\left(\int_{|x-x_0| \geq 2d} |x - x_0|^s |T(a)(x)|^2 \, dx \right)^{1/2} \leq C'd^{(s-n)/2}$$

as required. The "integral on the interior" $\int_{|x-x_0| \leq 2d} |T(a)(x)|^2 \, dx$ is estimated by ignoring the range of integration and using the continuity of T on $L^2(\mathbb{R}^n)$. Lastly, $\int Ta(x)\,dx = 0$ because ${}^tT(1) = 0$.

To define T on $H^1(\mathbb{R}^n)$ correctly (we must take care because of the non-uniqueness of atomic decompositions), we again use Lemma 1.

We may re-use the above proof to establish, under the same hypotheses, the continuity of T on $H^p(\mathbb{R}^n)$ when $n(1/p - 1) < \gamma \le 1$ and $0 < p \le 1$. The molecules in H^p are defined by condition (3.5) of *Wavelets and Operators*, Chapter 6. For $m \le n(1/p - 1) < m + 1$, (2.3) is replaced by the more precise condition: for all multi-indices $\alpha \in \mathbb{N}^n$, such that $|\alpha| \le m + 1$,

$$(4.1) \qquad |\partial_y^\alpha K(x, y)| \le C|x - y|^{-n - |\alpha|}.$$

Further, the condition ${}^tT(1) = 0$ has to be replaced by ${}^tT(x^\alpha) = 0$, modulo the polynomials of degree not greater than $|\alpha|$, for each multi-index $\alpha \in \mathbb{N}^n$ such that $|\alpha| \le m$. The precise meaning of this condition is that, if f is a square-integrable function of compact support and if $\int x^\alpha f(x)\,dx = 0$, for all multi-indices α with $|\alpha| \le m$, then the same holds for $\int x^\alpha Tf(x)\,dx$. The latter integral is defined because $Tf(x) = O(|x|^{-n-m-1})$ as $|x| \to \infty$. This property is obtained by replacing $K(x, y)$ on the support of f by its Taylor series in powers of $y - x_0$.

Given the above, Theorem 3 can be generalized in the following way.

Proposition 4. *Let $T : L^2(\mathbb{R}^n) \to L^2(\mathbb{R}^n)$ be a continuous linear operator whose kernel satisfies (4.1). If $0 < p \le 1$ and if $m \le n(1/p - 1) < m+1$, then T extends to a continuous linear operator on the Hardy space $H^p(\mathbb{R}^n)$ if and only if ${}^tT(x^\alpha) = 0$, modulo the polynomials of degree not greater than $|\alpha|$, for each multi-index $\alpha \in \mathbb{N}^n$ such that $|\alpha| \le m$.*

The condition ${}^tT(x^\alpha) = 0$ is automatically satisfied when T is a convolution operator, so this condition never appeared when the theory dealt only with convolution operators.

Corollary. *A Calderón-Zygmund operator T defines a continuous linear operator on BMO if and only if $T(1) = 0$.*

We extend T to BMO by writing $\langle T(f), g \rangle = \langle f, {}^tT(g) \rangle$, where $f \in$ BMO and $g \in H^1$. The corollary follows immediately.

Let us return to the invariance of the definition of the Stein and Weiss space H^1, under a change of wavelet basis. We must show that, if ψ_λ, $\lambda \in \Lambda$, and $\tilde{\psi}_\lambda$, $\lambda \in \Lambda$, are two such bases, then, for any sequence of coefficients $\alpha(\lambda)$, $\lambda \in \Lambda$, the conditions $\sum \alpha(\lambda)\psi_\lambda(x) \in H^1$ and $\sum \alpha(\lambda)\tilde{\psi}_\lambda(x) \in H^1$ are equivalent. To do this, we consider the unitary operator $U : L^2(\mathbb{R}^n) \to L^2(\mathbb{R}^n)$ defined by $U(\psi_\lambda) = \tilde{\psi}_\lambda$. The kernel of U is the distribution $S(x, y) = \sum_{\lambda \in \Lambda} \tilde{\psi}_\lambda(x) \bar{\psi}_\lambda(y)$ and it follows immediately (by using the regularity and localization properties of wavelets) that $S(x, y)$, restricted to $y \neq x$, satisfies conditions (2.1), (2.2) and (2.3) of Definition 1, with $\gamma = 1$. The operator U is thus a Calderón-Zygmund operator.

We now use Lemma 3 to show that $U(1) = 0$ and $U^*(1) = 0$. We require approximations U_m, $m \in \mathbb{N}$, of U, defined as follows.

Let Λ_m be an increasing sequence of finite subsets of Λ whose union is Λ. We define U_m by the distribution $S_m(x, y) = \sum_{\lambda \in \Lambda_m} \tilde{\psi}_\lambda(x) \bar{\psi}_\lambda(y)$ which is its kernel. Then the $S_m(x, y)$ satisfy conditions (2.1), (2.2) and (2.3), uniformly in m, while the U_m converge strongly to U. Each $U_m(1) = 0$, because $\int \psi_\lambda(y)\, dy = 0$, and, similarly, $U_m^*(1) = 0$. It follows that $U(1) = 0$ and $U^*(1) = 0$. Theorem 3 gives the required result.

This line of proof has many variants. For example, it shows (but we already know the result) that the wavelets ψ_λ, $\lambda \in \Lambda$, form an unconditional basis of $H^1(\mathbb{R}^n)$. For every bounded sequence $m(\lambda)$, $\lambda \in \Lambda$, we consider the operator M whose eigenfunctions are ψ_λ, $\lambda \in \Lambda$, and whose eigenvalues are $m(\lambda)$. Once again, M is a Calderón-Zygmund operator and $M(1) = {}^t M(1) = 0$, so M can be extended as a continuous linear operator of $H^1(\mathbb{R}^n)$ into $H^1(\mathbb{R}^n)$. Thus $\sum \alpha(\lambda)\psi_\lambda(x) \in H^1$ implies that $\sum m(\lambda)\alpha(\lambda)\psi_\lambda(x) \in H^1$, which shows that the $\psi_\lambda(x)$ form an unconditional basis of $H^1(\mathbb{R}^n)$.

This can be further generalized by replacing $H^1(\mathbb{R}^n)$ by other spaces B of functions or distributions. If the Calderón-Zygmund operators T satisfying $T(1) = {}^t T(1) = 0$ can be extended as a continuous linear operator from B to itself and if the subspace $S_0(\mathbb{R}^n)$, of those functions in $S(\mathbb{R}^n)$ whose moments all vanish, is dense in B, then the wavelets ψ_λ, $\lambda \in \Lambda$, form an unconditional basis of B.

Following the same pattern, we could show (although we already know it) that the wavelets ψ_λ arising from an r-regular multiresolution approximation formed an unconditional basis of the Hardy space $H^p(\mathbb{R}^n)$ when $n(1/p-1) < r$. The condition ${}^t M(x^\alpha) = 0$, $|\alpha| \leq n(1/p-1)$ follows from the cancellation properties of wavelets (*Wavelets and Operators*, Chapter 3, Section 7). The operator M is defined by $M(\psi_\lambda) = m(\lambda)\psi_\lambda$, with $m(\lambda) \in l^\infty(\Lambda)$.

5 Pointwise estimates for Calderón-Zygmund operators

One of the purposes of this section is, if possible, to define a Calderón-Zygmund operator as the principal value of a singular integral. That is, we want to know whether

(5.1) $$Tf(x) = \lim_{\varepsilon \downarrow 0} \int_{|x-y| \geq \varepsilon} K(x,y)f(y)\,dy$$

for every function $f \in L^2(\mathbb{R}^n)$ and almost all $x \in \mathbb{R}^n$.

Experience shows that (5.1) is proved by first establishing (5.1) for a dense linear subspace V of $L^2(\mathbb{R}^n)$. Usually, $V = \mathcal{D}(\mathbb{R}^n)$, but, in the case of the Cauchy kernel $(z(x) - z(y))^{-1}$, where $z(x) = x + ia(x)$ is such that $-M \leq a'(x) \leq M$, we must take $V = \{f(x)z'(x) : f \in \mathcal{D}(\mathbb{R}^n)\}$.

Once we know that (5.1) holds in a particular instance, we look at the corresponding maximal operator ([219]).

To construct that maximal operator, we define the truncated kernel $K_\varepsilon(x,y)$, for each $\varepsilon > 0$, by $K_\varepsilon(x,y) = K(x,y)$, when $|x - y| \geq \varepsilon$, and $K_\varepsilon(x,y) = 0$ otherwise. The operator T_ε is given by $T_\varepsilon f(x) = \int K_\varepsilon(x,y)f(y)\,dy$ and then, for all $x \in \mathbb{R}^n$, we define the maximal operator $T_\star$ by $(T_\star f)(x) = \sup_{\varepsilon > 0} |T_\varepsilon f(x)|$.

The operator $T_\star$, thus defined, is sublinear. More precisely, if f is a function in $L^2(\mathbb{R}^n)$, $T_\star f$ is a positive measurable function and we have $T_\star(\lambda f) = |\lambda| T_\star(f)$ and $T_\star(f_1 + f_2) \leq T_\star(f_1) + T_\star(f_2)$.

The operator $T_\star$ "dominates" the operator T in the following sense.

Lemma 4. *Let T be a Calderón-Zygmund operator. Then there exists a constant C such that, for each function $f \in L^2(\mathbb{R}^n)$,*

(5.2) $$|Tf(x)| \leq T_\star f(x) + C|f(x)| \qquad \text{for almost all } x \in \mathbb{R}^n.$$

To prove this, we need to define the Hardy-Littlewood maximal function $f^\star$ of a locally integrable function f. We put

(5.3) $$f^\star(x) = \sup_{B \ni x} \frac{1}{|B|} \int_B |f(y)|\,dy,$$

where the upper bound is taken over all balls B, of volume $|B|$, containing x. We similarly consider the centred maximal function defined by

$$f_c^\star(x) = \sup_{\varepsilon > 0} \frac{1}{c(n)\varepsilon^n} \int_{|y-x| \leq \varepsilon} |f(y)|\,dy,$$

where $c(n)$ is the volume of the unit ball. Clearly, $f^\star(x) \geq f_c^\star(x)$, but it is easy to see that $f^\star(x) \leq 2^n f_c^\star(x)$, so the two definitions are equivalent from the point of view of integration theory.

We shall assume the following classical results ([126], [127]).

Theorem 4. *The sublinear operator* $M : f \mapsto f^*$ *has the following properties:*

$$(5.4) \quad |\{x \in \mathbb{R}^n : f^*(x) > \lambda\}| \leq \frac{C(n)}{\lambda}\|f\|_1 \quad \text{if } f \in L^1(\mathbb{R}^n) \text{ and } \lambda > 0,$$

$$(5.5) \quad \|f^*\|_p \leq \frac{C(n)}{p-1}\|f\|_p \quad \text{if } 1 < p \leq 2 \text{ and } f \in L^p(\mathbb{R}^n),$$

and

$$(5.6) \quad \|f^*\|_p \leq C(n)\|f\|_p \quad \text{if } 2 \leq p \leq \infty \text{ and } f \in L^p(\mathbb{R}^n),$$

where $C(n)$ *is a constant depending only on* n.

Now for the proof of Lemma 4. As in Proposition 3, we let $\phi \in \mathcal{D}(\mathbb{R}^n)$ denote a radial function, which is 1, when $|x| \leq 1$, and is 0, when $|x| \geq 2$. We put $K_\varepsilon(x,y) = K(x,y)(1 - \phi(\varepsilon^{-1}(x-y)))$ and let S_ε denote the corresponding operator.

Let us estimate the norm of $T_\varepsilon - S_\varepsilon$ as an operator on $L^2(\mathbb{R}^n)$. We have $|(T_\varepsilon - S_\varepsilon)f(x)| \leq C\varepsilon^{-n} \int_{\varepsilon \leq |x-y| \leq 2\varepsilon} |f(y)|\, dy \leq Cf^*(x)$. From Theorem 4, we deduce that $\|T_\varepsilon - S_\varepsilon\| \leq C$.

Now, by Proposition 3, $\|S_\varepsilon\| < C$. It follows that $\|T_\varepsilon\| \leq C$, a constant independent of ε. Hence, there is a bounded linear operator $L : L^2(\mathbb{R}^n) \to L^2(\mathbb{R}^n)$ and a sequence $\varepsilon_j \to 0$ such that , for any $f, g \in L^2(\mathbb{R}^n)$, $\langle T_{\varepsilon_j} f, g\rangle \to \langle Lf, g\rangle$.

We conclude, as in the preamble to Proposition 3, that there is a function $m(x) \in L^\infty(\mathbb{R}^n)$ such that, $T - L$ is the operation of pointwise multiplication by m. Thus, if $f, g \in L^2(\mathbb{R}^n)$,

$$\langle Tf, g\rangle = \lim_{j\to\infty} \langle T_{\varepsilon_j} f, g\rangle + \langle mf, g\rangle.$$

Finally, this gives, for all $f, g \in L^2(\mathbb{R}^n)$,

$$(5.7) \quad \left|\int Tf(y)g(y)\, dy\right| \leq \int T_\star f(y)|g(y)|\, dy + C\int |f(y)||g(y)|\, dy,$$

where $C = \|m\|_\infty$. Since, for fixed f, (5.7) holds for arbitrary $g \in L^2(\mathbb{R}^n)$, we get the inequality (5.2), and Lemma 4 has been proved.

We now establish a pointwise inequality which is deep because it goes in the other direction. It is the fundamental result of this section: Cotlar's theorem.

Theorem 5. *Let* T *be a Calderón-Zygmund operator such that (2.1) to (2.4) are satisfied with* $C_0 = C_1 = C_2 = C_3 = 1$. *Then, for every* $f \in L^2(\mathbb{R}^n)$ *and every* $x \in \mathbb{R}^n$, *the following inequality holds:*

$$(5.8) \quad T_\star f(x) \leq 2(Tf)^\star(x) + C(n,\gamma)f^*(x).$$

Without loss of generality, we may assume $x = 0$. The proof is based on the following two lemmas.

Lemma 5. *For every function $f \in L^2(\mathbb{R}^n)$, each $\gamma > 0$ and all $\varepsilon > 0$,*

(5.9)
$$\varepsilon^\gamma \int_{|y| \geq \varepsilon} |y|^{-n-\gamma} |f(y)| \, dy \leq C f^\star(0) \,,$$

where the constant $C = C(n, \gamma)$ depends only on $\gamma > 0$ and the dimension.

To establish this inequality, we split the range of integration into the dyadic shells $\Gamma_k = \{2^k \varepsilon \leq |y| < 2^{k+1} \varepsilon\}$, $k \in \mathbb{N}$. In carrying out the estimate, we observe that Γ_k is a subset of the ball $B_k = \{y : |y| \leq 2^k \varepsilon\}$ and apply the definition of the maximal function to the ball B_k. The calculation is straightforward and is similar to one carried out earlier (proof of (6.7) in *Wavelets and Operators* Chapter 2).

Lemma 6. *For every Calderón-Zygmund operator T, there exists a constant C such that, for each function $f \in L^2(\mathbb{R}^n)$, all $x \in \mathbb{R}^n$, every $\varepsilon > 0$ and each $x' \in \mathbb{R}^n$ satisfying $|x' - x| \leq \varepsilon/2$, we have*
$$|T_\varepsilon f(x') - T_\varepsilon f(x)| \leq C f^\star(x) \,.$$

We first find an upper bound for $|\int_{|x-y| \geq \varepsilon} (K(x', y) - K(x, y)) f(y) \, dy|$. This is done by using (2.2) and (5.9).

We then estimate the difference between $\int_{|x'-y| \geq \varepsilon} K(x', y) f(y) \, dy$ and $\int_{|x-y| \geq \varepsilon} K(x', y) f(y) \, dy$. Since $|x' - x| \leq \varepsilon/2$, the symmetric difference of the sets $|x' - y| \geq \varepsilon$ and $|x - y| \geq \varepsilon$ is a subset of the shell $\varepsilon/2 \leq |x' - y| \leq 3\varepsilon/2$ (and, by symmetry, of the shell $\varepsilon/2 \leq |x - y| \leq 3\varepsilon/2$). Thus, in that region, $|K(x', y)|$ may be bounded above by $2^n C_0 \varepsilon^{-n}$, and the difference between our two integrals is less than $C f^\star(x)$.

We return to the proof of Cotlar's inequality. Suppose that $\varepsilon > 0$ and let T_ε be the corresponding truncated operator. Let B be the ball $|x| \leq \varepsilon/2$ and let $\tilde{B}$ be the ball $|x| \leq \varepsilon$. We then write f_1 for the product of f by the characteristic function of $\tilde{B}$ and put $f_2 = f - f_1$.

We intend to show that, for every $\varepsilon > 0$,

(5.10)
$$|T_\varepsilon f(0)| \leq 2(Tf)^\star(0) + C f^\star(0) \,.$$

Taking the supremum over $\varepsilon > 0$ of the left-hand side of (5.10), will give Cotlar's inequality.

To prove (5.10), we observe that $T_\varepsilon f(0) = Tf_2(0)$ which we shall compare with the other values taken by Tf_2 on B.

Lemma 6 shows that, for each $x \in B$,

(5.11)
$$|Tf_2(x) - Tf_2(0)| \leq C(n, \gamma) f^\star(0) \,.$$

In order to estimate $Tf(x)$, we note that, for almost all $x \in B$, $Tf_2(x) = Tf(x) - Tf_1(x)$ and (5.11) gives, for almost all $x \in B$,

(5.12)
$$|Tf_2(0)| \leq |Tf(x)| + |Tf_1(x)| + C(n, \gamma) f^\star(0) \,.$$

The last part of the proof consists of estimating the "weak L^1 norms" of both sides of the inequality (5.12), with respect to the uniformly distributed measure μ on the ball B, that is, the measure of constant density $|B|^{-1}$ on B and whose support is B. If $f : B \to \mathbb{C}$ is a measurable function, we put $N(f) = \sup_{\lambda > 0}(\lambda \mu\{x \in B : |f(x)| > \lambda\})$. If $|f| \le |g|$ on B, then $N(f) \le N(g)$. If $f(x)$ is identically equal to α on B, then $N(f) = \alpha$. Further, if f_1, f_2 and f_3 are three measurable functions on B, then

$$N(f_1 + f_2 + f_3) \le 2N(f_1) + 4N(f_2) + 4N(f_3).$$

These remarks let us change (5.12) to

$$|Tf_2(0)| \le 2N(Tf) + 4N(Tf_1) + 4C(n, \gamma)f^\star(0).$$

A simple application of the Bienaymé-Chebyshev inequality gives

$$N(Tf) \le \|Tf\|_{L^1(d\mu)} = |B|^{-1} \int_B |Tf(y)| \, dy \le (Tf)^\star(0).$$

To estimate Tf_1, we use Theorem 1 and get

$$N(Tf_1) = \sup_{\lambda > 0} \lambda(\mu\{x \in B : |Tf_1| > \lambda\}) \le C|B|^{-1}\|f_1\|_1 \le 2^n C f^\star(0).$$

The proof of Cotlar's theorem is complete.

Corollary. *If T is a Calderón-Zygmund operator and $f \in L^p(\mathbb{R}^n)$, for $1 < p < \infty$, then $T_\star f \in L^p(\mathbb{R}^n)$.*

For certain Calderón-Zygmund operators T, we now show how this result allows us to prove that $\lim_{\varepsilon \downarrow 0} \int_{|x-y| \ge \varepsilon} K(x, y)f(y) \, dy = Lf(x)$ exists, for almost all $x \in \mathbb{R}^n$, when $f \in L^p(\mathbb{R}^n)$, $1 < p < \infty$. The connection between T and L is then given by $Tf = Lf + mf$, where $m \in L^\infty(\mathbb{R}^n)$.

Suppose there is a dense vector subspace V of $L^p(\mathbb{R}^n)$, $1 < p < \infty$, such that, for $f \in V$, $Lf(x)$ exists for almost all $x \in \mathbb{R}^n$. Let us show that $Lf(x)$ exists almost everywhere, for every $f \in L^p(\mathbb{R}^n)$. For that, we set $g_\varepsilon(x) = \int_{|x-y| \ge \varepsilon} K(x, y)f(y) \, dy$ and

$$\omega(f; x) = \lim_{\varepsilon \downarrow 0} \left(\sup_{0 < s < t < \varepsilon} |g_t(x) - g_s(x)| \right),$$

for any $f \in L^p(\mathbb{R}^n)$. To show that $Lf(x)$ exists almost everywhere, for each function $f \in L^p(\mathbb{R}^n)$, reduces to proving that $\omega(f; x) = 0$ almost everywhere, or rather, that $|\{x \in \mathbb{R}^n : \omega(f; x) > \alpha\}| = 0$, for each $\alpha > 0$.

The three properties of $\omega(f; x)$ that we shall use are the following obvious remarks:

$$\omega(f_1 + f_2; x) \le \omega(f_1; x) + \omega(f_2; x),$$
$$\omega(f; x) \le 2T_\star f(x)$$

and, lastly,

$$\omega(f; x) = 0 \quad \text{almost everywhere} \quad \text{if } f \in V.$$

Let us suppose that $f \in L^p(\mathbb{R}^n)$. We fix $\alpha > 0$ and verify that $|\{x \in \mathbb{R}^n : \omega(f; x) > \alpha\}| = 0$. Indeed, let $\beta > 0$ be a real number, which will tend to 0, and let $g \in V$ denote a function such that $\|f - g\|_p \le \beta$. Then

$$\omega(f; x) \le \omega(f - g; x) + \omega(g; x) = \omega(f - g; x) \quad \text{almost everywhere}$$

and, thus,

$$|\{x : \omega(f; x) > \alpha\}| \le \alpha^{-p}\|\omega(f - g)\|_p^p$$
$$\le 2^p \alpha^{-p}\|T_*(f - g)\|_p^p \le C^p \alpha^{-p} \beta^p.$$

To finish, we just let β tend to 0.

We give some examples.

Let us begin by supposing that the kernel $K(x, y)$ has mean 0 over each sphere with centre x. This property holds for convolution kernels of the first generation Calderón-Zygmund operators (see the introduction). Then $\lim_{\varepsilon \downarrow 0} \int_{|x-y| \ge \varepsilon} K(x, y) f(y)\, dy$ exists, for every C^1 function f of compact support. Indeed, the integral can be written as $\int_{\varepsilon \le |x-y| \le R} K(x, y) f(y)\, dy$, for R sufficiently large, and then rewritten as $\int_{\varepsilon \le |x-y| \le R} K(x, y)(f(y) - f(x))\, dy$. Finally, we can let ε tend to 0, because the last integral is absolutely convergent.

A more subtle example is given, in dimension 1, by Calderón's first commutator, which will be studied systematically in Chapter 9. We start with a Lipschitz function $A : \mathbb{R} \to \mathbb{C}$ and form the antisymmetric kernel $K(x, y) = (A(x) - A(y))/(x - y)^2$. Then, for a C^1 function f of compact support,

$$T_\varepsilon f(x) = \int_{|x-y| \ge \varepsilon} K(x, y) f(y)\, dy$$
$$= \int_{-\infty}^{x-\varepsilon} K(x, y) f(y)\, dy + \int_{x+\varepsilon}^{\infty} K(x, y) f(y)\, dy.$$

We next integrate each of these integrals by parts, observing that the derivative with respect to y of $(x - y)^{-1}$ is $(x - y)^{-2}$. If we let $a \in L^\infty$ denote the derivative of A, this gives

$$T_\varepsilon f(x) = \frac{A(x) - A(x - \varepsilon)}{\varepsilon} f(x - \varepsilon) - \frac{A(x + \varepsilon) - A(x)}{\varepsilon} f(x + \varepsilon)$$
$$+ \int_{|x-y| \ge \varepsilon} \frac{a(y) f(y)}{x - y}\, dy - \int_{|x-y| \ge \varepsilon} \frac{A(x) - A(y)}{x - y} f'(y)\, dy.$$

Since $a \in L^\infty$,

$$\frac{A(x + \varepsilon) - A(x)}{\varepsilon} = \frac{1}{\varepsilon} \int_{\varepsilon}^{x+\varepsilon} a(t)\, dt \to a(x)$$

almost everywhere. To arrive at this conclusion, it is enough to suppose that $a(x)$ is locally integrable.

The Hilbert transform H is a first generation Calderón-Zygmund operator. If $a \in L^\infty$, since f is continuous and of compact support, af belongs to $L^2(\mathbb{R})$ and the convergence of the first integral above to $H(af)$ is a consequence of the conclusion of the first example. Finally, the last integral is absolutely convergent, because $|(A(x) - A(y))/(x - y)| \le \|a\|_\infty$.

All that is necessary to conclude is to prove that the operator $Tf = \lim_{\varepsilon \downarrow 0} T_\varepsilon f$ is L^2 continuous.

Our third and last example is given by the Cauchy kernel on a Lipschitz graph. In the calculation to follow, $\log z$ denotes the branch of the logarithm which is 0 at 1 and defined on the complex plane excluding the half-line joining 0 to $-i\infty$. We suppose that Γ is the graph of a Lipschitz function $a : \mathbb{R} \to \mathbb{R}$ and put $z(x) = x + ia(x)$.

We intend to verify that $\lim_{\varepsilon \downarrow 0} \int_{|x-y| \ge \varepsilon} (z(x) - z(y))^{-1} z'(y) f(y) \, dy$ exists, for every C^1 function f of compact support. To do this, we introduce the auxiliary function $F(y) = \log(z(y) - z(x))$, which is defined on the graph Γ except at $z(x)$. Now, $F'(y) = z'(y)/(z(y) - z(x))$, so

$$\int_{|x-y| \ge \varepsilon} (z(x) - z(y))^{-1} z'(y) f(y) \, dy = -\int_{|x-y| \ge \varepsilon} F'(y) f(y) \, dy$$

$$= \log(z(x + \varepsilon) - z(x)) f(x + \varepsilon) - \log(z(x - \varepsilon) - z(x)) f(x - \varepsilon)$$

$$+ \int_{|x-y| \ge \varepsilon} \log(z(y) - z(x)) f'(y) \, dy \, .$$

For almost all $x \in \mathbb{R}$,

$$\lim_{\varepsilon \downarrow 0} \frac{z(x + \varepsilon) - z(x)}{\varepsilon} = z'(x) = \lim_{\varepsilon \downarrow 0} \frac{z(x) - z(x - \varepsilon)}{\varepsilon} \, .$$

Hence, the integrated terms converge almost everywhere to $-\pi i f(x)$. The integral $\int_{|x-y| \ge \varepsilon} \log(z(y) - z(x)) f'(y) \, dy$ is absolutely convergent.

In this last example, the kernel $K(x, y)$ defines the operator T by the bias of the distribution $\mathrm{PV}(z(x) - z(y))^{-1}$. The dense subspace V of $L^2(\mathbb{R})$ is the set of products $z'f$, where f is a C^1 function of compact support. Of course, Cotlar's inequality can only be applied once we have shown that the operators in question are L^2 continuous.

6 Calderón-Zygmund operators and singular integrals

Cotlar's inequality will let us give a complete answer to the following question: if T is a Calderón-Zygmund operator with kernel $K(x, y)$, can we construct the operator T just using the kernel K?

One answer has already been given. We let $\theta \in C^\infty(\mathbb{R}^n)$ be a radial function, equal to 1 when $|x| \geq 2$ and to 0 when $|x| \leq 1$. For each $\varepsilon > 0$, we form the truncated kernel $K_\varepsilon(x,y) = \theta((x-y)/\varepsilon)K(x,y)$ and denote by T_ε the operator defined by this kernel. Then the T_ε, $\varepsilon > 0$, form a bounded set of Calderón-Zygmund operators. Thus, there is a sequence ε_j tending to 0, such that the operators T_{ε_j} converge weakly to a Calderón-Zygmund operator L. Here $T = L + M$, where M is the operator of pointwise multiplication by a function $m(x) \in L^\infty(\mathbb{R}^n)$. The kernels $K(x,y)$ of T and L are identical and it is obvious that $K(x,y)$ can give us no information about $m(x)$.

We have to pass to a subsequence even in the simplest examples. Consider the operator $M_\gamma = (-\Delta)^{i\gamma/2}$, $\gamma \in \mathbb{R}$, which is defined, via the Fourier transform, by $(M_\gamma f)\widehat{}(\xi) = |\xi|^{i\gamma}\hat{f}(\xi)$. If $\gamma \neq 0$, the kernel $K(x,y)$ of M_γ is $c(n,\gamma)|x-y|^{-n-i\gamma}$. For a C^1 function f of compact support, $\lim_{\varepsilon_j \downarrow 0} \int_{|x-y|\geq\varepsilon_j} |x-y|^{-n-i\gamma}f(y)\,dy$ exists if and only if the limit of $\varepsilon_j^{-i\gamma}$ does. We see this by reducing the range of integration to $\varepsilon_j \leq |x-y| \leq R$, for large enough R, and then writing $f(y) = f(y) - f(x) + f(x)$.

We conclude that the operators T_{ε_j} converge weakly if and only if $\varepsilon_j^{-i\gamma}$ tends to a limit, that is, if $(\gamma/2\pi)\log\varepsilon_j$ converges modulo 1.

A more elaborate example will be useful in what follows. We consider the pseudo-differential operator $\sigma(x,D)$ defined by the symbol $\sigma(x,\xi) = (1+\xi^2)^{i\alpha(x)/2}$, where $\alpha(x) = 2 + \sin x$ and $x,\xi \in \mathbb{R}$. This symbol belongs to all the Hörmander classes $S_{1,\rho}^0$ ([68], Chapter 2) and the corresponding operator $\sigma(x,D)$ is thus a Calderón-Zygmund operator. The kernel $K(x,y)$ of $\sigma(x,D)$ is the sum of a principal term $K_0(x,y)$ and an error term $K_1(x,y)$. The principal term is given by

$$K_0(x,y) = \gamma(x)|x-y|^{-1-i\alpha(x)},$$

where

$$\gamma(x) = -2i\Gamma(1 + 2i\alpha(x))\sinh\pi\alpha(x),$$

Γ being Euler's Gamma function.

The error term $K_1(x,y)$ belongs to $L^\infty(\mathbb{R} \times \mathbb{R})$ and thus does not give rise to a singular integral.

Suppose there is a sequence $\varepsilon(j) \to 0$, with $\varepsilon(j) > 0$, such that $\lim_{j\to\infty} \int_{|x-y|\geq\varepsilon(j)} K(x,y)f(y)\,dy$ exists almost everywhere, whenever $f(x)$ is a C^1 function of compact support. The existence of this limit is equivalent to that of $\lim_{j\to\infty} e^{-i\alpha(x)\log\varepsilon(j)}$. However, the latter cannot exist almost everywhere, by the following lemma ([239], p.316).

Lemma 7. *Let λ_j be a sequence of real numbers whose absolute values tend to infinity. Then the set E of real numbers t, such that $e^{it\lambda_j}$ tends to a limit, is of measure zero.*

To see this, we let $f(t)$ be the limit of the functions $e^{it\lambda_j}$ on E. Then, for every compact interval $[a, b]$, on applying Lebesgue's dominated convergence theorem, we get

$$(6.1) \qquad \int_a^b e^{it\lambda_j} \chi_E(t)\, dt \ \to\ \int_a^b f(t)\chi_E(t)\, dt,$$

where χ_E is the characteristic function of E.

Next we observe that the left-hand side of (6.1) is the Fourier transform at $-\lambda_j$ of the characteristic function of $E \cap [a, b]$. This function is in $L^1(\mathbb{R})$, so the Riemann-Lebesgue lemma shows that the limit in (6.1) is zero. So, for every $a \in \mathbb{R}$ and every $b > a$, $\int_a^b f(t)\chi_E(t)\, dt = 0$. As a consequence, $f(t)\chi_E(t) = 0$ almost everywhere. But $|f(t)| = 1$ on E. We conclude that $|E| = 0$.

We return to the basic problem of calculating the action of a Calderón-Zygmund operator T on a function $f \in L^2(\mathbb{R}^n)$ via a generalization of PV $\int K(x, y)f(y)\, dy$, where K is the kernel corresponding to T.

Let T_ε be the operator given by $T_\varepsilon f(x) = \int_{|x-y|\geq\varepsilon} K(x, y)f(y)\, dy$, for $\varepsilon > 0$. Theorem 5 tells us that the operators T_ε form a bounded subset of $\mathcal{L}(L^2(\mathbb{R}^n), L^2(\mathbb{R}^n))$. It is therefore possible to find a subsequence ε_j, tending to 0, such that T_{ε_j} converges weakly to an operator L. Then L is a Calderón-Zygmund operator whose kernel is precisely $K(x, y)$. Finally, $T = L + M$, where M is the operator of pointwise multiplication by a function $m(x) \in L^\infty(\mathbb{R}^n)$.

We thus have the following integral representation formula, where f and g belong to $L^2(\mathbb{R}^n)$:

$$(6.2) \quad \langle Tf, g \rangle = \lim_{\varepsilon_j \downarrow 0} \int_{|x-y|\geq\varepsilon_j} K(x, y)f(y)g(x)\, dy\, dx + \int m(x)f(x)g(x)\, dx.$$

We can ask for more, namely, that $T_{\varepsilon_j}f(x)$ should converge to $Lf(x)$ almost everywhere, as ε_j tends to 0. We have already seen that we then need to replace the constants ε_j by functions $\varepsilon_j(x)$ tending to 0. These remarks lead to the following result.

Theorem 6. *Let* $T : L^2(\mathbb{R}^n) \to L^2(\mathbb{R}^n)$ *be a Calderón-Zygmund operator. Then there exist a sequence* $\varepsilon_j(x)$ *of strictly positive measurable functions on* $\mathbb{R}^n$, *such that* $\lim_{j\to\infty} \varepsilon_j(x) = 0$, *and a function* $m(x) \in L^\infty(\mathbb{R}^n)$ *such that, for every function* $f \in L^2(\mathbb{R}^n)$,

$$(6.3) \qquad Tf(x) = m(x)f(x) + \lim_{j\to\infty} \int_{|x-y|\geq\varepsilon_j(x)} K(x, y)f(y)\, dy,$$

for almost all $x \in \mathbb{R}^n$.

The right-hand side of (6.3) also converges to Tf *in* L^2 *norm, for* f *belonging to* $L^2(\mathbb{R}^n)$.

As we have seen, Theorem 6 is the best possible result. It is basically a reformulation of Cotlar's theorem.

To prove Theorem 6, we use the sequence $m_j(x)$, $j \in \mathbb{N}$, of continuous functions on $\mathbb{R}^n$, defined by $m_j(x) = \int_{(j+1)^{-1} \leq |x-y| \leq 1} K(x,y)\, dy$.

If T is a convolution operator, the $m_j(x)$ are constant functions and the sequence so formed is bounded, by Cotlar's theorem.

Theorem 6 follows quite easily from the following lemma.

Lemma 8. *There exists a sequence* $j_q : \mathbb{R}^n \to \mathbb{N}$, $q \in \mathbb{N}$, *of measurable functions such that* $\lim_{q \to \infty} j_q(x) = \infty$, *for all* $x \in \mathbb{R}^n$, *and such that, restricted to any ball* B, *centre the origin, the function* $s(x) = \lim_{q \to \infty} m_{j_q(x)}(x)$ *belongs to* $L^2(B)$. *In particular,* $s(x)$ *exists almost everywhere and, where the limit exists,* $|m_{j_q(x)} - s(x)| < \sqrt{2}q^{-1}$.

Let us assume the lemma, for the moment, and prove Theorem 6.

We put $\varepsilon_q(x) = (j_q(x)+1)^{-1}$ and begin by showing that the sequence

$$(6.4) \qquad g_q(x) = \int_{|x-y| \geq \varepsilon_q(x)} K(x,y)f(y)\, dy$$

converges, almost everywhere and in $L^2(\mathbb{R}^n)$, to a function g, when f is a C^1 function of compact support. The operator L will be defined by $g = Lf$ and $T - L$ will be the operator of pointwise multiplication by $m(x) \in L^\infty(\mathbb{R}^n)$.

To prove convergence, we write

$$g_q(x) = f(x) \int_{\varepsilon_q(x) \leq |x-y| \leq 1} K(x,y)\, dy$$

$$+ \int_{\varepsilon_q(x) \leq |x-y| \leq 1} K(x,y)(f(y) - f(x))\, dy$$

$$+ \int_{|x-y| \geq 1} K(x,y)f(y)\, dy.$$

The first integral is exactly $f(x)m_{j_q(x)}(x)$, whose convergence, pointwise almost everywhere and in $L^2(\mathbb{R}^n)$, is guaranteed by Lemma 8. The second is absolutely convergent, because f is a C^1 function. The third converges because f has compact support.

The passage to arbitrary functions $f \in L^2(\mathbb{R}^n)$ is done by the methods of section 5.

We still have to prove Lemma 8.

Let us first show that $\sup_{j \geq 1} |m_j(x)| < \infty$ almost everywhere.

We let $R \geq 1$ be an arbitrary integer and restrict to $|x| \leq R$. The union, as R varies, of the exceptional sets of measure 0, where $|x| \leq R$ and $\sup_{j \geq 1} |m_j(x)| = \infty$, will again be a set of measure 0.

Let $\chi(y)$ be the characteristic function of the ball $|y| \leq R + 1$. Then

$m_j(x) = T_{1/(j+1)}\chi(x) - r(x)$, where $r(x) = \int_E K(x,y)\,dy$, with $E = \{y : |x - y| \geq 1$ and $|y| \leq R + 1\}$. Hence, $r(x)$ is bounded for fixed R. An application of Cotlar's theorem now gives $\sup_{j \geq 1} |m_j(x)| < \infty$, almost everywhere.

The same theorem, together with Theorem 4, shows that $T_*\chi(x) \in L^2(\mathbb{R}^n)$. Now $\sup_{j \geq 1} |m_j(x)|$ differs from $T_*\chi(x)$ by a bounded function on the ball of radius R, centre the origin. We conclude that $\sup_{j \geq 1} |m_j(x)|$, restricted to that ball, belongs to $L^2(\mathbb{R}^n)$.

Having got to this point, we can completely forget how the functions $m_j(x)$ were defined and apply the following lemma.

Lemma 9. *Let $m_j(x)$, $j \in \mathbb{N}$, be a sequence of real-valued or complex-valued, measurable functions on $\mathbb{R}^n$ such that $\sup_{j \geq 1} |m_j(x)|$ is finite almost everywhere. Then there exists a sequence $j_q : \mathbb{R}^n \to \mathbb{N}$, $q \in \mathbb{N}$, of measurable functions such that $\lim_{q \to \infty} j_q(x) = \infty$, for all $x \in \mathbb{R}^n$, and such that $\lim_{q \to \infty} m_{j_q(x)}(x)$ exists almost everywhere.*

To show this, we first consider the real part of the $m_j(x)$, which will lead to a first subsequence. Then we extract a second subsequence from the first, in order to take account of the imaginary part.

If the $m_j(x)$ are real, we put $m_j^*(x) = \sup_{k \geq 0} m_{j+k}(x)$ and we let $j_q(x)$ denote the smallest integer $j \geq q$ such that $m_j(x) \geq \limsup m_k(x) - 1/q$. Then $m_q^*(x) \geq m_{j_q}(x) \geq \limsup m_k(x) - 1/q$, which ensures that the functions $m_{j_q(x)}(x)$ converge to $\limsup m_k(x)$.

7 A more detailed version of Cotlar's inequality

We devote this section to an improvement of Theorem 5, which will be used in the proof of the "good λ" inequalities of section 8.

To do this, we need a variant of the Hardy-Littlewood maximal function, defined, for $0 < \delta \leq 1$, by

$$M_\delta f(x) = \sup_{B \ni x} \left(\frac{1}{|B|} \int_B |f(y)|^\delta \, dy \right)^{1/\delta}.$$

With this notation, we have

Theorem 7. *For $0 < \gamma \leq 1$ and $0 < \delta < 1$, there exists a constant $C(n, \gamma, \delta)$ such that, if T is a Calderón-Zygmund operator for which (2.1), (2.2), (2.3) and (2.4) are satisfied with $C_0 = C_1 = C_2 = C_3 = 1$, then, for all $x \in \mathbb{R}^n$,*

(7.1) $T_*f(x) \leq 3M_\delta(Tf)(x) + C(n, \gamma, \delta)f^*(x).$

The proof of Theorem 7 conforms to the same general pattern as that of Theorem 5, but the "weak L^1 norm" estimate is improved by the following inequality due to Kolmogorov.

Lemma 10. *Let $f : \mathbb{R}^n \to \mathbb{C}$ be a weak L^1 function. Put*

(7.2)
$$\theta = \sup_{\lambda > 0} \lambda |\{x : |f(x)| > \lambda\}| .$$

Then, for each Borel set $E \subset \mathbb{R}^n$, of finite measure $|E|$,

(7.3)
$$\int_E |f|^\delta \, dx \le C(n, \delta) |E|^{1-\delta} \theta^\delta .$$

Let us first prove (7.1). We follow the proof of Theorem 5 as far as (5.12) and raise that inequality to the power δ, then take the mean on the ball B (of radius $\varepsilon/2$, centre 0.) This gives

(7.4) $|Tf_2(0)| \le 3[M_\delta(Tf)](0) + 3C\varepsilon^{-n/\delta} \left(\int_B |Tf_1|^\delta \right)^{1/\delta} + C'f^\star(0) .$

We estimate the integral on the right-hand side of (7.4), using Kolmogorov's inequality and the continuity of $T : L^1(\mathbb{R}^n) \to \text{weak-}L^1$. This gives

$$\int_B |Tf_1|^\delta \, dx \le C(n, \delta)\varepsilon^{n-n\delta}\|f_1\|_1^\delta \le C'(n, \delta)\varepsilon^n (f^\star(0))^\delta ,$$

which concludes the proof.

To prove Kolmogorov's inequality, we can clearly restrict to the case $\theta = 1$. We set $E_k = \{x \in \mathbb{R}^n : |f(x)| > 2^k\}$, $k \in \mathbb{Z}$, and let k_0 denote the largest integer k such that $2^{-k} \ge |E|$.

For $k < k_0$, we use $|E|$ as an upper bound for $|E_k|$ and use 2^{-k} if $k \ge k_0$ (the second case is where we use $\theta = 1$). Finally

$$\int_E |f|^\delta \, dx \le \sum_{-\infty}^{\infty} 2^{(k+1)\delta}|E_k|$$

$$\le 2^\delta |E| \sum_{-\infty}^{k_0-1} 2^{k\delta} + 2^\delta \sum_{k_0}^{\infty} 2^{-k(1-\delta)} \le C_\delta 2^{-k_0(1-\delta)} .$$

Now for our first application of the more detailed form of Cotlar's inequality.

Theorem 8. *Let T be a Calderón-Zygmund operator and $T_\star$ the corresponding maximal operator. Then, for each function $f \in L^1(\mathbb{R}^n)$, $T_\star(f)$ belongs to weak L^1.*

As before, we first suppose $f \in L^1 \cap L^2$ and then extend to general $f \in L^1$ by the denseness argument used in the proof of Theorem 1. We suppose, as in Theorem 7, that $C_0 = C_1 = C_2 = C_3 = 1$ and $\|f\|_1 = 1$. We then use (7.1) and the classical Hardy-Littlewood theorem (if $f \in L^1$

then $f^\star$ belongs to weak L^1). So it is enough to verify that $M_\delta(Tf)$ is in weak L^1, for $0 < \delta < 1$.

To do this, we observe that $M_\delta(Tf) \in L^2$, if $f \in L^2$ and $0 < \delta < 1$. The set $E = \{x \in \mathbb{R}^n : M_\delta(Tf) > \lambda\}$ thus has finite measure. For the moment, let us assume the following lemma.

Lemma 11. *Let g be a function in $L^1(\mathbb{R}^n)$, let $\lambda > 0$ and put $E = \{x \in \mathbb{R}^n : g^\star(x) > \lambda\}$. Then*

$$(7.5) \qquad |E| \le C_n \lambda^{-1} \int_E |g(x)| \, dx$$

where C_n is a constant which depends only on the dimension.

Inequality (7.5) is a more precise form of the Hardy-Littlewood theorem, in that the right-hand side of (7.5) uses $\int_E |g(x)| \, dx$ instead of $\|g\|_1$. In our situation,

$$|E| = |\{x : (|Tf|^\delta)^\star > \lambda^\delta\}| \le C_n \lambda^{-\delta} \int_E |Tf|^\delta \, dx \,.$$

Once again, we use Kolmogorov's inequality and get $\int_E |Tf|^\delta \, dx \le C|E|^{1-\delta}$ (we use the assumptions that Tf belongs to weak L^1 and that $\|f\|_1 = C_0 = C_1 = C_2 = 1$).

Thus, $|E| \le C' \lambda^{-\delta} |E|^{1-\delta}$ and we can conclude the proof, since $|E| < \infty$.

To prove Lemma 11, we put $h(x) = g(x)$, if $x \in E$, and $h(x) = 0$ otherwise. This gives $\{x : g^\star(x) > \lambda\} = \{x : h^\star(x) > \lambda\}$. Indeed, it is obvious that $g^\star(x) \ge h^\star(x)$ and thus $h^\star(x) > \lambda$ implies $g^\star(x) > \lambda$. To show the converse, we let B denote a ball containing x such that $\int_B |g(y)| \, dy > \lambda|B|$. For all $x' \in B$, we have $g^\star(x') > \lambda$, so B is contained in E. Thus $h(x) = g(x)$ on B, which gives $h^\star(x) > \lambda$. Finally,

$$|E| = |\{x : h^\star(x) > \lambda\}| \le C_n \lambda^{-1} \|h\|_1 = C_n \lambda^{-1} \int_E |g(x)| \, dx \,.$$

Corollary. *Let T be a Calderón-Zygmund operator. Suppose that there exists a dense subspace V of $L^1(\mathbb{R}^n)$ such that, for each function $f \in V$,*

$$(7.6) \qquad \lim_{\varepsilon \downarrow 0} \int_{|x-y| \ge \varepsilon} K(x,y) f(y) \, dy = Tf(x) \,,$$

for almost all $x \in \mathbb{R}^n$.

Then (7.6) is also satisfied, for almost all $x \in \mathbb{R}^n$, if f is an arbitrary function in $L^1(\mathbb{R}^n)$.

The proof is exactly the same as that given in section 5.

8 The good λ inequalities and the Muckenhoupt weights

A further remarkable consequence of Cotlar's theorem is the possibility of "controlling" the maximal operator $T_* f$ "in measure" by the Hardy-Littlewood maximal function f^*, when T is a Calderón-Zygmund operator.

In order to make the theorem more useful still, we shall change the ambient measure dx to $\omega(x)\, dx$, where $\omega(x) > 0$ is a Borel-measurable function satisfying Muckenhoupt's A_∞ condition.

Definition 4. *Let $\omega : \mathbb{R}^n \to (0, \infty)$ be a Borel-measurable function which is locally integrable. For every Borel set $E \subset \mathbb{R}^n$, we put $\omega(E) = \int_E \omega(x)\, dx$. The function ω is said to satisfy Muckenhoupt's A_∞ condition if there exist an exponent $\delta \in (0, 1]$ and a constant C such that, for every cube $Q \subset \mathbb{R}^n$ and every Borel set $E \subset Q$,*

$$(8.1) \qquad \frac{\omega(E)}{\omega(Q)} \le C \left(\frac{|E|}{|Q|} \right)^\delta .$$

The next result is the Calderón-Zygmund operator version of the well-known "good λ" inequalities introduced by Burkholder and used by Burkholder and Gundy in the context of martingales ([31], [32]).

Theorem 9. *Let T be a Calderón-Zygmund operator and let ω be an A_∞ weight. Then there exist an exponent $\delta > 0$ (the same as in (8.1)), a constant $C > 0$, and a number $\gamma_0 > 0$ such that the following property holds. For every continuous function f of compact support on $\mathbb{R}^n$, every $\lambda > 0$ and every $\gamma \in (0, \gamma_0)$,*

$$(8.2) \quad \omega\{x \in \mathbb{R}^n : T_* f(x) > 2\lambda \text{ and } f^*(x) \le \gamma\lambda\}$$
$$\le C\gamma^\delta \omega\{x \in \mathbb{R}^n : T_* f(x) > \lambda\} .$$

We shall see what conclusions should be drawn from (8.2). But let us first establish this "good λ" inequality.

We first remark that $T_* f(x) = \sup_{\varepsilon > 0} |T_\varepsilon f(x)|$ is a lower semi-continuous function. The set $\Omega = \{x \in \mathbb{R}^n : T_* f(x) > \lambda\}$ is thus open. Since $T_* f(x) = O(|x|^{-n})$, as $|x| \to \infty$ (due to the conditions we have imposed on f), Ω is a bounded open set. Because of this, up to a set of measure 0, Ω is a disjoint union of dyadic cubes $Q_j \subset \Omega$, which are maximal with respect to the inclusion relation. We let $\tilde{Q}_j$ denote the "parent" of Q_j, that is, the dyadic cube containing Q_j with double the diameter. Then $\tilde{Q}_j$ is not contained in Ω and there exists a point $\alpha_j \in \tilde{Q}_j$ outside Ω.

We put $|x| = \sup(|x_1|, \dots, |x_n|)$, $x \in \mathbb{R}^n$, so that Q_j is defined by $|x - \beta_j| \le d_j/2$, where β_j is the centre of Q_j. Hence, $T_* f(\alpha_j) \le \lambda$ and $|\alpha_j - \beta_j| \le 2d_j$.

We now partition the set of Q_j into two classes. We let $J \subset \mathbb{N}$ be the set of indices j such that Q_j contains a point ξ_j satisfying $f^*(\xi_j) \leq \gamma\lambda$. For j not in J, the set E, defined by $T_*f(x) > 2\lambda$ and $f^*(x) \leq \gamma\lambda$, does not intersect Q_j. Since E is a subset of Ω, we have $E \subset \bigcup Q_j$ and, hence, $\omega(E) = \sum_{j \in J} \omega(E \cap Q_j)$.

For each $j \in J$, let E_j be the set of $x \in Q_j$ such that $T_*f(x) > 2\lambda$. We shall show that, for every $j \in J$,

$$(8.3) \qquad |E_j| \leq C\gamma|Q_j| \qquad \text{if } 0 < \gamma < \gamma_0\,.$$

Then, on applying (8.1), we get

$$(8.4) \qquad \omega(E_j) \leq C'\gamma^\delta \omega(Q_j)$$

and the proof concludes by summing the inequalities over $j \in J$.

To prove (8.3), we let Q_j^* be the cube with the same centre as Q_j, but with quadrupled diameter. Then we split f as $f_1 + f_2$, where f_1 is the product of f by the characteristic function of Q_j^* and f_2 is zero on Q_j^*.

By the same calculations as those of Lemma 6, we check that, for all $x \in Q_j$,

$$(8.5) \qquad |T_*f_2(x) - T_*f_2(\alpha_j)| \leq Cf^*(\xi_j) \leq C\gamma\lambda\,.$$

Similar considerations give

$$(8.6) \qquad T_*f_2(\alpha_j) \leq T_*f(\alpha_j) + Cf^*(\xi_j) \leq \lambda + C\gamma\lambda\,.$$

So, if $x \in Q_j$, we have $T_*f_2(x) \leq \lambda + 2C\gamma\lambda$. The set E_j is thus a subset of $E_j' = \{x \in Q_j : T_*f_1(x) > \lambda(1 - 2C\gamma)\}$. We now set $\gamma_0 = 1/4C$, so that if $0 < \gamma < \gamma_0$, then $1 - 4C\gamma > 0$. We then use Theorem 8 to get

$$|E_j| \leq \frac{C'}{\lambda(1 - 2C\gamma)}\|f_1\|_1 \leq C''\gamma|Q_j|\,.$$

So we have proved Theorem 9. We can now provide two corollaries.

Corollary 1. *Let T be a Calderón-Zygmund operator and let ω be an A_∞ weight. Then, for $0 < p < \infty$, there exists a constant C such that, for every locally integrable function $f(x)$,*

$$(8.7) \qquad \int (T_*f)^p \omega\, dx \leq C \int (f^*)^p \omega\, dx\,.$$

If $0 < p \leq 1$, we cannot, in general, replace the right-hand side of (8.7) by $\int |f|^p \omega\, dx$. For example, if $p = 1$ and $\omega(x) = 1$, the consequent inequality would lead, in particular, to the continuity of the operator T on the space L^1, which is not the case.

On the other hand, if $p > 1$, it is possible, for certain weights, to get the more precise inequality

$$(8.8) \qquad \int (T_*f)^p \omega\, dx \leq C \int |f|^p \omega\, dx\,.$$

For $1 < p < \infty$, we define the class $A_p \subset A_\infty$ as the set of positive, locally integrable functions ω on $\mathbb{R}^n$ such that there is a constant C for which, on each cube $Q \subset \mathbb{R}^n$, we have

$$(8.9) \qquad \left(\frac{1}{|Q|}\int_Q \omega(x)\,dx\right)\left(\frac{1}{|Q|}\int_Q (\omega(x))^{-1/(p-1)}\,dx\right)^{p-1} \le C.$$

Muckenhoupt has shown that this condition is equivalent to the continuity on $L^p(\mathbb{R}^n, \omega\,dx)$ of the sublinear operator $f \to f^*$. We refer the reader to the excellent book by J. Garcia-Cuerva and J. L. Rubio de Francia for the proof of this statement ([115], Theorem 2.8, p. 400).

In this situation, we have

Corollary 2. *Let $1 < p < \infty$ and let $\omega \in A_p$. Then every Calderón-Zygmund operator T can be extended to a continuous linear operator from $L^p(\mathbb{R}^n, \omega\,dx)$ to itself.*

We shall give the details of the calculations which establish the inequality (8.7). We begin by proving the identity

$$(8.10) \qquad \int |f|^p\,d\mu = p\int_0^\infty \mu\{x \in \mathbb{R}^n : |f(x)| > \lambda\}\lambda^{p-1}\,d\lambda,$$

where $p > 0$ and μ is an arbitrary positive regular Borel measure. To verify (8.10), we introduce the auxiliary function $F(x,\lambda) = \lambda^{p-1}$, when $|f(x)| > \lambda$, and $F(x,\lambda) = 0$, otherwise. We then apply Fubini's theorem to calculate the double integral $\int_{\mathbb{R}^n}\int_0^\infty F(x,\lambda)\,d\lambda\,d\mu(x)$.

This lets us use the following lemma.

Lemma 12. *Let μ be a positive regular Borel measure, let $p > 0$ and let u, v be Borel-measurable functions on $\mathbb{R}^n$ such that $\|u\|_p = (\int |u|^p\,d\mu)^{1/p}$ and $\|v\|_p$ are finite.*

Suppose that $u(x)$ and $v(x)$ are related by the following "good λ" inequality: there exist an exponent $\delta > 0$, a real number $\gamma_0 > 0$ and a constant C_0, such that, for every $0 < \gamma < \gamma_0$ and every $\lambda > 0$,

$$(8.11) \quad \mu\{x \in \mathbb{R}^n : |u(x)| > 2\lambda \text{ and } |v(x)| \le \gamma\lambda\}$$
$$\le C_0\gamma^\delta\mu\{x \in \mathbb{R}^n : |u(x)| > \lambda\}.$$

Then, if $C_0\gamma^\delta < 2^{-p}$, it follows that

$$(8.12) \qquad \|u\|_p \le (2^{-p} - C_0\gamma^\delta)^{-1/p}\gamma^{-1}\|v\|_p.$$

Indeed, from (8.11) we obtain the inequality

$$(8.13) \qquad \mu\{|u(x)| > 2\lambda\} \le \mu\{|v(x)| > \gamma\lambda\} + C_0\gamma^\delta\mu\{|u(x)| > \lambda\}.$$

We multiply both sides of (8.13) by λ^{p-1} and integrate with respect to λ, using (8.10). This gives

$$2^{-p}\|u\|_p^p \le \gamma^{-p}\|v\|_p^p + C_0\gamma^\delta\|u\|_p^p.$$

For $2^{-p} > C_0 \gamma^\delta$ and $\|u\|_p < \infty$, the conclusion of the lemma follows.

In the application to Calderón-Zygmund operators, we suppose that f is continuous, of compact support and such that $T_* f(x) = O(|x|^{-n})$ as $|x| \to \infty$. If $1 < p < \infty$, then $T_* f \in L^p(\mathbb{R}^n, \omega\, dx)$, when $\omega \in A_p$. If $0 < p \le 1$, we suppose that $\int_{|x| \ge 1} |x|^{-np} \omega(x)\, dx < \infty$ and that ensures that $\int (f^*)^p \omega\, dx < \infty$ and $\int_{|x| \ge R} (T_* f)^p \omega\, dx < \infty$, for sufficiently large R. To deal with $\int_{|x| \le R} (T_* f)^p \omega\, dx$, it is convenient to use the following lemma.

Lemma 13. *If ω is an A_p weight, then, for every $m \ge 1$, $\omega_m(x) = \inf(m, \omega(x))$ satisfies (8.9), with a constant C which does not depend on m.*

Let us assume this result. We still suppose that f is continuous and of compact support, but we replace ω by ω_m. Then $\int_{|x| \le R} (T_* f)^p \omega_m\, dx$ is finite: indeed, we can bound ω_m above by m and

$$\int_{|x| \le R} (T_* f)^p\, dx \le C R^{n(1-(p/2))} \int_{|x| \le R} (T_* f)^2\, dx$$

$$\le C' R^{n(1-(p/2))} \|f\|_2^2 .$$

Lemma 12 and Theorem 9 let us write (8.7) with ω_m instead of ω. After this, it is enough to pass to the limit in order to get (8.7) without having to impose the condition that $\int (T_* f)^p \omega\, dx < \infty$.

Returning to the lemma, we write $\omega(x) = e^{\beta(x)}$ and, for each ball B, we let β_B denote the mean of β on B. The A_p condition can thus be written in the form

$$(8.14) \quad \left(\frac{1}{|B|} \int_B e^{\beta(x) - \beta_B}\, dx \right) \left(\frac{1}{|B|} \int_B e^{-(\beta(x) - \beta_B)/(p-1)}\, dx \right)^{p-1} \le C .$$

We then note that, if $f(x)$ is a real-valued funtion on B,

$$\frac{1}{|B|} \int_B e^{f(x)}\, dx \ge \exp\left(\frac{1}{|B|} \int_B f(x)\, dx \right)$$

by the convexity of the exponential function.

It follows that each of the two terms forming the product in (8.14) is not less than 1. Each is thus bounded and the condition $\omega \in A_p$ can be written as the two simultaneous conditions

$$\frac{1}{|B|} \int_B e^{\beta(x) - \beta_B}\, dx \le C'$$

and

$$\frac{1}{|B|} \int_B e^{-(\beta(x) - \beta_B)/(p-1)}\, dx \le C' .$$

We can carry out two final manoeuvres with these estimates. Firstly, we replace $\beta(x) - \beta_b$ by $(\beta(x) - \beta_b)^+$ in the first of them and by $(\beta(x) - \beta_b)^-$

in the second, because the only values we have to take into account, on integrating, are those greater than 1. Then, we may replace β_B (the mean of β on B) by a real constant $\lambda(B)$, whose relationship with $\beta(x)$ we do not know, a priori.

Indeed, if we simultaneously have

(8.15) $$\frac{1}{|B|} \int_B e^{(\beta(x)-\lambda(B))^+} \, dx \le C'$$

and

(8.16) $$\frac{1}{|B|} \int_B e^{(-(\beta(x)-\lambda(B))/(p-1))^+} \, dx \le C' \,,$$

then, on adding the inequalities, we get

$$\frac{1}{|B|} \int_B \cosh(\tau(\beta(x) - \lambda(B))) \, dx \le C' \,,$$

where $\tau = \inf(1, (p-1)^{-1})$. It follows that $\frac{1}{|B|} \int_B (\beta(x)-\lambda(B))^2 \, dx \le C''$ and, finally, that $|\lambda(B) - \beta_B| \le C_1$. This allows us to work back, replacing $\lambda(B)$ throughout by β_B, which only changes the constants.

Now Lemma 13 becomes obvious. We simultaneously replace $\beta(x)$ by $\beta_m(x) = \inf(\beta(x), \log m)$ and $\lambda(B)$ by $\lambda_m(B) = \inf(\lambda(B), \log m)$. Then $(\beta_m(x) - \lambda_m(B))^+ \le (\beta(x) - \lambda(B))^+$ and the same is true for $(\beta_m(x) - \lambda_m(B))^-$.

A final remark is necessary. Muckenhoupt's class A_∞ is the union of all the A_p's. If $\omega \in A_\infty$, then $\log \omega$ will belong to the space BMO of John and Nirenberg. We see this straight away by returning to the conditions (8.15) and (8.16). Conversely, if $\beta(x)$ belongs to BMO, there exists a sufficiently small $\varepsilon > 0$ so that $\omega(x) = e^{\varepsilon \beta(x)}$ is a weight in A_∞.

9 Notes and additional remarks

The theory which we have described is only one aspect of the research in progress on singular integrals. After the pioneering work of Nagel, Rivière and Wainger ([195]), Stein and Phong systematically studied the operators $T : L^2(\mathbb{R}^n) \to L^2(\mathbb{R}^n)$ whose kernels have stronger singularities, in a certain sense, than those described by (2.2) and (2.3). Their starting point was the Hilbert transform along a parabola. This is an operator $T : L^2(\mathbb{R}^n) \to L^2(\mathbb{R}^n)$ defined by

$$Tf(x,y) = \frac{1}{\pi} \operatorname{PV} \int_{-\infty}^{\infty} f(x - t, y - t^2) \, \frac{dt}{t} \,.$$

In other words, $Tf = f \star S$, where S is the distribution which is the image of $\pi^{-1} \operatorname{PV} t^{-1}$ under the mapping $t \mapsto (t, t^2)$.

The most recent results of Phong and Stein are about a generalization of the "Hilbert transform along a parabola", a generalization in which the singularities of the kernel $S(x, y)$ lie on a sub-manifold V_x containing x and whose translations or whose geometric form satisfies a certain curvature condition ([203]). These extensions of the idea of Calderón-Zygmund operator come close to Fourier integral operators.

Finally, a very active branch of research concerns the case of Banach-space valued functions (the kernel remaining scalar-valued). All of the theory of this chapter remains unchanged, but the difficulty that arises is that of proving the basic L^2 estimate. Even in the case of the Hilbert transform, this estimate is not automatic. J. Bourgain put the coping stone on work of D. Burkholder's: these authors characterized those Banach spaces B for which the Hilbert transform defines a continuous operator on $L^2(\mathbb{R}; B)$ (square-summable functions on $\mathbb{R}$ taking values in B). Those Banach spaces are the UMD spaces ([26], [30] and [33]).

The operators we have studied in this chapter are not convolution operators, which suggests that we can dispense with the group structure of the underlying space $\mathbb{R}^n$. This point of view has been developed systematically in [75]. The theory, extended to homogeneous spaces, applies, for example, to nilpotent Lie groups.

The main difficulty, which will be resolved in the next chapter, is to replace the (non-existing) Fourier transform by some other tool in order to give the L^2 estimate on which the theory is based.

A similar difficulty occurs in the work of Bourgain and Burkholder, where the Calderón-Zygmund operators are convolution operators, but the functions take their values in a Banach space. If the Banach space in question is not a Hilbert space, we do not have a Plancherel theorem at our command to give the basic L^2 estimate.

8

David and Journé's $T(1)$ theorem

1 Introduction

This chapter continues and completes the preceding one, correcting a weakness to be found there. The definition of Calderón-Zygmund operators that we have given includes four conditions, of which the first three can be verified directly. These involve the restriction $K(x, y)$ of the distribution-kernel $S(x, y)$ of the operator T to the open set $y \neq x$ in $\mathbb{R}^n \times \mathbb{R}^n$, and are

(1.1) $$|K(x, y)| \leq C_0 |x - y|^{-n},$$

(1.2) $$|K(x', y) - K(x, y)| \leq C_1 |x' - x|^\gamma |x - y|^{-n-\gamma},$$

if $|x' - x| \leq |x - y|/2$, and

(1.3) $$|K(x, y') - K(x, y)| \leq C_1 |y' - y|^\gamma |x - y|^{-n-\gamma},$$

if $|y' - y| \leq |x - y|/2$.

The other condition is the continuity of the operator T on $L^2(\mathbb{R}^n)$. The verification of this condition proves to be very difficult as soon as we go beyond the special case of convolution operators. Is it possible, with the help of (1.1), (1.2) and (1.3), to find an equivalent form of the L^2 continuity which is easier to use? G. David and J.L. Journé's "$T(1)$" theorem gives a preliminary response to this question. It can be applied very easily in certain cases, but is of no use in others, even though the condition it gives is necessary and sufficient.

In the special case where $K(y, x) = -K(x, y)$ and where the kernel of T is the distribution PV $K(x, y)$, the criterion of David and Journé becomes $T(1) \in$ BMO (hence the name of the theorem). The action of T on the function identically 1 is defined by $T(1) = \lim_{\varepsilon \downarrow 0}(T(\phi_\varepsilon) - c_\varepsilon)$, where $\phi \in \mathcal{D}(\mathbb{R}^n)$, $\phi(0) = 1$, $\phi_\varepsilon(x) = \phi(\varepsilon x)$, and where the c_ε are constants, which are chosen so that the limit exists in the sense of distributions.

The $T(1)$ criterion gives the continuity of the "Calderón commutators" defined by the antisymmetric kernels $(A(x) - A(y))^k (x - y)^{-k-1}$, where $x \in \mathbb{R}$, $y \in \mathbb{R}$ and $A : \mathbb{R} \to \mathbb{C}$ is a Lipschitz function. The criterion, however, does not apply directly to the L^2 continuity of the Cauchy kernel on a Lipschitz curve, because, in this case, we do not know how to identify $T(1)$ with a known BMO function. This is why we shall give an account of the "$T(b)$ theorem", an enhancement of the $T(1)$ theorem, which has the advantage of working for the case of the Cauchy kernel on Lipschitz curves.

At present, there is no universal criterion. For example, we do not know whether, for every Lipschitz function $A : \mathbb{R} \to \mathbb{R}$ and every Lipschitz function $F : \mathbb{R} \to \mathbb{C}$, the antisymmetric kernel $K(x, y) = (x-y)^{-1}F((A(x)-A(y))/(x-y))$ defines a bounded operator on $L^2(\mathbb{R}^n)$. None of the known criteria apply in this case. If, conversely, F is a C^2 function, the operator in question is bounded ([99]), even though this cannot be established directly from either of the $T(1)$ or $T(b)$ theorems.

In this chapter, we give two proofs of the $T(1)$ theorem. In both cases, we first examine the operators T such that $T(1) = 0$ (and $T^*(1) = 0$ in the non-antisymmetric case). The L^2 continuity of these operators is established by calculating their matrices with respect to the orthonormal wavelet basis. These matrices are characterized by the rate of decrease of the entries as the distance from the principal diagonal increases; the continuity of such matrices on the space of square-summable sequences is established using Schur's lemma.

We get the general case by "correcting" the operator T being investigated. This correction is made by means of "pseudo-products", that is, continuous linear actions of BMO on L^2: the usual product does not work, but a very simple variant, based on the expansion as a wavelet series, gives the required action. These "pseudo-products" gave rise to J.M. Bony's "paraproducts" which were the source of the "paradifferential operators", of which we shall speak again in Chapter 16.

The second proof uses the well-known near-orthogonality lemma of Cotlar and Stein and is based on the same reduction to the special case where $T(1) = T^*(1) = 0$, a case which we can always reach using corrections by suitable pseudo-products.

On the way, we shall construct certain noteworthy Banach algebras of Calderón-Zygmund operators, the $\mathcal{M}_\gamma$ algebras. The operators T in $\mathcal{M}_\gamma$, and their adjoints, are continuous on homogeneous Hölder spaces, as well as on Besov spaces, as long as the index of regularity is not greater than γ.

These continuity properties of Calderón-Zygmund operators as operators on spaces other than L^2 will be the subject of Chapter 10.

Finally, the general results of this chapter will enable us to construct various noteworthy examples of Calderón-Zygmund operators. This construction will be the subject of Chapter 9.

2 Statement of the $T(1)$ theorem

We shall say that an operator T is *weakly defined* if T is a continuous linear operator from $\mathcal{D}(\mathbb{R}^n) \to \mathcal{D}'(\mathbb{R}^n)$, where $\mathcal{D}(\mathbb{R}^n)$ is the topological vector space of infinitely differentiable functions of compact support and $\mathcal{D}'(\mathbb{R}^n)$ is the dual space of distributions. According to a well-known theorem of Schwartz, such an operator is always defined by a unique distribution $S \in \mathcal{D}'(\mathbb{R}^n \times \mathbb{R}^n)$, and the relation between T and S is given by

$$(2.1) \qquad \langle Tf, g \rangle = \langle S, g \otimes f \rangle,$$

where f and g belong to $\mathcal{D}(\mathbb{R}^n)$ and where the left-hand side of (2.1) is the bilinear form expressing the duality between $\mathcal{D}'(\mathbb{R}^n)$ and $\mathcal{D}(\mathbb{R}^n)$. On the right-hand side of (2.1), $g \otimes f$ is the function $g(x)f(y)$ of $2n$ variables and $\langle \cdot, \cdot \rangle$ is the bilinear form giving the duality between $\mathcal{D}'(\mathbb{R}^n \times \mathbb{R}^n)$ and $\mathcal{D}(\mathbb{R}^n \times \mathbb{R}^n)$.

The distribution S is called the *distribution-kernel* of T. It determines T completely.

In many applications, the natural domain of the operators we study is a topological vector space V sandwiched between $\mathcal{D}(\mathbb{R}^n)$ and $L^2(\mathbb{R}^n)$, that is, $\mathcal{D}(\mathbb{R}^n) \subset V \subset L^2(\mathbb{R}^n)$, where the two injections are continuous and have dense ranges. We then denote the topological dual of V by V' and get $\mathcal{D}(\mathbb{R}^n) \subset V \subset L^2(\mathbb{R}^n) \subset V' \subset \mathcal{D}'(\mathbb{R}^n)$, where all the inclusions are, again, continuous and of dense range. We shall see that we can always reduce to the case where V is the topological vector space of C^1 functions of compact support. We shall call the functions in V *test* functions and the elements of V' will be the corresponding distributions.

Throughout this chapter, we shall start from an operator $T : V \to V'$ which will naturally be supposed to be continuous and of dense range. We shall try to extend T to a continuous linear operator on $L^2(\mathbb{R}^n)$, or—what amounts to the same thing—show that there is a constant C_2

such that, for every $f \in V$, $\|T(f)\|_2 \leq C_2\|f\|_2$. Such a problem is too general to receive anything other than a tautological answer and we shall clearly have to limit ourselves to the special case where the restriction of the kernel of T to $x \neq y$ satisfies (1.1), (1.2) and (1.3).

We shall, as usual, denote the bilinear form between V and V' by $\langle \cdot, \cdot \rangle$.

Let us start by introducing an obviously necessary condition for the L^2 continuity of a linear operator $T : V \to V'$.

Definition 1. *Let $q \geq 1$ be an integer. For each ball $B \subset \mathbb{R}^n$, centre x_0 and radius R, and for every function $f \in \mathcal{D}(\mathbb{R}^n)$ whose support lies in B, we put*

$$(2.2) \qquad N_q^B(f) = R^{n/2} \sum_{|\alpha| \leq q} R^{|\alpha|} \|\partial^\alpha f\|_\infty .$$

If $0 < s < 1$, we similarly define $N_s^B(f)$ by

$$(2.3) \qquad N_s^B(f) = R^{(n/2)+s} \sup_{x \neq y} \frac{|f(x) - f(y)|}{|x - y|^s} .$$

We say that a continuous linear operator $T : V \to V'$ is *weakly continuous* on $L^2(\mathbb{R}^n)$ if there exist a constant C and an integer q such that, for every ball $B \subset \mathbb{R}^n$ and every pair (f, g) of functions in V with supports in B,

$$(2.4) \qquad |\langle Tf, g \rangle| \leq C N_q^B(f) N_q^B(g) .$$

The following result shows us that, subject to a very weak condition on the kernel of T, the value of q is not important.

Theorem 0. *Let $T : V \to V'$ be a continuous linear operator. Let $K(x, y)$ be the restriction of the distribution-kernel of T to the open subset of $\mathbb{R}^n \times \mathbb{R}^n$ defined by $x \neq y$. We suppose that $|K(x, y)| \leq C_0|x - y|^{-n}$, for a certain constant C_0. If, further, T satisfies property (2.4) above, for some integer $q \geq 1$, then, for every $s > 0$, there exists a constant $C(s)$ such that*

$$(2.5) \qquad |\langle Tf, g \rangle| \leq C(s) N_s^B(f) N_s^B(g) ,$$

for every ball B and every pair (f, g) of functions in $\mathcal{D}(\mathbb{R}^n)$ with supports in B.

Once we have proved it, Theorem 0 will let us extend T as a continuous linear operator from V_s to V_s', where V_s is the space of continuous functions $f(x)$ of compact support, whose moduli of continuity $\omega(h) = \sup_{|x-y| \leq h} |f(x) - f(y)|$ satisfy $\omega(h) = o(h^s)$ as h tends to 0. In other words an "ε of regularity" is enough to define T weakly.

Let us pass to the proof of the theorem. To simplify the notation, we shall write $\beta(f,g) = \langle Tf, g \rangle$, where f and g are test functions.

In outline, the proof consists of successive improvements of (2.4). We start by assuming that, for a certain $x_0 \in \mathbb{R}^n$, $r > 0$ and $R \geq 2r$, the supports of f and g are contained in $|x - x_0| \leq r$ and $|x - x_0| \leq R$, respectively. We shall exploit the fact that r is small compared with R and verify the inequality

$$(2.6) \quad |\beta(f,g)| \leq Cr^n \log \frac{R}{r} \|f\|_\infty \|g\|_\infty$$

$$+ Cr^n \left(\sum_{|\alpha| \leq q} r^{|\alpha|} \|\partial^\alpha f\|_\infty \right) \left(\sum_{|\alpha| \leq q} R^{|\alpha|} \|\partial^\alpha g\|_\infty \right).$$

From this estimate, we shall easily get an "intermediate inequality" of type

$$(2.7) \quad |\beta(f,g)| \leq CR^n N_s^B(f) N_q^B(g),$$

where f and g now have supports in the ball B, centre x_0 and radius R.

Then we shall repeat the proof, but start from (2.7), to get (2.5).

To prove (2.6), we let B' denote the ball $|x - x_0| \leq 3r/2$ and we suppose that $\theta \in \mathcal{D}(\mathbb{R}^n)$ is a function which is identically equal to 1 on B' and is 0 when $|x - x_0| \geq 2r$. We further suppose that $\|\partial^\alpha \theta\|_\infty \leq C(\alpha) r^{-|\alpha|}$, for every $\alpha \in \mathbb{N}^n$. Putting $\eta = 1 - \theta$, we can write $\beta(f,g) = \beta(f, \theta g) + \beta(f, \eta g)$. We now use the hypothesis on $K(x,y)$ to estimate the second term, getting

$$|\beta(f, \eta g)| \leq C_0 \|f\|_\infty \|g\|_\infty \int_{E(r,R)} |x - y|^{-n} \, dx \, dy,$$

where $E(r,R) = \{|x - x_0| \leq r, \ 3r/2 \leq |y - x_0| \leq R\}$. An upper bound for the double integral is given by $C(n) r^n \log(R/r)$, and we see that we have the first term on the right-hand side of (2.6).

The second term $\beta(f, \theta g)$ is estimated directly by using (2.4) for the ball $|x - x_0| \leq 2r$ and this gives

$$|\beta(f, \theta g)| \leq Cr^n \left(\sum_{|\alpha| \leq q} r^{|\alpha|} \|\partial^\alpha f\|_\infty \right) \left(\sum_{|\alpha| \leq q} R^{|\alpha|} \|\partial^\alpha g\|_\infty \right).$$

Let us now move to the proof of the intermediate estimate (2.7). We begin by dealing with the case where B is the unit ball $|x| \leq 1$. The general case is a consequence of this one, as we shall see shortly.

Using the basis of wavelets of compact support of *Wavelets and Operators*, Chapter 3, section 8, we can expand f, if $N_s^B(f) \leq 1$, as a series $\sum_{j \geq 0} \sum_k \alpha(j,k) u_{jk}(x)$, where the functions $u_{jk}(x)$ are C^∞ functions, with supports in the balls $B(j,k) = \{x \in \mathbb{R}^n : |x - 2^{-j}k| \leq C2^{-j}\}$, satisfying

$$\|\partial^\gamma u_{jk}(x)\|_\infty \le C 2^{j|\gamma|} \qquad \text{for } |\gamma| \le q,$$

and where $|\alpha(j,k)| \le C 2^{-js}$.

By virtue of (2.6), we then have $|\beta(u_{jk}, g)| \le C 2^{-nj} \log(2 + j)$ and we just need to sum these estimates, noting that, for fixed j, the number of values of k is $O(2^{nj})$. The series converges by comparison with the convergent series $\sum_0^\infty 2^{-js} \log(2 + j)$.

For the general case of (2.7), where B is an arbitrary ball, it is enough to observe that (2.7) and the hypotheses on T are invariant under the action of the affine unitary group. More explicitly, if we put $g(x) = ax + b$, where $a > 0$ and $b \in \mathbb{R}^n$, and let $U_g f(x) = a^{-n/2} f((x - b)/a)$, we can see that the conjugates $U_g T U_g^{-1}$ of T satisfy (2.4) with the same constants as T.

As we remarked above, to establish (2.5), we now repeat the proof of (2.7) word for word, but start with the intermediate estimate (2.7). This lets us replace $N_q^B(g)$ by $N_s^B(g)$ in the right-hand side of (2.7).

We now approach the heart of the matter by defining a class of operators prefiguring the class of Calderón-Zygmund operators. Let Ω denote the subset of $\mathbb{R}^n \times \mathbb{R}^n$ defined by $y \ne x$, $x, y \in \mathbb{R}^n$.

Definition 2. *A continuous linear operator $T : V \to V'$ corresponds to a singular integral if there are an exponent $\gamma \in (0, 1]$, two constants C_0 and C_1 and a function $K : \Omega \to \mathbb{C}$, such that*

$$(2.8) \qquad |K(x,y)| \le C_0 |x - y|^{-n},$$

$$(2.9) \qquad |K(x',y) - K(x,y)| \le C_1 |x' - x|^\gamma |x - y|^{-n-\gamma},$$

if $|x' - x| \le |x - y|/2$,

$$(2.10) \qquad |K(x,y') - K(x,y)| \le C_1 |y' - y|^\gamma |x - y|^{-n-\gamma},$$

if $|y' - y| \le |x - y|/2$, and, finally,

$$(2.11) \qquad Tf(x) = \int K(x,y) f(y)\, dy,$$

for every function $f \in V$ and every x not in the support of f.

In other words, the restriction of the distribution Tf to the complement of the support of f is, in fact, a continuous function, given by the integral representation (2.11). Another way of expressing (2.11) is to say that the kernel $K(x,y)$ corresponding to T is the restriction to Ω of the distribution-kernel $S(x,y)$ of T.

This definition is different from Definition 1 of Chapter 7 in that we do not require T to be continuous on $L^2(\mathbb{R}^n)$. It is similar, however, in that, if (2.9) and (2.10) are satisfied for a particular exponent γ, they will hold for every exponent $\gamma' \in (0, \gamma)$.

We come now to the definition of $T(1)$. In fact we shall give two definitions of $T(1)$. Naturally, the difficulty is that 1 is not a test function. This will lead to $T(1)$ being not a distribution, but an equivalence class of tempered distributions, modulo the constant functions.

The first definition is based on the following, fairly obvious, remark.

Lemma 1. *Let $S \in \mathcal{D}'(\mathbb{R}^n)$ be a distribution. Suppose that there exists $R > 0$ such that the restriction of S to the open set $|x| > R$ is a continuous function and such that $S(x) = O(|x|^{-n-\gamma})$ as $|x| \to \infty$. If $\gamma > 0$, then the integral $\int_{\mathbb{R}^n} S(x)\,dx = \langle S, 1 \rangle$ converges.*

To see this and to define the integral, we write $1 = \phi_0(x) + \phi_1(x)$, where $\phi_0 \in \mathcal{D}(\mathbb{R}^n)$ and $\phi_0 = 1$ in a neighbourhood of $|x| \leq R$. Then $\langle S, 1 \rangle$ is defined by $\langle S, \phi_0 \rangle + \int S(x)\phi_1(x)\,dx$. The integral converges absolutely. It follows immediately that $\langle S, 1 \rangle$ is independent of the decomposition.

The transpose of $T : V \to V'$ is the continuous linear operator ${}^t T : V \to V'$ given by $\langle {}^t Tf, g \rangle = \langle f, Tg \rangle$, for all $f, g \in V$. The hypotheses (2.8), (2.9) and (2.10) are all invariant under transposition, the kernel corresponding to the transpose ${}^t T$ being $L(x, y) = K(y, x)$.

Now, if $f \in \mathcal{D}(\mathbb{R}^n)$ and $\int f(x)\,dx = 0$, we can define $\langle T(1), f \rangle$ by $\langle {}^t Tf, 1 \rangle$. Indeed, if the support of f is contained in $|x| \leq R$, then

$${}^t Tf(x) = \int (K(y, x) - K(0, x))f(y)\,dy = O(|x|^{-n-\gamma}) \qquad \text{for } |x| > R.$$

As a mathematical object, $T(1)$ is thus a continuous linear form on the subspace $\mathcal{D}_0 \subset \mathcal{D}(\mathbb{R}^n)$ defined by $\int f(x)\,dx = 0$. We extend $T(1)$ to a distribution $S \in \mathcal{D}'(\mathbb{R}^n)$ as follows. Let $\phi \in \mathcal{D}$ be a function of mean 1: then every $f \in \mathcal{D}$ can be written uniquely as $f = \lambda\phi + g$, where $\lambda = \int f(x)\,dx$ and $g \in \mathcal{D}_0$. We make an arbitrary choice for the value of $\langle S, \phi \rangle$ and put $\langle S, f \rangle = \lambda\langle S, \phi \rangle + \langle T(1), g \rangle$.

As a mathematical object $T(1)$ is now a distribution modulo the constants.

Another possible definition of $T(1)$ is given by the following direct approach. We start with a function $\phi \in \mathcal{D}(\mathbb{R}^n)$ which equals 1 at 0 and put $\phi_\varepsilon(x) = \phi(\varepsilon x)$, for every $\varepsilon > 0$.

We then consider the distribution $S_\varepsilon = T(\phi_\varepsilon)$.

Lemma 2. *We can "correct" the distributions S_ε by "renormalization constants" $c(\varepsilon)$ such that $\lim_{\varepsilon \downarrow 0}(S_\varepsilon - c(\varepsilon))$ exists in the sense of distributions. This limit coincides with $T(1)$.*

Indeed, if $f \in \mathcal{D}_0$, we have $\langle T(\phi_\varepsilon), f \rangle = \langle \phi_\varepsilon, {}^t Tf \rangle$. We must verify that the right-hand side converges to $\langle 1, {}^t Tf \rangle$. Putting $1 = u + v$, where $u \in \mathcal{D}$ and $u = 1$ in a neighbourhood of the support of f, we can handle

$\lim_{\varepsilon\downarrow 0}\langle\phi_\varepsilon, u\,{}^tTf\rangle$, using the definition of a distribution. The Lebesgue dominated convergence theorem then shows that $\lim_{\varepsilon\downarrow 0}\langle\phi_\varepsilon, v\,{}^tTf\rangle$ exists.

Let $w \in \mathcal{D}$ be a function of mean 1 and put $c(\varepsilon) = \langle S_\varepsilon, w\rangle$. Then, for $f \in \mathcal{D}$ and $\lambda = \int f(x)\,dx$,

$$\langle S_\varepsilon - c(\varepsilon), f\rangle = \langle S_\varepsilon, f\rangle - \lambda\langle S_\varepsilon, w\rangle = \langle S_\varepsilon, g\rangle \qquad \text{where } g = f - \lambda w.$$

Then $g \in \mathcal{D}_0$, so the required limit exists.

Having made these definitions, we are in a position to state the theorem of David and Journé.

Theorem 1. *Let $T : V \to V'$ be a continuous linear operator corresponding to a singular integral as in Definition 2. Then a necessary and sufficient condition for the extension of T as a continuous linear operator on $L^2(\mathbb{R}^n)$ is that the three following properties are all satisfied:*

(a) *$T(1)$ belongs to $\mathrm{BMO}(\mathbb{R}^n)$;*
(b) *${}^tT(1)$ belongs to $\mathrm{BMO}(\mathbb{R}^n)$;*
(c) *T is weakly continuous on $\mathbb{R}^n$.*

We first note that an equivalent form of condition (b) is $T^\star(1) \in \mathrm{BMO}$. Here, $T^\star(1)$ is the complex conjugate of ${}^tT(1)$.

Let us use a few simple examples to show that conditions (a), (b) and (c) are independent.

If T is defined by pointwise multiplication by a function $m(x) \in \mathrm{BMO}$, we clearly have $K(x,y) = 0$ and $T(1) = {}^tT(1) = m(x) \in \mathrm{BMO}$. Property (c) is only satisfied if $m(x) \in L^\infty(\mathbb{R}^n)$. Another example in the same direction is that of convolution by the distribution $S = \mathrm{fp}\,|x|^{-n}$. The kernel $K(x,y) = |x-y|^{-n}$ satisfies (2.8) to (2.10) and $T(1) = {}^tT(1) = 0$, because T is a convolution operator. But, once again, (c) is not satisfied.

An interesting example where (a) and (c) are satisfied, but (b) is not, is that of the pseudo-differential operators $\sigma(x, D)$ whose symbols $\sigma(x, \xi)$ satisfy the "forbidden" conditions

$$(2.12) \qquad |\partial_\xi^\alpha \partial_x^\beta \sigma(x,\xi)| = C(\alpha,\beta)(1 + |\xi|)^{|\beta|-|\alpha|}.$$

It is not hard to show that the corresponding kernels $K(x,y)$ satisfy

$$(2.13) \qquad |\partial_x^\alpha \partial_y^\beta K(x,y)| \leq C'(\alpha,\beta)|x - y|^{-n-|\alpha|-|\beta|}.$$

In fact, (c) is just a consequence of the hypothesis $|\sigma(x,\xi)| \leq C$. In general, however, such an operator is not bounded on $L^2(\mathbb{R}^n)$. Nevertheless, $T(1) = \sigma(x,0) \in L^\infty(\mathbb{R}^n)$. What is missing is condition (b). We refer the reader to [96] for a more detailed discussion of such examples. We also pursue these matters further in Chapter 9.

A noteworthy special case of Theorem 1 is when the kernel $S(x,y)$ of T is PV $K(x,y)$, where $K(x,y) = -K(x,y)$ and $K(x,y)$ satisfies (2.8) and (2.9). Then ${}^tT = -T$ and (c) is trivially satisfied. For such anti-symmetric kernels, the L^2 continuity of T is equivalent to the condition $T(1) \in$ BMO.

For example, consider the kernels

$$K_m(x, y) = (A(x) - A(y))^m / (x - y)^{m+1},$$

where $A : \mathbb{R} \to \mathbb{C}$ is a Lipschitz function and $m \in \mathbb{N}$. The corresponding operators T_m are the "Calderón commutators" and a simple integration by parts gives the following remarkable identity:

(2.14) $T_m(1) = T_{m-1}(A')$ for $m \geq 1$.

Assuming Theorem 1, we show how the continuity of the T_m follows. T_0 is obviously continuous, since $T_0 = \pi H$, where H is the Hilbert transform. Then we argue by induction on $m \geq 1$. If we know that T_{m-1} is a Calderón-Zygmund operator, it follows that T_{m-1} is continuous as a mapping from L^∞ to BMO (*Wavelets and Operators*, Chapter 5, section 3). But $A'(x)$ is in $L^\infty(\mathbb{R}^n)$ and thus $T_m(1) \in$ BMO. This gives the L^2 continuity of T_m and the induction proceeds.

A closer examination of this recursive proof gives the estimate

(2.15) $\|T_m(f)\|_2 \leq \pi X^m \|A'\|_\infty \|f\|_2 \,,$

where $X > 1$ is a constant whose value is not given by this method of proof.

Before approaching the proof of Theorem 1, we observe that (a), (b) and (c) are indeed necessary for the L^2 continuity of T. As far as (a) is concerned, we have already remarked that every Calderón-Zygmund operator is also continuous as a mapping from L^∞ to BMO. Of course, the constant function 1 is in L^∞. The transpose of a Calderón-Zygmund operators is still a Calderón-Zygmund operator, so (b) is satisfied. Finally, any operator which is L^2 continuous is automatically weakly continuous.

3 The wavelet proof of the $T(1)$ theorem

The $T(1)$ theorem is a statement which becomes more powerful as the exponent $\gamma > 0$ decreases. We may therefore suppose that $0 < \gamma < 1$ in what follows.

To prove Theorem 1, we must establish the L^2 continuity of T, given the conditions (2.8), (2.9), (2.10), (a), (b), and (c). As we have seen, we may suppose that V is the topological vector space of C^1 functions of compact support. We now fix a multiresolution approximation V_j, $j \in \mathbb{Z}$, of $L^2(\mathbb{R}^n)$ giving a function ϕ and wavelets ψ_λ which are C^1 and have compact supports. Recall that $\phi(x - k)$, $k \in \mathbb{Z}^n$, is an orthonormal basis of V_0 and that $\psi_\lambda(x) = 2^{nj/2}\psi_\varepsilon(2^j x - k)$, where $\varepsilon \in \{0,1\}^n \setminus \{(0,\ldots,0)\}$ and $\lambda = 2^{-j}k + 2^{-j-1}\varepsilon$. We further let $Q(\lambda)$ denote the dyadic cube defined by $2^j x - k \in [0,1)^n$: the support of $\psi_\lambda(x)$ is contained in $mQ(\lambda)$ where $m \geq 1$ is a fixed integer and $mQ(\lambda)$ is defined by $2^j x - k \in [-m/2 + 1/2, m/2 + 1/2)^n$.

The functions ψ_λ are real-valued and the idea of the proof is to estimate the absolute values $|\tau(\lambda, \lambda')|$ of the entries $\tau(\lambda, \lambda') = \langle T\psi_\lambda, \psi_{\lambda'} \rangle = \langle \psi_\lambda, {}^t T\psi_{\lambda'} \rangle$ of the matrix representing T with respect to the wavelet basis of the Hilbert space. We try to show that these entries become very small when the cubes $Q(\lambda)$ and $Q(\lambda')$ differ in either position or magnitude. To do this, we exploit the trivial remark that an integral $\int f(x)g(x)\,dx$ is very small if one of the terms of the integrand varies very slowly whilst the other is sharply localized (on a very much smaller scale than that of the variations of the former) and has mean 0.

In our particular situation, we can always suppose that $|Q(\lambda)|$ is (much) larger than $|Q(\lambda')|$, by changing T to its transpose, if necessary. If $|Q(\lambda)|$ is larger than $|Q(\lambda')|$, the wavelet ψ_λ is "flat" where $\psi_{\lambda'}$ has zero mean. To conclude, we need to know that ${}^t T\psi_{\lambda'}$ also has zero mean, so we need $T(1) = 0$.

The condition ${}^t T(1) = 0$ is needed to deal with the case $|Q(\lambda')| \geq |Q(\lambda)|$.

These considerations suggest that the matrix M of T in the wavelet basis is almost diagonal, in the sense of Schur's lemma, when $T(1) = {}^t T(1) = 0$. This is what we shall now confirm, by establishing the next result.

Proposition 1. *Let $T : V \to V'$ be a weakly continuous operator on $L^2(\mathbb{R}^n)$ which corresponds to a singular integral, as in Definition 2. Suppose, further, that $T(1) = {}^t T(1) = 0$. Let ψ_λ, $\lambda \in \Lambda$, be an orthonormal C^1 wavelet basis of compact support. Then there exists a constant C such that, for $\lambda = 2^{-j}k + 2^{-j-1}\varepsilon$ and $\lambda' = 2^{-j'}k' + 2^{-j'-1}\varepsilon'$,*

$$(3.1) \quad |\tau(\lambda, \lambda')| \leq C 2^{-|j-j'|((n/2)+\gamma)} \left(\frac{2^{-j} + 2^{-j'}}{2^{-j} + 2^{-j'} + |2^{-j}k - 2^{-j'}k'|} \right)^{n+\gamma}.$$

We can simplify the proof of (3.1) by a few preliminary remarks. First of all, everything is symmetric in λ and λ' (interchanging T and ${}^t T$, if necessary). We may therefore suppose that $j \leq j'$.

If $j = j'$, (3.1) is proved as follows. When the supports of ψ_λ and $\psi_{\lambda'}$ intersect, (3.1) is an immediate consequence of the weak continuity of

T. If the supports are disjoint, we have

$$\tau(\lambda, \lambda') = \int\int K(x,y)\psi_{\lambda'}(x)\psi_{\lambda}(y)\,dx\,dy$$
$$= \int\int (K(x,y) - K(x,\lambda))\psi_{\lambda'}(x)\psi_{\lambda}(y)\,dx\,dy$$

and we get the upper bound from (2.10).

If $j' \leq j-1$, the wavelets $\psi_{\lambda'}$ are in V_j. Before computing $\langle T\psi_{\lambda}, \psi_{\lambda'} \rangle$, we may orthogonally project $T\psi_{\lambda}$ on V_j, which does not change the scalar product. We then replace the distribution $T\psi_{\lambda}$ by

$$(3.2) \qquad g(x) = \sum_{l \in \mathbb{Z}^n} c(l)2^{nj/2}\phi(2^j x - l),$$

where

$$c(l) = \langle T\psi_{\lambda}, \phi_{jl} \rangle, \qquad \phi_{jl}(x) = 2^{nj/2}\phi(2^j x - l).$$

The coefficients $c(l)$ are easily dealt with, either using the weak continuity of T, or using the estimate (2.10) of the kernel. In this way we get

$$(3.3) \qquad |c(l)| \leq C(1 + |k - l|)^{-n-\gamma}.$$

Further, $\sum_{l \in \mathbb{Z}^n} c(l) = 0$, since $\sum \phi(x-l) = 1$ and ${}^tT(1) = 0$.

Then, $\tau(\lambda, \lambda') = \int g(x)\psi_{\lambda'}(x)\,dx$, where $\psi_{\lambda'}$ is "flat", whereas $g(x)$ is well localized (about λ) and of mean 0, because of the properties of the coefficients $c(\lambda)$. The estimate (3.1) then follows from the next lemma.

Lemma 3. *Let $f(x)$ be a C^1 function whose support is contained in the unit ball of $\mathbb{R}^n$ and whose partial derivatives $\partial f/\partial x_j$, $1 \leq j \leq n$, satisfy $\|\partial f/\partial x_j\|_\infty \leq 1$.*

Let $\gamma \in (0,1)$ be an exponent and let $g \in L^1(\mathbb{R}^n)$ be a function satisfying $|g(x)| \leq (1 + |x|)^{-n-\gamma}$ and $\int g(x)\,dx = 0$. Then there is a constant $C = C(n, \gamma)$ such that, for all $x_0 \in \mathbb{R}^n$ and every real $R \geq 1$,

$$(3.4) \qquad \left| \int g(x)f\left(\frac{x - x_0}{R}\right)dx \right| \leq CR^{-\gamma} \qquad \text{if } |x_0| \leq R$$

and

$$(3.5) \qquad \left| \int g(x)f\left(\frac{x - x_0}{R}\right)dx \right| \leq CR^n|x_0|^{-n-\gamma} \qquad \text{if } |x_0| \geq R.$$

To establish (3.4) and (3.5), it is enough to write

$$g(x) = \frac{\partial}{\partial x_1}g_1(x) + \cdots + \frac{\partial}{\partial x_n}g_n(x),$$

where $|g_1(x)| \leq C(1 + |x|)^{-n+1-\gamma}, \ldots, |g_n(x)| \leq C(1 + |x|)^{-n+1-\gamma}$ and then integrate by parts.

The calculation of the matrix coefficients $\tau((\lambda, \lambda')) = \langle T\psi_{\lambda}, \psi_{\lambda'} \rangle$ involves two separate scales (2^{-j} and $2^{-j'}$). It is worth remarking that

the calculation can be done using a preliminary computation involving
only one of those scales (the smaller one). Writing ψ_{jk} instead of ψ_λ,
for $\lambda = 2^{-j}k + 2^{-j-1}\varepsilon$, where $\varepsilon \in \{0,1\}^n \setminus (0,\ldots,0)$, we define three
"non-standard" matrices corresponding to T by

$$A = \langle T\psi_{jk}, \psi_{jl} \rangle,$$
$$B = \langle T\phi_{jk}, \psi_{jl} \rangle,$$
$$C = \langle T\psi_{jk}, \phi_{jl} \rangle,$$

for $j \in \mathbb{Z}$, $k \in \mathbb{Z}^n$, and $l \in \mathbb{Z}^n$. Their entries are denoted, respectively,
by $\alpha(j,k,l)$, $\beta(j,k,l)$, and $\gamma(j,k,l)$. The singular integral operators of
Definition 2 are characterized by the simple conditions

$$|\alpha(j,k,l)| \le C(1 + |k - l|)^{-n-\gamma},$$
$$|\beta(j,k,l)| \le C(1 + |k - l|)^{-n-\gamma},$$

and

$$|\gamma(j,k,l)| \le C(1 + |k - l|)^{-n-\gamma}.$$

Once we have computed these three non-standard matrices, the stan-
dard matrix $\langle T\psi_\lambda, \psi_{\lambda'} \rangle$ may be obtained by a very simple manipulation,
consisting of writing $\psi_{\lambda'} \in V_j$, for $j' \le j - 1$, and decomposing $\psi_{\lambda'}$ with
respect to the orthogonal basis ϕ_{jk}, $k \in \mathbb{Z}^n$, of V_j.

In [E4], G. Beylkin, R. Coifman, and V. Rokhlin demonstrate that
the above remarks have a considerable significance in numerical analysis.
The reader may also refer to [E7] and [E19].

4 Schur's lemma

For the moment, we shall forget the problem of the continuity of
singular integral operators, and recall the statement of Schur's lemma.

Lemma 4. *Let $M = (m(p,q))_{p,q\in\mathbb{N}}$ be an infinite matrix and suppose
that $\omega(p) > 0$ is a sequence of positive real numbers. Suppose, further,
that, for every $p \in \mathbb{N}$,*

$$(4.1) \qquad \sum_{q\in\mathbb{N}} |m(p,q)|\omega(q) \le \omega(p)$$

and, symmetrically, for every $q \in \mathbb{N}$,

$$(4.2) \qquad \sum_{p\in\mathbb{N}} |m(p,q)|\omega(p) \le \omega(q).$$

*Then $M : l^2(\mathbb{N}) \to l^2(\mathbb{N})$ is bounded and the norm of M is not greater
than 1.*

Indeed, suppose that $\sum_0^\infty |x(q)|^2 \le 1$ and put $y(p) = \sum_0^\infty m(p,q)x(q)$.

We intend to show that $\sum_0^\infty |y(p)|^2 \leq 1$. To to do this, we write

$$|m(p,q)||x(q)| = |m(p,q)|^{1/2}\omega^{1/2}(q)\omega^{-1/2}(q)|m(p,q)|^{1/2}|x(q)|.$$

Applying the Cauchy-Schwarz inequality gives

$$|y(p)|^2 \leq \left(\sum_0^\infty |m(p,q)|\omega(q)\right)\left(\sum_0^\infty |m(p,q)|\omega^{-1}(q)|x(q)|^2\right)$$

$$\leq \omega(p)\left(\sum_0^\infty |m(p,q)|\omega^{-1}(q)|x(q)|^2\right).$$

We sum with respect to p, change the order of summation and (4.2) gives what we want.

The magic of Schur's lemma is that the order of the indices is of no account.

Let us return to the problem of the $l^2(\Lambda)$ continuity of the matrices M whose coefficients satisfy (3.1).

We apply Schur's lemma with $\omega(\lambda) = 2^{-nj/2}$. This means we must estimate

$$\sum_{j'}\sum_{k'} 2^{-nj'/2}2^{-|j-j'|((n/2)+\gamma)}\left(\frac{2^{-j}+2^{-j'}}{2^{-j}+2^{-j'}+|2^{-j}k-2^{-j'}k'|}\right)^{n+\gamma}.$$

We start with the case $j' \geq j$. Putting $d = j' - j$, we bound the quotient above by $2(1+|k-2^{-d}k'|)^{-1}$. Now, $\sum_{k'} 2^{-nd}(1+|k-2^{-d}k'|)^{-n-\gamma} = \sum_{k'} 2^{-nd}(1+2^{-d}|k'|)^{-n-\gamma} \leq C(n,\gamma)$, as the latter series can be interpreted as a Riemann sum. This leaves

$$\left(\sum_{j'\geq j} 2^{-\gamma(j'-j)}\right)2^{-nj/2} = C(\gamma)2^{-nj/2} = C(\gamma)\omega(\lambda).$$

Similarly, if $j' < j$, we put $d = j - j'$ and the quotient is bounded above by $2(1+|2^{-d}k-k'|)^{-1}$. This leads to a summation over k' which is bounded above, uniformly with respect to k and d. We are left with

$$\sum_{j'<j} 2^{-nj'/2}2^{-(j-j')(n/2+\gamma)} = 2^{-nj/2}\sum_{d>0} 2^{-d\gamma} = C(\gamma)2^{-nj/2}.$$

So we have proved Theorem 1 when $T(1) = {}^tT(1) = 0$.

Choosing $\omega(\lambda) = 2^{-nj/2}$ has the following significance. If we look back at section 8 in Chapter 6 of *Wavelets and Operators*, we can see that the operators T and ${}^tT$ are continuous on the homogeneous Besov space $\dot{B}_1^{0,1}$, when $T(1) = {}^tT(1) = 0$. These two estimates give the L^2 continuity by interpolation.

5 Wavelets and Vaguelets

Before leaving the special case of the $T(1)$ theorem, considered above, we shall give a corollary of the proof we have just presented.

Definition 3. *The continuous functions $f_{j,k}(x)$, $j \in \mathbb{Z}$, $k \in \mathbb{Z}^n$, on $\mathbb{R}^n$, are called vaguelets if there are two exponents α, β, satisfying $0 < \beta < \alpha < 1$, and a constant C such that*

(5.1) $$|f_{j,k}(x)| \le C2^{nj/2}(1+|2^j x - k|)^{-n-\alpha},$$

(5.2) $$\int f_{j,k}(x)\, dx = 0,$$

and

(5.3) $$|f_{j,k}(x') - f_{j,k}(x)| \le C2^{j(n/2+\beta)}|x' - x|^\beta.$$

We then have the following result.

Theorem 2. *If the functions $f_{j,k}(x)$, $j \in \mathbb{Z}$, $k \in \mathbb{Z}^n$, are vaguelets, then there is a constant C' such that, for every sequence $\alpha(j,k)$ of coefficients,*

(5.4) $$\left\| \sum\sum \alpha(j,k)f_{j,k}(x) \right\|_2 \le C'\left(\sum\sum |\alpha(j,k)|^2\right)^{1/2}.$$

To establish (5.4), we compute the square of the L^2 norm of the series $\sum\sum \alpha(j,k)f_{j,k}(x) = f(x)$. To do this, we must compute the integrals $I(j,k;j'k') = \int f_{j,k}(x)\bar{f}_{j',k'}(x)\, dx$.

By symmetry, we can always reduce to the case where $j' \ge j$. To use (5.1), we first note that

$$\int \frac{2^{nj'}}{(1+|2^j x - k|)^{n+\alpha}(1+|2^{j'} x - k'|)^{n+\alpha}}\, dx \le \frac{C(n,\alpha)}{(1 + 2^j|k'2^{-j'} - k2^{-j}|)^{n+\alpha}}.$$

From this, we deduce that the integral we are interested in is bounded above, in modulus, by

$$C'2^{-(n/2)|j-j'|}\left(\frac{2^{-j} + 2^{-j'}}{2^{-j} + 2^{-j'} + |k2^{-j} - k'2^{-j'}|}\right)^{n+\alpha}.$$

The second estimate we use is given by

$$\left| \int (f_{j,k}(x) - f_{j,k}(2^{-j'}k'))f_{j',k'}(x)\, dx \right|$$
$$\le C2^{j(n/2+\beta)}\int |x - 2^{-j'}k'|^\beta |f_{j',k'}(x)|\, dx$$
$$\le C'2^{-(n/2+\beta)|j-j'|}.$$

We denote these two upper bounds of $|I(j,k;j',k')|$ by M_1 and M_2. From them, we obtain a third: $M_3 = M_2^\theta M_1^{1-\theta}$, where $0 < \theta < 1$. Choosing θ sufficiently small, we find we have an estimate of the form (3.1) and we can finish by applying Schur's lemma.

Theorem 1 can be related to Theorem 2 by showing that every operator which satisfies the hypotheses of Theorem 1, and such that $T(1) = {}^tT(1) = 0$, transforms an orthonormal wavelet basis ψ_{jk}^ε, $\varepsilon \in \{0,1\}^n$, $\varepsilon \neq (0,\ldots,0)$, into vaguelets f_{jk}^ε. Inequality (5.4) then becomes a new proof of the L^2 continuity of T.

Corollary. If f_{jk}, $j \in \mathbb{Z}$, $k \in \mathbb{Z}^n$, are vaguelets and if $Q(j,k) = \{x \in \mathbb{R}^n : 2^j x - k \in [0,1)^n\}$, then there is a constant C such that, for every dyadic cube $Q_0 = Q(j_0, k_0)$ and every function $b \in \mathrm{BMO}(\mathbb{R}^n)$,

$$(5.5) \qquad \sum_{Q(j,k) \subset Q_0} \left| \int b(x) f_{jk}(x)\, dx \right|^2 \leq C \|b\|_{\mathrm{BMO}}^2 |Q_0| \,.$$

To establish (5.5), we start with the dual form of (5.4). For $f \in L^2(\mathbb{R}^n)$

$$(5.6) \qquad \left(\sum\sum |\langle f, f_{jk} \rangle|^2 \right)^{1/2} \leq C' \|f\|_2 \,.$$

We then write $b = b_1 + b_2 + b_3$ as follows. Let Q_1 denote the cube which is Q_0 doubled (same centre, double diameter) and let b_3 be the mean of b on Q_1. The function $b_1(x)$ is $b(x) - b_3$ on Q_1 and 0 otherwise. So $b_2(x)$ is $b(x) - b_3$, if $x \notin Q_1$, and 0 otherwise. Then

$$\langle b, f_{jk} \rangle = \langle b_1, f_{jk} \rangle + \langle b_2, f_{jk} \rangle \,.$$

Applying (5.6) gives

$$\sum\sum |\langle b_1, f_{jk} \rangle| \leq C'' \|b_1\|_2^2 \leq C'' 4^n \|b_1\|_{\mathrm{BMO}}^2 |Q_0| \,.$$

We then follow the proof in *Wavelets and Operators*, Chapter 5, of the corollary to the proof of Theorem 4. In this fashion, we get

$$\sum_{Q(j,k) \subset Q_0} \sum |\langle b_2, f_{jk} \rangle| \leq C'' \|b\|_{\mathrm{BMO}}^2 |Q_0| \,.$$

6 Pseudo-products and the rest of the proof of the $T(1)$ theorem

We have just proved Theorem 1 under the hypothesis that the integrals

$$(6.1) \qquad \alpha(\lambda) = \int {}^tT(\psi_\lambda)\, dx$$

and

$$(6.2) \qquad \beta(\lambda) = \int T(\psi_\lambda)\, dx$$

are zero for all $\lambda \in \Lambda$. But the conditions $T(1) \in \mathrm{BMO}$ and ${}^tT(1) \in \mathrm{BMO}$ indicate that, even if these integrals are non-zero, they are nonetheless sufficiently small. Indeed, for every dyadic cube Q,

$$(6.3) \qquad \sum_{Q(\lambda) \subset Q} |\alpha(\lambda)|^2 \leq C |Q| \|T(1)\|_{\mathrm{BMO}}^2$$

and

(6.3)
$$\sum_{Q(\lambda) \subset Q} |\beta(\lambda)|^2 \le C|Q| \|{}^tT(1)\|_{\mathrm{BMO}}^2 \,.$$

Conditions (6.3) and (6.4) are the characterization of BMO given by Theorem 4 of *Wavelets and Operators*, Chapter 5.

These two conditions will enable us to construct Calderón-Zygmund operators R and S such that

(6.5) $R(1) = T(1), \quad {}^tR(1) = 0, \quad S(1) = 0, \quad \text{and} \quad {}^tS(1) = {}^tT(1)\,.$

We then put $N = T - R - S$ and get:

(a) $N : V \to V'$ is weakly continuous and its kernel satisfies (2.8), (2.9) and (2.10),

because all three of R, S and T have these properties;

(b) $N(1) = {}^tN(1) = 0$,

which is the point of the construction.

Thus, from what we did before, N is continuous on $L^2(\mathbb{R}^n)$. Since the same is true for R and S, it follows that T is continuous on $L^2(\mathbb{R}^n)$, and the proof of the $T(1)$ theorem is complete.

We now forget all the foregoing and start with an arbitrary sequence of complex numbers $\alpha(\lambda)$, $\lambda \in \Lambda$. Suppose there is a constant C_0 such that $|\alpha(\lambda)| \le C_0 2^{-nj/2}$, where $\lambda = k2^{-j} + \varepsilon 2^{-j-1}$, $\varepsilon \in \{0,1\}^n \backslash \{(0,\ldots,0)\}$. Let $\theta(x)$ be a function in $\mathcal{D}(\mathbb{R}^n)$, of mean 1. Put

(6.6)
$$\theta_\lambda(x) = 2^{nj}\theta(2^j x - k)\,.$$

Then we consider the distribution $S \in \mathcal{D}'(\mathbb{R}^n \times \mathbb{R}^n)$, defined by

(6.7)
$$S(x,y) = \sum_{\lambda \in \Lambda} \alpha(\lambda)\psi_\lambda(x)\theta_\lambda(y)\,.$$

That this series converges, in the sense of distributions, follows immediately from the inequality $|\alpha(\lambda)| \le C_0 2^{-nj/2}$, the localization of θ_λ and ψ_λ, and the fact that $\int \psi_\lambda(x)\,dx = 0$.

It is equally easy to check that the restriction of the distribution $S(x,y)$ to $x \ne y$ is a function $K(x,y)$ satisfying

$$|\partial_x^\mu \partial_y^\nu K(x,y)| \le C|x - y|^{-n-|\mu|-|\nu|} \qquad \text{for } |\mu|,|\nu| \le q$$

(where the integer q defines the regularity of the wavelets).

Then we have

Proposition 2. *The operator* $T : \mathcal{D}(\mathbb{R}^n) \to \mathcal{D}'(\mathbb{R}^n)$ *whose distribution-kernel is* $S(x,y)$ *can be extended as a continuous linear operator on* $L^2(\mathbb{R}^n)$ *if and only if* $\sum_{\lambda \in \Lambda} \alpha(\lambda)\psi_\lambda(x) \in \mathrm{BMO}$, *or, equivalently, if there is a constant* C *such that, for every dyadic cube* Q,

(6.8)
$$\sum_{Q(\lambda) \subset Q} |\alpha(\lambda)|^2 \le C|Q|\,.$$

Since T satisfies (2.8), (2.9) and (2.10), we know that condition (6.8) is necessary, because it means that $T(1) \in$ BMO. In fact $T(1) = \sum_{\lambda \in \Lambda} \alpha(\lambda) \psi_\lambda(x)$, because $\int \theta_\lambda(y)\, dy = 1$.

To see that (6.8) is sufficient, we apply T to an arbitrary function $f \in L^2(\mathbb{R}^n)$. This gives $T(f) = \sum_{\lambda \in \Lambda} \alpha(\lambda) \theta_\lambda(f) \psi_\lambda(x)$, where $\theta_\lambda(f) = \int f(y)\theta_\lambda(y)\, dy$. Since the wavelets ψ_λ form an orthonormal basis of $L^2(\mathbb{R}^n)$, we have

$$\|T(f)\|_2^2 = \sum_{\lambda \in \Lambda} |\alpha(\lambda)|^2 |\theta_\lambda(f)|^2 \, .$$

To find an upper bound for this last sum, we use Carleson's well-known lemma.

Lemma 5. *Let $p(\lambda)$, $\lambda \in \Lambda$, be a sequence of positive numbers such that $\sum_{Q(\lambda) \subset Q} p(\lambda) \leq |Q|$, for every dyadic cube Q. Then, for every sequence $\omega(\lambda) \geq 0$, $\lambda \in \Lambda$, we have*

$$(6.9) \qquad \sum_{\lambda \in \Lambda} \omega(\lambda) p(\lambda) \leq \int_{\mathbb{R}^n} \omega(x)\, dx \, ,$$

where

$$\omega(x) = \sup_{Q(\lambda) \ni x} \omega(\lambda) \, .$$

In our case, if we ignore the constants, we have $p(\lambda) = |\alpha(\lambda)|^2$, $\omega(\lambda) = |\theta_\lambda(f)|^2$ and thus $\omega(x) = (f^\star(x))^2$, where $f^\star(x)$ is the Hardy-Littlewood maximal function of f. We finish by observing that $\int_{\mathbb{R}^n} (f^\star(x))^2\, dx \leq C\|f\|_2^2$.

We still need to prove Lemma 5. The characteristic function $\chi(\lambda, t)$ is defined by $\chi(\lambda, t) = 1$, if $0 < t < \omega(\lambda)$, and 0 otherwise. Then

$$\sum_{\lambda \in \Lambda} \omega(\lambda) p(\lambda) = \int_0^\infty \sum_{\lambda \in \Lambda} \chi(\lambda, t) p(\lambda)\, dt \, .$$

For $t > 0$, let Ω_t be the set of x such that $\omega(x) > t$: then Ω_t can also be expressed as the union of all the cubes $Q(\lambda)$ such that $\omega(\lambda) > t$. By the Bienaymé-Chebyshev inequality, $|\Omega_t| \leq t^{-1} \int \omega(x)\, dx$, and we may suppose that the integral is finite, because, if not, (6.9) loses all interest.

Let Q_k denote the maximal dyadic cubes contained in Ω_t. Ω_t is the union of the Q_k and, for $t > 0$, we get

$$\sum_{\lambda \in \Lambda} \chi(\lambda, t) p(\lambda) \leq \sum_{Q(\lambda) \subset \Omega_t} p(\lambda) = \sum_k \sum_{Q(\lambda) \subset Q_k} p(\lambda) \, .$$

We now use the assumption about $p(\lambda)$ and get $\sum_k |Q_k| = |\Omega_t|$ as an upper bound for the double sum. We finish by observing that $\int_0^\infty |\Omega_t|\, dt = \int_{\mathbb{R}^n} \omega(x)\, dx$.

Proposition 2 is now completely established and gives the operator R

we have been trying to construct. To define S, we replace $\alpha(\lambda)$ by $\beta(\lambda)$, and S is the transpose of the operator R constructed using the $\beta(\lambda)$.

7 Cotlar and Stein's lemma and the second proof of David and Journé's theorem

We state Cotlar and Stein's well-known lemma.

Lemma 6. *Let H be a Hilbert space and let $T_j : H \to H$, $j \in \mathbb{Z}$, be bounded linear operators with adjoints $T_j^\star$. Suppose that there exists a sequence $\omega(j) \geq 0$, $j \in \mathbb{Z}$, such that $\sum_{-\infty}^{\infty} (\omega(j))^{1/2} < \infty$,*

$$(7.1) \qquad \|T_j^\star T_k\| \leq \omega(j - k), \qquad \text{for all } j, k \in \mathbb{Z},$$

and

$$(7.2) \qquad \|T_j T_k^\star\| \leq \omega(j - k), \qquad \text{for all } j, k \in \mathbb{Z}.$$

Then, for all $x \in H$, the series $\sum_{-\infty}^{\infty} T_j(x)$ converges in H. Putting $T(x) = \sum_{-\infty}^{\infty} T_j(x)$ gives $\|T\| \leq \sum_{-\infty}^{\infty} (\omega(j))^{1/2}$.

We follow [108] and first consider the finite sums $S = S_N = \sum_{-N}^{N} T_j$. Of course, $\|S\|^2 = \|S^\star S\|$ and, since $S^\star S$ is a positive self-adjoint operator, $\|S^\star S\| = \|(S^\star S)^M\|^{1/M}$ for every integer $M \geq 1$. Now

$$(S^\star S)^M = \sum_{j_1} \sum_{k_1} \cdots \sum_{j_m} \sum_{k_m} T_{j_1}^\star T_{k_1} \cdots T_{j_M}^\star T_{k_M} .$$

A brute force upper bound for $\|(S^\star S)^M\|$ is given by

$$\sum_{j_1} \sum_{k_1} \cdots \sum_{j_m} \sum_{k_m} \|T_{j_1}^\star T_{k_1} \cdots T_{j_M}^\star T_{k_M}\|$$

and we then consider adjacent pairs of terms in two ways. We can either use (7.1) to get the estimate $\omega(j_1 - k_1) \cdots \omega(j_M - k_M)$ or separate off the first and last terms and employ (7.2) for the rest. By (7.1), $\|T_k\| \leq (\omega(0))^{1/2}$, so the second upper bound is $\omega(0)\omega(k_1 - j_2) \cdots \omega(k_{M-1} - j_M)$.

We then take the geometric mean of the two estimates to get

$$\|T_{j_1}^\star T_{k_1} \cdots T_{j_M}^\star T_{k_M}\| \leq (\omega(0))^{1/2}(\omega(j_1 - k_1))^{1/2}(\omega(k_1 - j_2))^{1/2} \cdots$$
$$\cdots (\omega(k_{M-1} - j_M))^{1/2}(\omega(j_M - k_M))^{1/2}.$$

Let $\sigma = \sum_{-\infty}^{\infty} (\omega(j))^{1/2}$. Summing successively over $j_1, k_1, \ldots, j_M$, we get $(\omega(0))^{1/2}\sigma^{2M-1}$. Summing over k_M multiplies the estimate by $2N + 1$.

Putting all this together gives $\|S\|^{2M} \leq (2N + 1)(\omega(0))^{1/2}\sigma^{2M-1}$, or $\|S\| \leq ((2N + 1)(\omega(0)^{1/2}\sigma^{-1})^{1/2M}\sigma$. It is now enough to let M tend to infinity to get $\|S\| \leq \sigma$.

To pass to the general case, we follow the argument of J.L. Journé by remarking that, if the scalars λ_j are such that $|\lambda_j| \leq 1$, the operators $\lambda_j T_j$ still satisfy (7.1) and (7.2) with the same ω. As a consequence, if we put $y_j = T_j(x)$, we get, for every $N \geq 1$,

$$(7.3) \qquad \left\| \sum_{-N}^{N} \lambda_j y_j \right\| \leq \sigma \|x\| .$$

We then put

$$\varepsilon(m) = \sup_{m' \geq m} \sup_{|\lambda_j| \leq 1} \| \sum_{m \leq |\lambda_j| \leq m'} \lambda_j y_j \|$$

and aim to show that $\varepsilon(m)$ tends to 0 as m tends to infinity. We observe that $\varepsilon(m)$ is decreasing and that $\varepsilon(m) > \varepsilon > 0$, if $\varepsilon(m)$ does not tend to 0. In that case, we can take disjoint intervals $[m_k, m'_k]$ and $z_k = \sum_{m_k \leq |j| \leq m'_k} \lambda_j z_j$, such that $\|z_k\| \geq \varepsilon$. We then consider the sums $\pm z_1 \pm z_2 \pm \cdots \pm z_k = Z(\theta, k)$, where $\theta \in \{-1, 1\}^k$. By (7.3), $\|Z(\theta, k)\|^2 \leq \sigma^2 \|x\|^2$. But if we now take the mean of $\|Z(\theta, k)\|^2$ over the 2^k sequences of ± 1's, we get $\|z_1\|^2 + \cdots + \|z_k\|^2 \geq k\varepsilon^2$. This is a contradiction.

We have shown that the series $\sum_{-\infty}^{\infty} T_j(x)$ converges unconditionally for all $x \in H$.

In the application we have in mind, $H = L^2(\mathbb{R}^n)$ and the conditions attaching to the operators T_j will be described by the weight $w(x, y) = (1 + |x - y|)^{-n-\gamma}$, for some exponent $\gamma > 0$. We put $w_j(x, y) = 2^{nj} w(2^j x, 2^j y)$ and we suppose that

$$T_j f(x) = \int T_j(x, y) f(y) \, dy ,$$

where

$$(7.4) \qquad |T_j(x, y)| \leq w_j(x, y) .$$

Furthermore, there will be an exponent $\beta \in (0, \gamma]$, such that
$$(7.5) \quad |T_j(x, y) - T_j(x, y')| \leq 2^{j\beta} |y - y'|^\beta (w_j(x, y) + w_j(x, y')) ,$$
$$(7.6) \quad |T_j(x, y) - T_j(x', y)| \leq 2^{j\beta} |x - x'|^\beta (w_j(x, y) + w_j(x', y)) ,$$
and the $T_j(x, y)$ will also satisfy

$$(7.7) \qquad \int T_j(x, y) \, dy = 0 \qquad \text{for all } x \in \mathbb{R}^n,$$

$$(7.8) \qquad \int T_j(x, y) \, dx = 0 \qquad \text{for all } y \in \mathbb{R}^n.$$

The lemma of Cotlar and Stein then assumes the following form.

Lemma 8. *Given the hypotheses (7.4) to (7.8), there exists a constant $C(\beta, \gamma, n)$ such that*

$$(7.9) \qquad \| \sum_{-\infty}^{\infty} T_j \| \leq C(\beta, \gamma, n) .$$

We prove this estimate by applying the lemma of Cotlar and Stein. To do this, we use the following continuous version of Schur's lemma to get an upper bound for the norms of the operators $T_j^* T_k$ and $T_j T_k^*$.

Lemma 9. *Let $A : \mathbb{R}^n \times \mathbb{R}^n \to \mathbb{C}$ be a Borel-measurable function. If $\int_{\mathbb{R}^n} |A(x,y)|\, dy \le 1$, for all $x \in \mathbb{R}^n$, and $\int_{\mathbb{R}^n} |A(x,y)|\, dx \le 1$, for all $y \in \mathbb{R}^n$, then the operator L, given by $Lf(x) = \int A(x,y) f(y)\, dy$, is a continuous operator on $L^p(\mathbb{R}^n)$, $1 \le p \le \infty$, of norm not greater than 1.*

Assuming this result, we continue the proof of Lemma 8. The kernel of $T_k^* T_j$ is

$$A_{jk}(x,y) = \int_{\mathbb{R}^n} \overline{T}_k(z,x) T_j(z,y)\, dz\,,$$

while that of $T_j T_k^*$ is

$$B_{jk}(x,y) = \int_{\mathbb{R}^n} T_k(x,z) \overline{T}_j(y,z)\, dz\,.$$

The hypotheses on $T_j(x,y)$ are symmetric in the variables x and y. For that reason, we can restrict our discussion to $I(j,k) = \int |A_{jk}(x,y)|\, dx$ and $J(j,k) = \int |A_{jk}(x,y)|\, dy$, taking $j \le k$. Writing

$$A_{jk}(x,y) = \int_{\mathbb{R}^n} (\overline{T}_k(z,x) - \overline{T}_k(y,x)) T_j(z,y)\, dz\,,$$

we get

$$|A_{jk}(x,y)| \le 2^{k\beta} \int_{\mathbb{R}^n} |y - z|^\beta (w_k(z,x) + w_k(y,x)) w_j(z,y)\, dz\,.$$

To estimate $I(j,k)$, we evaluate the double integral by first integrating with respect to x. Since $\int w_k(z,x)\, dx = C$, we are left with

$$2^{k\beta} \int_{\mathbb{R}^n} |y - z|^\beta w_j(z,y)\, dz = C' 2^{-(\gamma-\beta)|j-k|}\,.$$

We leave the reader to conclude the proof of Lemma 8 by a similar estimate of $J(j,k)$.

We return to the particular case of Theorem 2, in which $T(1) = 0$ and ${}^t T(1) = 0$. To establish the continuity of T, using the lemma of Cotlar and Stein, we must expand T appropriately as $T = \sum_{-\infty}^{\infty} T_j$.

To this end, we let $\theta(x)$, $x \in \mathbb{R}^n$, denote a C^1 function depending only on $|x|$, whose support is contained in $|x| \le 1$ and whose mean is 1. Let S_j be the operator defined by convolution with $\theta_j(x) = 2^{nj} \theta(2^j x)$ and put $\Delta_j = S_{j+1} - S_j$.

The series decomposition of T to which the lemma of Cotlar and Stein applies can then be written as

$$T = \sum_{-\infty}^{\infty} (S_{j+1} T S_{j+1} - S_j T S_j) = \sum_{-\infty}^{\infty} \Delta_j T S_j + \sum_{-\infty}^{\infty} S_j T \Delta_j + \sum_{-\infty}^{\infty} \Delta_j T \Delta_j\,.$$

The details of the proof are very simple. We give them for the convenience of the reader.

Definition 4. *Let $L : V \to V'$ be a continuous linear operator. We say that L is regular if*

$$\lim_{j \to -\infty} \langle LS_j u, S_j v \rangle = 0 \qquad \text{for } u, v \in V.$$

The significance of the definition is that a regular operator L is the sum of its telescoping series $\sum_{-\infty}^{\infty}(S_{j+1}LS_{j+1} - S_j LS_j)$. It is not hard to see that every weakly continuous operator on $L^2(\mathbb{R}^n)$ is regular. On the other hand, it is easy to give examples of operators which are not regular.

The theorem of David and Journé then follows from

Proposition 3. *If T satisfies the hypotheses of Theorem 2 and if, further, $T(1) = {}^tT(1) = 0$, then the lemma of Cotlar and Stein applies to each of the three series*

$$\sum_{-\infty}^{\infty} \Delta_j TS_j, \qquad \sum_{-\infty}^{\infty} S_j T\Delta_j, \qquad \text{and} \qquad \sum_{-\infty}^{\infty} \Delta_j T\Delta_j.$$

In fact, a more precise result is valid: the condition $T(1) = 0$ together with the weak continuity of T is enough to deal with the first series; ${}^tT(1) = 0$ is what is needed for the second; the final series requires only the weak continuity of T.

We concentrate our efforts on the first series. Let $\eta_j = \theta_{j+1} - \theta_j$. The kernel $T_j(x, y)$ of the operator $\Delta_j TS_j$ is calculated by the same rule as the coefficients of a product of three matrices. We have

$$(7.10) \qquad T_j(x, y) = \sum\sum \eta_j(x - u)S(u, v)\theta_j(v - y) \, du \, dv$$

where $S(u, v)$ is the distribution-kernel of T. We put $\eta_j^{(x)}(u) = \eta_j(x - u)$ and $\theta_j^{(y)}(v) = \theta_j(v - y)$, so that we also have

$$(7.11) \qquad T_j(x, y) = \langle T\theta_j^{(y)}, \eta_j^{(x)} \rangle.$$

If the distance from x to y is not greater than $4 \cdot 2^{-j}$, the estimates are a consequence of the weak continuity of T, whereas, if $|x - y| > 4 \cdot 2^{-j}$, we can replace $S(u, v)$ by the kernel $K(u, v)$ corresponding to T. From here, (7.4) can be obtained by the usual calculations.

The verification of (7.5) and (7.6) is simple and left to the reader. In equation (7.7), the integral is $T_j(1) = \Delta_j TS_j(1) = \Delta_j T(1) = \Delta_j(0) = 0$. This formal calculation is easy to justify and we leave that task to the reader. Verifying (7.8) is even more straightforward.

8 Other formulations of the $T(1)$ theorem

To deal with certain applications, it is convenient to have several different ways of expressing the necessary and sufficient conditions which appear in the $T(1)$ theorem.

We suppose, once and for all, that conditions (2.8), (2.9) and (2.10) are satisfied by an operator $T : V \to V'$ and the corresponding kernel $K(x,y)$. We shall examine, in greater depth, the relationship between $T(1)$, ${}^tT(1)$ and the property of weak continuity.

Proposition 4. *Suppose that T and $K(x,y)$ satisfy (2.8), (2.9) and (2.10) and that T is weakly continuous. Then $T(1)$ and ${}^tT(1)$ belong to the homogeneous Besov space $\dot{B}_\infty^{0,\infty}$. Conversely, if u and v are arbitrary elements of $\dot{B}_\infty^{0,\infty}$, there exists an operator $T : V \to V'$ satisfying properties (2.8), (2.9) and (2.10), which is weakly continuous and such that $T(1) = u$ and ${}^tT(1) = v$.*

To prove the first part, we observe that $\dot{B}_\infty^{0,\infty}$ is the dual of $\dot{B}_1^{0,1}$. Moreover, the elements of $\dot{B}_1^{0,1}$ have an atomic decomposition using special atoms. The special atoms are constructed from basic functions $\psi \in \mathcal{D}(\mathbb{R}^n)$ whose supports lie in the unit ball $|x| \leq 1$ and which satisfy $\int \psi(x)\, dx = 0$ Then the "special atoms" are the functions of the form $a^{-n}\psi(a^{-1}(x - x_0))$, where $a > 0$ and $x_0 \in \mathbb{R}^n$ are arbitrary.

After using the isometric invariance of the hypotheses under the action of the group $ax + b$, everything boils down to showing that the estimate $|\langle T(1), \psi \rangle| \leq C_0$ holds, where C_0 depends only on the constants which appear in the hypotheses made about T and on $\sum_{|\alpha|=q} \|\partial^\alpha \psi\|_\infty$, for a certain integer $q \geq 1$. This estimate is obtained by repeating, word for word, the argument used to define $T(1)$. The property of weak continuity appears in the term $\langle T(\phi_0), \psi \rangle$, where $\phi_0 \in \mathcal{D}(\mathbb{R}^n)$, $\phi_0(x) = 1$, for $|x| \leq 2$, and $\phi(x) = 0$, for $|x| \geq 3$.

To show the converse, we look at the construction of the pseudo-products used in the proof of Theorem 2. The details are left to the reader.

In the opposite direction, it is sometimes useful to be able to deduce the weak continuity from another condition.

Proposition 5. *Suppose that property (2.8) is satisfied by the kernel $K(x,y)$ corresponding to an operator $T : V \to V'$. Then the weak continuity of T follows from the continuity of $T : \dot{B}_1^{0,1} \to L^1$.*

Here we can again reduce the problem to estimating $|\langle Tg, f \rangle|$, where f and g are sufficiently regular functions with supports in the unit ball

$|x| \leq 1$. Consider $h(x) = f(x) - f(x - x_0)$, where $|x_0| = 3$. Then $h \in \dot{B}_1^{0,1}$ (in fact, h is a special atom) and we get

(8.1) $\qquad \langle Tf, g \rangle = \langle Th, g \rangle + \iint K(x,y)f(y - x_0)g(x)\, dx\, dy$.

We estimate the first term, using the continuity of $T : \dot{B}_1^{0,1} \to L^1$, and the second term, using (2.8).

The continuity of $T : \dot{B}_1^{0,1} \to L^1$ is a consequence of the condition

(8.2) $\qquad\qquad\qquad\qquad \|T(\psi_\lambda)\|_2 \leq C$,

where ψ_λ, $\lambda \in \Lambda$, is the orthonormal wavelet basis of compact support of *Wavelets and Operators* Chapter 3, section 8. Since $\|\psi_\lambda\|_2 = 1$, condition (8.2) expresses a weak form of the L^2 continuity of T. To check our assertion, observe first that (8.2) and (2.10) imply (Chapter 7, section 3) that

(8.3) $\qquad\qquad\qquad\qquad \|T(\psi_\lambda)\|_1 \leq C 2^{-nj/2}$

and then that (8.3) is equivalent to the continuity of $T : \dot{B}_1^{0,1} \to L^1$, because of the atomic decomposition of $\dot{B}_1^{0,1}$.

Proposition 6. *Suppose that (2.8) and (2.9) hold. For each $\xi \in \mathbb{R}^n$, let χ_ξ denote the character $x \mapsto e^{ix \cdot \xi}$. Then the weak continuity of T is implied by the existence of a constant C_1, such that, for each $\xi \in \mathbb{R}^n$, the $\dot{B}_\infty^{0,\infty}$ norm of $T(\chi_\xi)$ is not greater than C_1.*

We need to use (2.9) to define $T(\chi_\xi)$. To prove the weak continuity, we once again reduce the problem to $\langle Tf, g \rangle$, where f and g are sufficiently regular and have supports contained in the unit ball. This time, we put $h(x) = g(x) - g(x - x_0)$ and everything comes down to estimating $\langle Tf, h \rangle$. To do this, we write

$$ f(x) = (2\pi)^{-n} \int \hat{f}(\xi) e^{ix \cdot \xi}\, d\xi $$

which gives

$$ \|Tf\|_{\dot{B}_\infty^{0,\infty}} \leq (2\pi)^{-n} \int |\hat{f}(\xi)| \|T(\chi_\xi)\|_{\dot{B}_\infty^{0,\infty}}\, d\xi \leq C_1 . $$

This gives an upper bound for $|\langle Tf, h \rangle|$, because h is a special atom.

9 Banach algebras of Calderón-Zygmund operators

We recall that the Riesz transforms R_j, $1 \leq j \leq n$, are defined by $R_j = D_j(-\Delta)^{-1/2}$, where $D_j = -i\partial/\partial x_j$. The R_j are first generation Calderón-Zygmund operators. We have $R_j(1) = {}^t R_j(1) = 0$ and this property is shared by all Calderón-Zygmund operators which are convolution operators, in other words, those that commute with translations. The collection $\mathcal{B}$ of these operators is a commutative algebra of Calderón-Zygmund operators and $T(1) = {}^t T(1) = 0$ for every $T \in \mathcal{B}$.

For a result in the opposite direction, we let $\mathcal{A}$ denote the collection of all Calderón-Zygmund operators T satisfying $T(1) = {}^t T(1) = 0$.

We shall show that $\mathcal{A}$ is an algebra of operators. In fact, $\mathcal{A}$ is the largest algebra of operators which consists of Calderón-Zygmund operators and contains $\mathcal{B}$. More precisely, if T is a Calderón-Zygmund operator and if the same is true for $R_1 T$, $T R_1, \ldots, R_n T$ and $T R_n$, then T necessarily belongs to $\mathcal{A}$.

Let CZO denote the vector space of all Calderón-Zygmund operators. Then the proof of the $T(1)$ theorem gives the isomorphism

(9.1) $\text{CZO} = \mathcal{A} \oplus \text{BMO} \oplus \text{BMO} .$

This decomposition amounts to putting $T(1) = \beta$, ${}^t T(1) = \gamma$, for $T \in \text{CZO}$, and using these two BMO functions to construct the pseudo-products L_β and ${}^t L_\gamma$, by the technique of Section 6. By these means, we get $T = L_\beta + {}^t L_\gamma + N$, where $N \in \mathcal{A}$.

In conclusion, $\mathcal{A}$ gives a good approximation to the entire collection CZO of all Calderón-Zygmund operators and is the largest algebra containing the usual Calderón-Zygmund operators.

We still have to pose and resolve the basic problem of the symbolic calculus for $\mathcal{A}$. We shall show that $\mathcal{A}$ is the increasing union of the Banach algebras $Op\mathcal{M}_\gamma$, $\gamma > 0$, which are isomorphic to explicit matrix algebras. This will enable us to construct an operator $T \in \mathcal{A}$ which is an isomorphism on $L^2(\mathbb{R}^n)$ and whose inverse does not belong to $\mathcal{A}$.

The time has come to prove these assertions.

Proposition 7. If the $2n + 1$ operators T, $R_1 T$, $T R_1, \ldots, R_n T$ and $T R_n$ belong to CZO, then $T(1) = 0$ and ${}^t T(1) = 0$.

To see this, we use the definition of $H^1(\mathbb{R}^n)$ given by Stein and Weiss, namely, $f \in H^1(\mathbb{R}^n)$ if and only if $f \in L^1(\mathbb{R}^n)$ and $R_j f \in L^1(\mathbb{R}^n)$, for $1 \leq j \leq n$.

So let f be a function which is in the unit ball of $H^1(\mathbb{R}^n)$ and in L^2, so that all the operations we are about to perform make sense. Since T, $R_1 T, \ldots, R_n T$ are all Calderón-Zygmund operators, we have $Tf \in L^1$, $R_1 Tf \in L^1, \ldots, R_n Tf \in L^1$, and thus $Tf \in H^1$. This means that T extends to a continuous linear operator on H^1, which implies that ${}^t T(1) = 0$.

We repeat the same line of argument, but starting with ${}^t T$ instead of T, to get $T(1) = 0$.

We now intend to verify that $\mathcal{A}$ is an algebra of operators by establishing the indentity $\mathcal{A} = \bigcup_{0 < \gamma \leq 1} \mathcal{O}p\mathcal{M}_\gamma$, where:

(9.2) $\mathcal{O}p\mathcal{M}_\gamma$ is a non-commutative Banach algebra of Calderón-Zygmund operators;

(9.3) $\mathcal{O}p\mathcal{M}_\gamma \subset \mathcal{O}p\mathcal{M}_{\gamma'}$, if $0 < \gamma' < \gamma$.

We start from a Banach algebra $\mathcal{M}_\gamma$, $\gamma > 0$, of matrices acting on $l^2(\Lambda)$: remember that Λ is the disjoint union of the Λ_j, where $\Lambda_j = 2^{-j-1}\mathbb{Z}^n \setminus 2^{-j}\mathbb{Z}^n$. Every $\lambda \in \Lambda$ can thus be written uniquely as $\lambda = 2^{-j}(k + r)$, where $r = \varepsilon/2$, for $\varepsilon \in \{0,1\}^n \setminus \{(0,\ldots,0)\}$.

We fix $\gamma > 0$ and define $\omega_\gamma : \Lambda \times \Lambda \to (0, \infty)$ by

$$\omega_\gamma = \frac{2^{-|j'-j|(n/2+\gamma)}}{1 + (j' - j^2)} \left(\frac{2^{-j} + 2^{-j'}}{2^{-j} + 2^{-j'} + |k2^{-j} - k'2^{-j'}|} \right)^{n+\gamma}.$$

Definition 5. *A matrix* $A = (\alpha(\lambda, \lambda'))_{(\lambda,\lambda') \in \Lambda \times \Lambda}$ *belongs to* $\mathcal{M}_\gamma$ *if there exists a constant* $C > 0$ *such that, for all* $(\lambda, \lambda') \in \Lambda \times \Lambda$,

(9.4) $$|\alpha(\lambda, \lambda')| \leq C\omega_\gamma(\lambda, \lambda').$$

Proposition 8. *For every* $\gamma > 0$, $\mathcal{M}_\gamma$ *is an algebra.*

The proof is a simple verification. We just need to prove the existence of a constant $C(n, \gamma)$ such that, for each $\lambda_0 \in \Lambda$ and $\lambda_1 \in \Lambda$,

(9.5) $$\sum_{\lambda \in \Lambda} \omega_\gamma(\lambda_0, \lambda)\omega_\gamma(\lambda, \lambda_1) \leq C(n, \gamma)\omega_\gamma(\lambda_0, \lambda_1).$$

With the obvious notation, we can distinguish three partial sums in (9.5), corresponding to $j \geq j_1 \geq j_0$, $j_1 \geq j \geq j_0$ and $j_1 \geq j_0 \geq j$ (by symmetry, we may suppose that $j_1 \geq j_0$).

In the first case, we note that, if $0 < \eta \leq \varepsilon \leq 1$ and $x, y \in \mathbb{R}^n$, then

$$\sum_{k \in \mathbb{Z}^n} (1 + |x - \varepsilon k|)^{-n-\gamma}(1 + |y - \eta k|)^{-n-\gamma}$$

$$\leq C'(n, \gamma)\varepsilon^{-n}(1 + |y - x\eta\varepsilon^{-1}|)^{-n-\gamma},$$

which immediately leads to the estimate we want. For this case, the terms $(1 + (j - j_0)^2)^{-1}$ and $(1 + (j - j_1)^2)^{-1}$ could have been left out.

The case $j_1 \geq j_0 \geq j$ can be dealt with similarly.

The situation where $j_1 \geq j \geq j_0$ is more interesting: we then need to observe that

$$\sum_{j_1 \geq j \geq j_0} (1 + (j - j_0)^2)^{-1}(1 + (j - j_1)^2)^{-1} \leq C(1 + (j_1 - j_0)^2)^{-1}.$$

Omitting these terms would have introduced a "logarithmic" singularity.

Let ψ_λ, $\lambda \in \Lambda$, be a wavelet basis arising out of a multiresolution approximation of regularity $r > \gamma$. We define $\mathcal{O}p\mathcal{M}_\gamma$ to be the set of operators T whose matrices, with respect to the (orthonormal) wavelet basis, belong to $\mathcal{M}_\gamma$. We need to show that this definition does not depend on the particular choice of wavelet basis. To do this, let $\tilde{\psi}_\lambda$ be another orthonormal wavelet basis of regularity $r > \gamma$. Then the matrix $M = (m(\lambda, \lambda'))$ of the unitary operator defined by $U(\psi_\lambda) = \tilde{\psi}_{\lambda'}$ belongs to $\mathcal{M}_\gamma$, because $m(\lambda, \lambda') = (\tilde{\psi}_\lambda, \psi_{\lambda'})$ and the verification of (9.4) is immediate (and similar to that carried out in section 5).

From now on, we consider only the case where $0 < \gamma \leq 1$. We let $\mathcal{A}_\gamma \subset \mathcal{A}$ denote the collection of Calderón-Zygmund operators satisfying (2.8), (2.9) and (2.10) with exponent γ and such that $T(1) = {}^t T(1) = 0$. Then we have

Theorem 3. *If $0 < \gamma \leq 1$ and $T \in \mathcal{O}p\mathcal{M}_\gamma$, then $T \in \mathcal{A}_\gamma$. Conversely, if $T \in \mathcal{A}_\gamma$ and if $0 < \gamma' < \gamma$, then $T \in \mathcal{O}p\mathcal{M}_{\gamma'}$.*

In one direction, (3.1) gives $T \in \mathcal{O}p\mathcal{M}_{\gamma'}$, for $0 < \gamma' < \gamma$. In the other, we must verify, for $T \in \mathcal{O}p\mathcal{M}_\gamma$,

$$(9.6) \qquad |K(x,y)| \leq C_0 |x - y|^{-n},$$

$$(9.7) \qquad |K(x',y) - K(x,y)| \leq C_1 |x' - x|^\gamma |x - y|^{-n-\gamma},$$

for $|x' - x| \leq |x - y|/2$, and

$$(9.8) \qquad |K(x,y') - K(x,y)| \leq C_1 |y' - y|^\gamma |x - y|^{-n-\gamma},$$

for $|y' - y| \leq |x - y|/2$.

Since the definition of $\mathcal{O}p\mathcal{M}_\gamma$ is independent of the choice of wavelet basis, we may do all the computations with a basis consisting of real-valued wavelets of compact support.

We intend to establish (9.8). For that, we decompose $K(x,y)$ as $K_1(x,y) + K_2(x,y) + K_3(x,y)$, where

$$K_1(x,y) = \sum_{j' > j} \sum \alpha(\lambda, \lambda') \psi_\lambda(x) \psi_{\lambda'}(y),$$

$$K_2(x,y) = \sum_{j' = j} \sum \alpha(\lambda, \lambda') \psi_\lambda(x) \psi_{\lambda'}(y),$$

and

$$K_3(x,y) = \sum_{j' < j} \sum \alpha(\lambda, \lambda') \psi_\lambda(x) \psi_{\lambda'}(y).$$

We shall see that these sums all satisfy (9.8). For the sake of brevity, we shall only discuss the first. We define $j_0, j_1 \in \mathbb{Z}$ by $2^{-j_0} \leq |x - y| <$

2^{-j_0+1} and $2^{-j_1} \leq |y' - y| < 2^{-j_1+1}$. We then write

$$|K(x,y') - K(x,y)| \leq C|y' - y| \sum_{\{j \leq j_0, j < j' < j_1\}} \frac{2^{(n+\gamma)j} 2^{(1-\gamma)j'}}{1 + (j' - j)^2}$$

$$+ C \sum_{\{j \leq j_0, j_1 \leq j'\}} \frac{2^{(n+\gamma)j} 2^{-\gamma j'}}{1 + (j' - j)^2}$$

$$+ C|y' - y| \sum_{\{j_0 < j < j' \leq j_1\}} \frac{2^{(n+\gamma)j} 2^{(1-\gamma)j'}}{(1 + (j' - j)^2)(1 + |k - l|)^{n+\gamma}}$$

$$+ C \sum_{\{j_0 < j, j_1 < j'\}} \frac{2^{(n+\gamma)j} 2^{-\gamma j'}}{(1 + (j' - j)^2)(1 + |k - l|)^{n+\gamma}},$$

where (j, k) corresponds to λ, (j', k') to λ' and where $l \in \mathbb{Z}^n$ is defined by $2^{-j'} k' \in Q(j, l)$.

A few words of explanation are, perhaps, necessary. The first series involves the "large" cubes $Q(\lambda)$ and the "large" cubes $Q(\lambda')$, where we exploit the regularity (in y) of the wavelets $\psi_{\lambda'}(y)$. Because the wavelets are localized, we can forget the sums over $k \in \mathbb{Z}^n$ and $k' \in \mathbb{Z}^n$, since, for fixed x, y and y', these sums are over sets of pairs (k, k'), where the size of the sets is uniformly bounded. The second series involves large cubes $Q(\lambda)$ and small cubes $Q(\lambda')$. The regularity of the wavelets $\psi_{\lambda'}$ does not help at the scale of $|y' - y|$, and we just use $\|\psi_{\lambda'}\|_\infty$ as a bound for $\psi_{\lambda'}$.

In the last two series, the cubes $Q(\lambda)$ and $Q(\lambda')$ are sufficiently small to be far apart: the order of magnitude of their distance is $2^{-j}(1 + |l| + |k - l|)$. For $j' \leq j_1$, we can use the regularity of the wavelets $\psi_{\lambda'}$.

There is no difficulty in making the required estimate for the first two series, if $0 < \gamma < 1$; the factor $(1 + (j' - j)^2)^{-1}$ is only involved if $\gamma = 1$.

Let $m \geq 1$ be an integer such that the support of the wavelet ψ_λ is contained in $mQ(\lambda) = mQ(j, k)$. In the last two series, j, k and l are connected by the conditions $x \in mQ(j, k)$ and $y \in mQ(j, l)$. The product $2^{-j}|k - l|$ is thus of order of magnitude $|y - x|$, while $2^{(m+\gamma)j}(1 + |k - l|)^{-n-\gamma}$ is essentially constant and equal to $|x - y|^{-n-\gamma}$. Once this estimate has been done, we sum over j, which gets rid of the term $(1 + (j' - j)^2)^{-1}$. Finally, the sum over j' gives $|y' - y|^\gamma$.

We have established (9.8). The estimate (9.7) follows by symmetry.

We now show that $T(1) = {}^tT(1) = 0$. Let Λ_m be an increasing sequence of finite subsets of Λ whose union is Λ and let T_m be the corresponding truncated operators, defined by $\alpha_m(\lambda, \lambda') = \alpha(\lambda, \lambda')$, if $(\lambda, \lambda') \in \Lambda_m \times \Lambda_m$, and $\alpha_m(\lambda, \lambda') = 0$, otherwise. Then $T_m(1) = 0$ and ${}^tT_m(1) = 0$: for once, these equations can be interpreted naively. The operators T_m form a bounded set of Calderón-Zygmund operators, to which we apply Lemma 3 of Chapter 7 (section 3). Passing to the limit, we get $T(1) = {}^t T(1) = 0$.

The operators $T \in \mathcal{O}p\mathcal{M}_\gamma$ have some noteworthy continuity properties which we would not normally expect of Calderón-Zygmund operators. For quite general $\gamma > 0$ we have

Theorem 4. *For all $\gamma > 0$ and each $s \in [-\gamma, \gamma]$, every $T \in \mathcal{O}p\mathcal{M}_\gamma$ is a bounded operator on the homogeneous Besov space $\dot{B}_p^{s,q}$.*

Once again, the verification is purely numerical. We work with the wavelet basis ψ_λ, $\lambda \in \Lambda$, which comes from the Littlewood-Paley multiresolution approximation (*Wavelets and Operators*, Chapter 3, section 2). We then use the characterization of the Besov spaces given by (10.5) of *Wavelets and Operators*, Chapter 6, namely, $\sum \xi(\lambda)\psi_\lambda(x) \in \dot{B}_p^{s,q}$ if and only if

$$(9.9) \qquad \left(\sum_{\lambda \in \Lambda_j} |\xi(\lambda)|^p 2^{nj(p/2-1)} \right)^{1/p} = 2^{-sj}\varepsilon_j \qquad \text{and} \qquad \varepsilon_j \in l^q(\mathbb{Z}) .$$

So it is enough to show that $\eta(\lambda) = \sum_{\lambda' \in \Lambda} \alpha(\lambda, \lambda')\xi(\lambda')$ still fulfils condition (9.9), when $\alpha(\lambda, \lambda')$ satisfies (9.4) and (9.9) holds for $\xi(\lambda)$, $\lambda \in \Lambda$.

To achieve this, we use the following lemma.

Lemma 10. *For $\gamma > 0$ and $n \geq 1$, there exists a constant $C(\gamma, n)$ such that, if $0 < \varepsilon \leq 1$, $1 \leq p \leq \infty$, and $1/p + 1/p' = 1$, then the norm of the matrix $A_\varepsilon = (1 + |k - \varepsilon l|)^{-n-\gamma}$ as an operator on $l^p(\mathbb{Z}^n)$ is not greater than $C(\gamma, n)\varepsilon^{-n/p'}$.*

In order to prove Lemma 10, we first consider the case $p = 1$ and $p' = \infty$. Then the matrices A_ε are uniformly bounded on $l^1(\mathbb{Z}^n)$, because $\sum_k (1 + |k - \varepsilon l|)^{-n-\gamma} \leq C(\gamma, n)$. The other extreme case is $p = \infty$ and $p' = 1$. This time, we observe that $\sum_l \varepsilon^n (1 + |k - \varepsilon l|)^{-n-\gamma} \leq C(\gamma, n)$, because the series is a Riemann sum of a convergent integral.

The general case of Lemma 10 is obtained by interpolation between the two extreme cases.

We return to Theorem 4. Define the weighted coefficients $x_\varepsilon(j, k)$ by

$$x_\varepsilon(j, k) = \xi(\lambda) 2^{nj(1/2 - 1/p)} 2^{sj} ,$$

where $\lambda = 2^{-j}(k + \varepsilon)$ and $\varepsilon = (\varepsilon_1/2, \ldots, \varepsilon_n/2)$, with $(\varepsilon_1, \ldots, \varepsilon_n) \in \{0, 1\}^n \setminus \{(0, \ldots, 0)\}$. With this notation, the condition for belonging to the Besov space becomes $(\sum_k |x_\varepsilon(j, k)|^p)^{1/p} \in l^q(\mathbb{Z}^n)$. Similarly, we define $y_\varepsilon(j, k)$, using $\eta(\lambda)$ and, to simplify the notation, we omit the subscript ε. All this gives

$$y(j,k) = \sum_{j'}\sum_{k'} \beta(j,j',k,k') x(j',k')$$

$$= \sum_{j'\geq j}\sum_{k'}(\cdots) + \sum_{j'<j}\sum_{k'}(\cdots) = u(j,k) + v(j,k).$$

To evaluate $(\sum_k |u(j,k)|^p)^{1/p}$, we apply Lemma 10 with $\varepsilon = 2^{-(j'-j)}$ and get

$$\left(\sum_k |u(j,k)|^p\right)^{1/p} \leq \frac{C(\gamma,n)}{2^{(\gamma+s)(j'-j)}(1+(j'-j)^2)} \left(\sum_{k'} |x(j',k')|^p\right)^{1/p}.$$

Since $\gamma + s \geq 0$, we only need to observe that convolution with $(1+j^2)^{-1}$ defines an operator which is bounded on $l^q(\mathbb{Z}^n)$, for $1 \leq q \leq \infty$.

To deal with $(\sum_k |v(j,k)|^p)^{1/p}$, note that the norm of $A_\varepsilon : l^p(\mathbb{Z}^n) \to l^p(\mathbb{Z}^n)$ is uniformly bounded for $1 \leq \varepsilon$ and $1 \leq p \leq \infty$. We conclude, since $\gamma - s + n/p \geq 0$.

10 Banach spaces of Calderón-Zygmund operators

Let H be a separable Hilbert space. A bounded sequence T_m of continuous operators on H *converges weakly* to $T : H \to H$ if $(T_m x, y) \to (Tx, y)$, for all $x, y \in H$. It is well-known that, given a bounded sequence $T_m : H \to H$, we can extract a subsequence T_{m_k} which converges weakly to some operator $T : H \to H$. This weak compactness theorem is a special case of the following general result.

Lemma 11. *Let E be a separable Banach space. Then the unit ball B of the dual space E^* of E is a topological space which is metrizable and compact for the weak topology $\sigma(E^*, E)$.*

For the case of the unit ball B of $\mathcal{L}(H, H)$, we let E denote the tensor product $H \widehat{\otimes}_\pi H$. Then E^* is exactly $\mathcal{L}(H, H)$, which is how the two statements are connected.

In Chapter 7, we showed that, for a bounded sequence T_m of Calderón-Zygmund operators (that is, T_m satisfy (2.8), (2.9) and (2.10), with the same exponent $\gamma > 0$ and uniform constants C_0 and C_1), the limit T is also a Calderón-Zygmund operator.

Will Lemma 11 help us to interpret this remark? We shall see that the answer is "yes", by showing that the Calderón-Zygmund operators sort themselves, in a natural way, into a family of Banach spaces $\mathcal{L}_\gamma$, $0 < \gamma \leq 1$, satisfying the following conditions:

(10.1) if (2.8), (2.9) and (2.10) are satisfied by T, then $T \in \mathcal{L}_{\gamma'}$, for

$0 < \gamma' < \gamma$, and the norm of T in $\mathcal{L}_\gamma$ only depends on the constants C_0, C_1, on the norm of $T : L^2 \to L^2$ and on γ and γ';

(10.2) if $0 < \gamma' < \gamma$, then $\mathcal{L}_\gamma \subset \mathcal{L}_{\gamma'}$;

(10.3) if $T \in \mathcal{L}_\gamma$, then (2.8), (2.9) and (2.10) are satisfied with the same γ;

(10.4) $\mathcal{L}_\gamma$ is the dual of a separable space $\mathcal{E}_\gamma$.

To define $\mathcal{E}_\gamma$ and $\mathcal{L}_\gamma$, we go back to the decomposition $T = R + S + N$ used in the proof of Theorem 1. We recall that R is the pseudo-product with the function $T(1) \in \text{BMO}$, that S is the transpose of the pseudo-product with ${}^tT(1)$ and that N satisfies $N(1) = {}^tN(1) = 0$.

We write $T \in \mathcal{L}_\gamma$ if (and only if) $N \in \mathcal{O}p\mathcal{M}_\gamma$. The norm of T in $\mathcal{L}_\gamma$ is

(10.5) $$\|T(1)\|_{\text{BMO}} + \|{}^tT(1)\|_{\text{BMO}} + \|N\|_{\mathcal{O}p\mathcal{M}_\gamma},$$

where the last norm is defined to be the infimum of the constants appearing in (9.4).

As a Banach space, $\mathcal{O}p\mathcal{M}_\gamma$ is isomorphic to l^∞. Indeed, $\|N\|_{\mathcal{O}p\mathcal{M}_\gamma}$ is the norm in $l^\infty(\Lambda^2)$ of $\alpha(\lambda, \lambda')/\omega_\gamma(\lambda, \lambda')$ and the bounded sequences in question are entirely arbitrary (because ψ_λ is a Hilbert basis). The Banach space $\mathcal{L}_\gamma$ is thus isomorphic to $\text{BMO}(\mathbb{R}^n) \oplus \text{BMO}(\mathbb{R}^n) \oplus l^\infty(\Lambda^2)$, which is itself the dual of $H^1(\mathbb{R}^n) \oplus H^1(\mathbb{R}^n) \oplus l^1(\Lambda^2)$.

What we have done proves the statements (10.1) to (10.4), but it might be useful to give the relevant isomorphisms explicitly. To do that, we use the following notation. If E is a Banach space whose dual is E^*, if $u \in E$ and $v \in E^*$, then $u \otimes v \in \mathcal{L}(E, E)$ is defined by $(u \otimes v)(x) = v(x)u$, for all $x \in E$.

We also need the idea of unconditional basis of a dual space E^*. Let E be a separable Banach space with an unconditional basis e_j, $j \in \mathbb{N}$. Then every vector $x \in E$ can be written uniquely as $x = \alpha_0 e_0 + \alpha_1 e_1 + \cdots$ and this series converges to x commutatively. The scalars α_j are given by $\alpha_j = \langle e_j^*, x \rangle$, where $\langle \cdot, \cdot \rangle$ denotes the bilinear form which encapsulates the duality between E^* and E.

Let $x^* \in E^*$. We define $\beta_j = \langle x^*, e_j \rangle$ and it follows that, for all $y \in E$,

(10.6) $$\langle x^*, y \rangle = \beta_0 \langle e_0^*, y \rangle + \cdots + \beta_j \langle e_j^*, y \rangle + \cdots.$$

In other words, the series $\beta_0 e_0^* + \cdots + \beta_j e_j^* + \cdots$ converges to x^* in the $\sigma(E^*, E)$ topology. Further, there exists a constant $C \geq 1$, such that all the series $\lambda_0 \beta_0 e_0^* + \cdots + \lambda_j \beta_j e_j^* + \cdots$, with $|\lambda_j| \leq 1$, converge, in the $\sigma(E^*, E)$ topology, to elements $y^* \in E^*$ satisfying $\|y^*\|_{E^*} \leq C \|x^*\|_{E^*}$.

We can then state

Theorem 5. *If θ_λ is given by (6.6), then the union of the three families $\psi_\lambda \otimes \theta_\lambda$, $\lambda \in \Lambda$, $\theta_\lambda \otimes \psi_\lambda$, $\lambda \in \Lambda$, and $\psi_\lambda \otimes \psi_{\lambda'}$, $(\lambda, \lambda') \in \Lambda^2$, forms an unconditional basis of $\mathcal{L}_\gamma$, where the norm is defined by (10.5).*

In other words, every operator $T \in \mathcal{L}_\gamma$ can be written uniquely as

$$(10.7) \quad T = \sum_{\lambda \in \Lambda} \alpha(\lambda)\psi_\lambda \otimes \theta_\lambda + \sum_{\lambda \in \Lambda} \beta(\lambda)\theta_\lambda \otimes \psi_\lambda + \sum_\lambda \sum_{\lambda'} \gamma(\lambda, \lambda')\psi_\lambda \otimes \psi_{\lambda'} \,.$$

Even more explicitly, (10.7) means that the distribution-kernel $S(x, y)$ of T is given by

$$(10.8) \quad S(x, y) = \sum_{\lambda \in \Lambda} \alpha(\lambda)\psi_\lambda(x)\theta_\lambda(y) + \sum_{\lambda \in \Lambda} \beta(\lambda)\theta_\lambda(x)\psi_\lambda(y)$$

$$+ \sum_\lambda \sum_{\lambda'} \gamma(\lambda, \lambda')\psi_\lambda(x)\psi_{\lambda'}(y) \,.$$

Further, if Λ_m, $m \in \mathbb{N}$, is an increasing sequence of finite subsets of Λ whose union is Λ, then the operators T_m, defined by (10.7), but with Λ replaced by Λ_m, form a bounded sequence of Calderón-Zygmund operators whose limit is T.

Given this information alone, it is easy to identify the three series which appear in (10.7). Indeed, we have $T_m(1) = \sum_{\lambda \in \Lambda_m} \alpha(\lambda)\psi_\lambda(x)$ and thus, by a simple passage to the limit, $T(1) = \sum_{\lambda \in \Lambda} \alpha(\lambda)\psi_\lambda(x)$. Thus, the first series on the right-hand side of (10.7) is the pseudo-product with $T(1)$. The second is the transpose of the pseudo-product with ${}^t T(1)$ and the last series is the operator $N \in \mathcal{O}p\mathcal{M}_\gamma$.

It is worth noting that the conditions on the coefficients of (10.7) apply to their absolute values (this is a general consequence of the definition of an unconditional basis). Conditions (6.3) and (6.4) apply to $\alpha(\lambda)$ and $\beta(\lambda)$ and we have condition (9.4) for $\gamma(\lambda, \lambda')$.

11 Variations on the pseudo-product

The proof we have given of the $T(1)$ theorem is based on a bilinear operation $\pi : \mathrm{BMO} \times L^2 \to L^2$ such that, for every function $b \in \mathrm{BMO}$, the operator $f \mapsto \pi(b, f)$ is a Calderón-Zygmund operator. Explicitly,

$$(11.1) \qquad \pi(b, f) = \sum_{\lambda \in \Lambda} \langle b, \psi_\lambda \rangle \langle f, \theta_\lambda \rangle \psi_\lambda \,.$$

We have called this operation a pseudo-product. This is why. If $b \in L^p(\mathbb{R}^n)$ and $f \in L^q(\mathbb{R}^n)$, $1 < p, q < \infty$, then $\pi(b, f)$ belongs to L^r, where $1/r = 1/p + 1/q$. Moreover, we can show that, when b and f are both in $L^2(\mathbb{R}^n)$, then $\pi(b, f)$ belongs to the Stein and Weiss space H^1. To see this, we let g be a function in BMO and try to show that $C\|b\|_2\|f\|_2\|g\|_{\mathrm{BMO}}$ is an upper bound for $|\langle \pi(b, f), g \rangle|$. In fact, we can see immediately that $\langle \pi(b, f), g \rangle = \langle b, \pi(g, f) \rangle$ and we then apply the initial description of π.

The meaning of the above is that the pseudo-product improves on the usual product as far as Hölder's inequality is concerned: the product of two functions in $L^2(\mathbb{R}^n)$ is an L^1 and not an H^1 function. Similarly, the product of a BMO function by an L^2 function does not, in general, belong to L^2.

These new algebraic operations on functions are one of the revolutionary aspects of Calderón's achievements. Let us recall how Calderón's argument goes. We let $\mathbb{H}^p$, $0 < p \leq \infty$, denote the space of functions which are holomorphic in the upper half-plane and satisfy

$$\sup_{y>0} \int_{-\infty}^{\infty} |f(x+iy)|^p \, dx < \infty.$$

If $f \in \mathbb{H}^p$ and $g \in \mathbb{H}^p$, then Calderón defines a third function $h(z)$ holomorphic on $\Im m\, z > 0$, by $h'(z) = f(z)g'(z)$ and $h(i\infty) = 0$.

Since the derivative of fg is given by $fg' + f'g$, we can think of $h(z)$ as one "half" of the product of f and g: the other half would be given by defining $h'(z) = f'(z)g(z)$.

In 1965, Calderón showed that his pseudo-product behaved like the usual product. In particular, his proof of the L^2 continuity of the first commutator (whose distribution kernel is $\mathrm{PV}((A(x) - A(y))/(x - y)^2)$, $A' \in L^\infty(\mathbb{R})$) relies on the inequality $\|h\|_1 \leq C\|f\|_2\|g\|_2$. This procedure can be found again in the proof of the $T(1)$ theorem, which relies on the continuity of the pseudo-product.

Another operation related to the pseudo-product is J.M. Bony's well-known *paraproduct* ([16]) which we shall encounter once more towards the end of this volume.

To define the paraproduct, we let $\gamma(x)$ denote a radial function, in the Schwartz space $S(\mathbb{R}^n)$, whose Fourier transform satisfies $\hat{\gamma}(\xi) = 1$ if $|\xi| \leq 1$ and $\hat{\gamma}(\xi) = 0$ if $|\xi| \geq 3/2$. For all $j \in \mathbb{Z}$, we let $S_j : L^2 \to L^2$ be the operator of convolution with $2^{nj}\gamma(2^j x)$ and we put $\Delta_j = S_{j+1} - S_j$.

The decomposition $1 = \sum_{-\infty}^{\infty} \Delta_j$ is the Littlewood-Paley decomposition. If f belongs to L^2, the support of the Fourier transform of $f_j = \Delta_j(f)$ is contained in the annulus Γ_j defined by $2^j \leq |\xi| \leq 3 \cdot 2^j$. Bony's paraproduct is the bilinear operation defined by

$$B(f,g) = \sum_{-\infty}^{\infty} S_{j-1}(f)\Delta_j(g).$$

The difference between the subscripts is no accident. Its effect is that the product $h_j = S_{j-1}(f)\Delta_j(g)$ has a good "frequency localization". To see this, we note that, if the spectrum (that is, the support of the Fourier transform) of a function u is a compact set A and if that of v is a closed set B, then that of the product uv is contained in $A + B$. On applying this remark to h_j, we find that the support of $\hat{h}_j$ is contained in $(1/4)2^j \leq |\xi| \leq (15/4)2^j$. The frequency localization ensures that the h_j are orthogonal when $j \equiv 0$ (mod 4), or ..., or $j \equiv 3$ (mod 4). Thus

$$\|B(f,g)\|_2 \leq 4 \left(\sum_{-\infty}^{\infty} \|S_{j-1}(f)\Delta_j(g)\|_2^2 \right)^{1/2}$$

and it is easy to show that $C\|f\|_2\|g\|_{\text{BMO}}$ is an upper bound for this expression, by using Carleson's lemma. Bony's paraproduct might just as well have been used to prove Theorem 2, since

$$B(1,g) = \sum_{-\infty}^{\infty} \Delta_j(g) = g \qquad \text{if } g \in \text{BMO}.$$

Let us return to the bilinear operation given by (11.1). Journé has been able to relate the continuity of this operation directly to the fact that the wavelets form an orthonormal basis of the space BMO. Journé's proof is based on the following observation, which we shall prove below.

Proposition 9. *Let* $T : \mathcal{D}(\mathbb{R}^n) \to \mathcal{D}'(\mathbb{R}^n)$ *be a continuous linear operator. Suppose that* T *extends as a continuous operator of* L^∞ *into* BMO *(to do this, we use the denseness of* $\mathcal{D}$ *in* L^∞ *with the weak topology* $\sigma(L^\infty, L^1)$*). If, moreover, the distribution-kernel of* T*, restricted to* $x \neq y$*, satisfies the estimate* $|K(x,y)| \leq C_0|x-y|^{-n}$ *and*

$$(11.2) \qquad \int_{|x-y|\geq 2|y'-y|} |K(x,y') - K(x,y)|\, dx \leq C$$

then T *extends as a continuous linear operator on* $L^2(\mathbb{R}^n)$*.*

The operator to which we apply this remark is $f \mapsto \pi(b,f)$. The proof of (11.2) is an immediate consequence of

$$|\langle b, \psi_\lambda \rangle| \leq \|b\|_{\text{BMO}}\|\psi_\lambda\|_{H^1} \leq C2^{-nj/2}\|b\|_{\text{BMO}}.$$

It is thus enough to check that $\|\pi(b,f)\|_{\text{BMO}} \leq C\|f\|_\infty\|b\|_{\text{BMO}}$ in order to conclude. But we have $|\langle f, \theta_\lambda \rangle| \leq C\|f\|_\infty$. We interpret the right-hand side of (11.1) as a modification of the series $b = \sum_{\lambda \in \Lambda} \langle b, \psi_\lambda \rangle \psi_\lambda$, which is unconditionally convergent in BMO. The modification consists of multiplying by a series of coefficients which belongs to $l^\infty(\Lambda)$.

We still have to prove Proposition 9. We first verify that T is a continuous mapping from H^1 to L^1. We then obtain the required result by interpolation ([75] or [109]).

To prove the continuity of T on H^1, we use the atomic decomposition of $H^1(\mathbb{R}^n)$ given by atoms normalized with respect to the L^∞ norm. Let $a(x)$ be such an atom, with support in a ball B, centre x_0 and radius $R > 0$. The atom is normalized by $\|a\|_\infty \leq \|B\|^{-1}$ and satisfies $\int a(x)\, dx = 0$.

Let B' be the "twin" of B: the radius of B' is R and its centre x_1 satisfies $|x_1 - x_0| = 3R$. Further, we let $3B$ denote the ball with centre x_0 and of radius $3R$.

If $|x - x_0| \geq 3R$, we write $Ta(x) = \int (K(x,y) - K(x,x_0))a(y)\,dy$, so that (11.2) applies and we get

$$\int_{|x-x_0|\geq 3R} |Ta(x)|\,dx \leq C\,.$$

It remains to estimate $\int_{3B} |Ta(x)|\,dx$. Since $a \in L^\infty$, it follows that $Ta \in \mathrm{BMO}$. Of course, this does not of itself allow us to estimate $\int_{3B} |Ta(x)|\,dx$ unless we have some information about the mean of Ta on $3B$. But, since $Ta \in \mathrm{BMO}$, the mean can equally well be estimated by doing the calculation on B'. There, we can use the estimate $|K(x,y)| \leq C_0|x - y|^{-n}$ and the result follows immediately, by a computation that we leave to the reader.

12 Additional remarks

One can try to find the minimal conditions on the regularity (with respect to x and y) of the kernel $K(x,y)$ of T, under which the conclusions of Theorem 1 still hold.

In order that $T(1)$ and ${}^tT(1)$ can be defined, it is sufficient that $K(x,y)$ satisfies the Hörmander conditions

$$(12.1) \qquad \int_{|x-y|\geq 2|y'-y|} |K(x,y') - K(x,y)|\,dx \leq C$$

and

$$(12.2) \qquad \int_{|x-y|\geq 2|x'-x|} |K(x',y) - K(x,y)|\,dy \leq C < \infty\,.$$

We do not know whether Theorem 1 is still valid under these conditions. The delicate part of the proof is that of the continuity of T under conditions (12.1), (12.2), $T(1) = 0$, ${}^tT(1) = 0$ and the weak continuity of T.

No longer are the operators of this chapter convolution operators. It thus seems natural that the $T(1)$ theorem should work in geometric frameworks other than that of locally compact abelian groups. A natural geometric framework has been defined by Coifman and Weiss: that of spaces of homogeneous type. These authors show that most of the results of Chapter 7 remain valid in this context. Finally, David, Journé and Semmes have proved the $T(1)$ theorem for such spaces ([76], [97], and [98]).

9

Examples of Calderón-Zygmund operators

1 Introduction

Initially, there was no difference between Calderón-Zygmund operators and pseudo-differential operators.

The former class consisted of kernels $K(x,y) = L(x, x - y)$, where $L(x,z) \in C^\infty(\mathbb{R}^n \times \mathbb{R}^n \setminus \{0\})$ satisfied

$$|\partial_x^\alpha \partial_z^\beta L(x,z)| \leq C(\alpha,\beta), \qquad \text{if } |z| = 1,$$
$$L(x, \lambda z) = \lambda^{-n} L(x,z), \qquad \text{if } \lambda > 0,$$

and, lastly,

$$\int L(x,z)\, d\sigma(z) = 0,$$

where $d\sigma$ was the usual surface measure of the unit sphere $|z| = 1$ in $\mathbb{R}^n$. The operator T was then defined by $Tf(x) = \text{PV} \int L(x, x - y)f(y)\, dy$.

The latter class consists of those operators which correspond to symbols $\sigma(x,\xi) \in C^\infty(\mathbb{R}^n \times \mathbb{R}^n \setminus \{0\})$ satisfying

$$|\partial_\xi^\alpha \partial_x^\beta \sigma(x,\xi)| \leq C(\alpha,\beta), \qquad \text{for } |\xi| = 1,$$

and

$$\sigma(x, \lambda\xi) = \sigma(x,\xi), \qquad \text{for } \lambda > 0.$$

It is not necessary to suppose that the mean of the symbol on $|\xi| = 1$ is zero, as long as we allow the inclusion of symbols which do not depend on ξ (these correspond to the operators of pointwise multiplication by bounded functions of x).

The relationship between the symbol $\sigma(x,\xi)$ and the operator T is given by $Tf(x) = (2\pi)^{-n} \int e^{ix\cdot\xi} \sigma(x,\xi)\hat{f}(\xi)\, d\xi$. Basically, $T(e^{ix\cdot\xi}) = \sigma(x,\xi)e^{ix\cdot\xi}$ and this operation is the analogue of amplitude modulation in radio detection. Finally, kernel and symbol are related by

$$(1.1) \qquad \text{PV} \int L(x,y)e^{-i\xi\cdot y}\, dy = \sigma(x,\xi)$$

(here, the Fourier transform is to be understood in the sense of distributions).

The relationship (1.1), discovered by Calderón and Zygmund in the

1950s, opened the way to all later developments in which the pseudo-differential operators were defined using algebras of symbols, without reference to any kernels. After the golden age just described, the two points of view diverged: Kohn and Nirenberg, for their part, and Hörmander, for his, systematically favoured the definition of pseudo-differential operators by symbols. Research on kernels remained very active in the school of Calderón and Zygmund and led to what we now call the "Calderón-Zygmund operators."

It remains to be seen whether the operators in question can be defined by symbols satisfying simple conditions of regularity and rate of growth at infinity.

Unfortunately, this is not the case for the set $\mathcal{C}$ of the operators of Calderón's programme. However, there does exist an algebra $\mathcal{A}_\infty \subset \mathcal{C}$ of operators which can be described, either by their kernels, or by their symbols, or by the matrices of their coefficients in a wavelet basis.

For example, suppose we start with symbols satisfying the illicit estimates

(1.2) $|\partial_\xi^\alpha \partial_x^\beta \sigma(x,\xi)| \leq C(\alpha,\beta)|\xi|^{|\alpha|-|\beta|}$.

We know that such conditions do not give operators which are bounded on $L^2(\mathbb{R}^n)$. In the light of the $T(1)$ theorem, however, we can see that it will be sufficient to require that the symbols of T and its adjoint T^* satisfy (1.2). The set $\mathcal{A}_\infty$ of these operators is a subalgebra of $\mathcal{L}(L^2(\mathbb{R}^n), L^2(\mathbb{R}^n))$. One of the purposes of this chapter is to show that the algebra $\mathcal{A}_\infty$ can be characterized just as well by conditions on the distribution-kernels of the operators.

We may also characterize the operators $T \in \mathcal{A}_\infty$ by their matrices in the orthonormal basis of Littlewood-Paley wavelets, and this property enables us to investigate the symbolic calculus. This algebra is unlike the usual algebras of pseudo-differential operators, in that there exists an operator $T \in \mathcal{A}_\infty$ whose inverse does not belong to $\mathcal{A}_\infty$, even though T is an isomorphism on $L^2(\mathbb{R}^n)$.

The operators $T \in \mathcal{A}_\infty$ arise when we use wavelets as unconditional bases in classical spaces of functions or distributions. If the orthonormal basis ψ_λ, $\lambda \in \Lambda$, is an unconditional basis of B, every bounded operator on $L^2(\mathbb{R}^n)$ which is diagonal with respect to the basis ψ_λ can be extended as a continuous linear operator on B. These are the operators which belong to $\mathcal{A}_\infty$.

The second group of examples that we deal with in this chapter cannot be described in the usual language of pseudo-differential operators. In the course of Chapter 13, we shall see that we can approach these examples by extending that language to multilinear operators. What we are discussing are the commutators $\Gamma_1 = [A_1, L_1]$, $\Gamma_2 = [A_1, [A_2, L_2]], \ldots, \Gamma_k = [A_1, [A_2, \ldots, [A_k, L_k]] \ldots]$, where the L_j are classical pseudo-differential operators of order j and the A_j are operators of pointwise multiplication by bounded functions $a_j(x)$ of limited regularity. Calderón's programme consists of exploring the algebras of operators containing the usual pseudo-differential operators and the operators of multiplication by functions which are only slightly regular.

We thus look for conditions of minimal regularity on the functions $a_j(x)$ so that the commutators $\Gamma_1, \ldots, \Gamma_k, \ldots$ are bounded on $L^2(\mathbb{R}^n)$, whatever the choice of operators $L_1, \ldots, L_k, \ldots$, of order $1, \ldots, k, \ldots$. By choosing L_k to be a differential operator of order k with constant coefficients, we see that the $a_j(x)$ must, necessarily, be Lipschitz functions.

The fact that this necessary condition is also sufficient is one of the nicest applications of the $T(1)$ theorem.

The final example of a Calderón-Zygmund operator that we present in this chapter is the Cauchy kernel on a Lipschitz curve and the operators that can be constructed therefrom by the method of rotations of Calderón and Zygmund.

We shall study the Cauchy kernel more systematically in Chapter 12, but here we want to show that its continuity is a consequence of the $T(1)$ theorem and some new real-variable methods introduced by G. David and then simplified by T. Murai ([193], [194]).

2 Pseudo-differential operators and Calderón-Zygmund operators

Let $\sigma(x, \xi) \in L^\infty(\mathbb{R}^n \times \mathbb{R}^n)$ and let $\sigma(x, D) : \mathcal{S}(\mathbb{R}^n) \to C^\infty(\mathbb{R}^n)$ be the operator defined by

$$(2.1) \qquad \sigma(x, D)f(x) = \frac{1}{(2\pi)^n} \int e^{ix \cdot \xi} \sigma(x, \xi) \hat{f}(\xi) \, d\xi.$$

If R is the translation operator defined by $(Rf)(x) = f(x - x_0)$, then $R^{-1}\sigma(x, D)R = \tau(x, D)$, where $\tau(x, \xi) = \sigma(x + x_0, \xi)$. If, now, δ_a, $a > 0$ is the dilation operator defined by $(\delta_a f)(x) = f(a^{-1}x)$, then $\delta_a^{-1}\sigma(x, D)\delta_a = \tau(x, D)$, where $\tau(x, \xi) = \sigma(ax, a^{-1}\xi)$.

The hypothesis $\sigma(x, \xi) \in L^\infty(\mathbb{R}^n \times \mathbb{R}^n)$ is isometrically invariant under these two operations. This enables us to make the following observation.

Lemma 1. *If $\sigma(x, \xi) \in L^\infty(\mathbb{R}^n \times \mathbb{R}^n)$, then the operator $\sigma(x, D)$ is weakly continuous on $L^2(\mathbb{R}^n)$.*

Indeed, let f be a C^q function, $q > n/2$, with compact support. We first show that $\hat{f}(\xi) \in L^1(\mathbb{R}^n)$. We suppose that q is an integer: then $x^\alpha \hat{f}(\xi)$ belongs to $L^2(\mathbb{R}^n)$ for $|\alpha| \le q$, since $\partial^\alpha f$ is continuous with compact support. So

$$(1 + |\xi|^q)\hat{f}(\xi) \in L^2(\mathbb{R}^n)$$

and it follows that

$$\hat{f}(\xi) \in L^1(\mathbb{R}^n),$$

by the Cauchy-Schwarz inequality.

Thus, if f is a C^q function, $q > n/2$, with support in the unit ball $|x| \le 1$, it follows from (2.1) that

$$|\sigma(x, D)f(x)| \le C\|f\|_{C^q}$$

which implies the weak continuity property. To pass to functions with support in an arbitrary ball, we use the remarks about invariance under the action of the group $ax + b$, $a > 0$, $b \in \mathbb{R}^n$.

Lemma 2. *Suppose that $\sigma(x, \xi)$ belongs to $C^\infty(\mathbb{R}^n \times \mathbb{R}^n \setminus \{0\})$ and that*

(2.2) $|\partial_\xi^\alpha \partial_x^\beta \sigma(x, \xi)| \le C(\alpha, \beta)|\xi|^{|\beta| - |\alpha|}$, $\alpha, \beta \in \mathbb{N}^n$.

Then the distribution-kernel $S(x, y)$ of $\sigma(x, D)$ satisfies

(2.3) $|\partial_x^\alpha \partial_y^\beta S(x, y)| \le C'(\alpha, \beta)|x - y|^{-n - |\alpha| - |\beta|}$, $\alpha, \beta \in \mathbb{N}^n$.

To see this, we use a technique of approximation by truncated symbols. Let $\psi_0(\xi)$ be a function in $\mathcal{D}(\mathbb{R}^n)$ which equals 1, if $|\xi| \le 1/2$, and equals 0, if $|\xi| \ge 1$. We put $\phi_1 = 1 - \phi_0$ and replace $\sigma(x, \xi)$ by $\sigma_j(x, \xi) = \sigma(x, \xi)\phi_0(j^{-1}\xi)$, $j \ge 1$. If $T_j = \sigma_j(x, D)$, it is an immediate consequence that, for every pair (f, g) of functions in $\mathcal{S}(\mathbb{R}^n)$,

(2.4) $\langle \sigma(x, D)f, g \rangle = \lim_{j \to \infty} \langle T_j f, g \rangle$.

The principle of the proof will be to show that the distribution-kernels S_j of T_j satisfy (2.3) uniformly. Since S is the weak limit of the S_j, we obtain (2.3) by passing to the limit.

In what follows, another truncation is going to be useful. The hypothesis on $\sigma(x, \xi)$ does not exclude a discontinuity at $\xi = 0$. If we want to avoid this problem, as we shall need to later, we replace $\sigma(x, \xi)$ by $\sigma(x, \xi)\phi_1(j\xi)\phi_0(j^{-1}\xi)$.

The essential fact is that the estimate (2.2) is not affected by these adjustments. We shall therefore forget about the subscript j in the calculations which follow. Moreover, by the previous considerations we can suppose that $\sigma(x, \xi) \in C^\infty(\mathbb{R}^n \times \mathbb{R}^n)$, $\sigma(x, 0) = 0, \ldots, \partial_\xi^\alpha \sigma(x, 0) = 0, \ldots$ and that, with respect to ξ, the support of $\sigma(x, \xi)$ is compact.

With these hypotheses in mind, the relationship between symbol and kernel is, evidently,

$$(2.5) \qquad S(x,y) = \frac{1}{(2\pi)^n} \int e^{i\xi \cdot (x-y)} \sigma(x,\xi) \, d\xi .$$

To prove (2.3), we begin by supposing that $|x - y| = 1$ and we first find a bound for $|S(x,y)|$.

We use the partition $1 = \phi_0(\xi) + \phi_1(\xi)$ once again by putting

$$S_0(x,y) = \frac{1}{(2\pi)^n} \int e^{i\xi \cdot (x-y)} \sigma(x,\xi) \phi_0(\xi) \, d\xi$$

and

$$S_1(x,y) = \frac{1}{(2\pi)^n} \int e^{i\xi \cdot (x-y)} \sigma(x,\xi) \phi_1(\xi) \, d\xi .$$

The required estimate of $|S_0(x,y)|$ is obvious and we integrate by parts $2m$ times to get

$$S_1(x,y) = \frac{1}{(2\pi)^n} (-1)^m \int e^{i\xi \cdot (x-y)} \Delta^m (\sigma(x,\xi) \phi_1(\xi)) \, d\xi .$$

We now exploit the hypotheses applying to the symbol and the fact that $\partial^\alpha \phi_1(\xi) = 0$, if $\alpha \neq 0$ and $|\xi| \geq 1$ or $|\xi| \leq 1/2$. From the latter considerations, we conclude that the term $(\Delta^m \sigma(x,\xi)) \phi_1(\xi)$ is the only one for which the range of integration is unbounded, and, for $2m > n+1$, the hypotheses allow us to conclude.

The estimates for $|\partial_x^\alpha \partial_y^\beta S(x,y)|$ are similar, when $|x - y| = 1$.

For the general case of $y \neq x$, we use the action of the group $ax + b$, $a > 0$, $b \in \mathbb{R}^n$. For $x_0 \in \mathbb{R}^n$ and $a > 0$, we observe that, if we conjugate the operator $T = \sigma(x, D)$ by the translation by x_0 and the dilation by a, we pass from the kernel $S(x,y)$ to the kernel $a^n S(ax + x_0, ay + y_0)$ and from the symbol $\sigma(x,\xi)$ to the symbol $\sigma(ax + x_0, a^{-1}\xi)$. Now, the hypotheses we have made on the symbol σ are not affected by these changes of variables. It is this observation that allows us deduce the general case from that in which $|x - y| = 1$.

The fundamental problem is whether the operator $T = \sigma(x, D)$ can be extended as a continuous linear operator on $L^2(\mathbb{R}^n)$. To this end, we apply the $T(1)$ theorem. The process starts with the following lemma.

Lemma 3. *With the hypotheses of Lemma 2, $T(1) = 0$.*

We establish this result by the method of approximating the symbol $\sigma(x,\xi)$ of T by the truncated symbols $\sigma_j(x,\xi)$ corresponding to the operators T_j.

Clearly we have $T_j(1) = 0$, in the simplest sense. Moreover, the distribution-kernels S_j of T_j satisfy (2.3) uniformly in j and are weakly continuous, uniformly in j. The T_j converge weakly to T in the sense of

(2.4). As a consequence, $T_j(1)$ converges to $T(1)$ in the space $\dot{B}^{0,\infty}_\infty$, in the $\sigma(\dot{B}^{0,\infty}_\infty, \dot{B}^{0,1}_1)$ topology. The proof of this remark is a straightforward adaptation of the proof of Lemma 3 in Chapter 7. Hence $T(1) = 0$.

We still have to calculate $\sigma = {}^tT(1)$.

We already know that σ belongs to $\dot{B}^{0,\infty}_\infty$. Because of this, σ is a continuous linear functional on $\dot{B}^{0,1}_1$. In order to proceed with the calculations, we need the following lemma.

Lemma 4. *The vector space V of functions $f \in \mathcal{S}(\mathbb{R}^n)$, whose Fourier transforms are zero in a neighbourhood of 0, is dense in $\dot{B}^{0,1}_1$.*

Here is a very simple proof of this remark.

We let S denote a continuous linear functional on $\dot{B}^{0,1}_1$ which satisfies $\langle S, f \rangle = 0$ for all $f \in V$. Then $S \in \dot{B}^{0,\infty}_\infty$ and we choose an arbitrary representative of S, modulo the affine functions, which is a tempered distribution. We still use S to denote this representative.

On passing to the Fourier transform, we get $\langle \hat{S}, g \rangle = 0$, whenever $g \in \mathcal{S}(\mathbb{R}^n)$ is zero in a neighbourhood of 0. Hence the support of the distribution $\hat{S}$ is 0 and thus S is a polynomial. Since S belongs to $\dot{B}^{0,\infty}_\infty$, it follows that S must be a constant. Hence the class of S in $\dot{B}^{0,\infty}_\infty$ is zero.

We can now proceed to the calculation of $\sigma = {}^tT(1)$.

Proposition 1. *Let $\sigma(x,\xi)$ be a symbol satisfying condition (2.2). Then the integral $(2\pi)^{-n} \int e^{ix\cdot\xi}\sigma(x,\xi)\,dx$ defines a distribution τ on $\mathbb{R}^n \setminus \{0\}$ and ${}^tT(1)$ is the generalized Fourier transform of τ in the sense that, for $f \in V$,*

(2.6) $\langle {}^tT(1), f \rangle = \dfrac{1}{(2\pi)^n} \displaystyle\int\int e^{ix\cdot\xi}\hat{f}(\xi)\sigma(x,\xi)\,dx\,d\xi.$

We first of all note that the convergence of the integral on the right-hand side of (2.6) poses no problem. Indeed, we consider the integral $I(x) = \int e^{ix\cdot\xi}\hat{f}(\xi)\sigma(x,\xi)\,d\xi$, observing that this function is continuous and $O(|x|^{-N})$ at infinity, for every integer N. This last property is established using integration by parts.

The distribution τ is thus well-defined, and (2.6) makes sense. The proof of the identity (2.6) is found by writing

$$\langle {}^tT(1), f \rangle = \int T(f)\,dx = \frac{1}{(2\pi)^n}\int I(x)\,dx.$$

We give an example. Let ψ_λ, $\lambda \in \Lambda$, be an orthonormal basis of infinitely differentiable wavelets (arising from the Littlewood-Paley multiresolution approximation). Let $\theta \in \mathcal{D}(\mathbb{R}^n)$ be a function of mean 1 and, for $\lambda = 2^{-j}(k + \varepsilon/2)$, $k \in \mathbb{Z}^n$, $\varepsilon = (\varepsilon_1,\ldots,\varepsilon_n) \in \{0,1\}^n \setminus \{(0,\ldots,0)\}$, let us put $\theta_\lambda(x) = 2^{nj}\theta(2^j x - k)$.

Consider the operator T defined by

(2.7)
$$Tf(x) = \sum_{\lambda \in \Lambda} \theta_\lambda(x) \langle a, \psi_\lambda \rangle \langle f \psi_\lambda \rangle,$$

where $a \in \dot{B}_\infty^{0,\infty}$.

The symbol $\sigma(x, \xi)$ of T can be calculated immediately and we get

(2.8)
$$\sigma(x, \xi) = \sum \sum \alpha(\lambda) \theta(2^j x - k) e^{-i 2^{-j} \xi \cdot (2^j x - k)} \hat{\psi}_\varepsilon(-2^{-j} \xi),$$

where $\alpha(\lambda) = 2^{nj/2} \langle a, \psi_\lambda \rangle \in l^\infty(\Lambda)$, because $a \in \dot{B}_\infty^{0,\infty}$.

The only values of j for which $\hat{\psi}_\varepsilon(-2^{-j}\xi) \neq 0$ are those for which $c_1 2^j \leq |\xi| \leq c_2 2^j$, $c_2 > c_1 > 0$ and, for fixed j and x, the sum over k is restricted by the compact support of θ. These two remarks let us find a bound for $|\partial_\xi^\alpha \partial_x^\beta \sigma(x, \xi)|$ and obtain the required estimates.

We have $T(1) = 0$ and ${}^t T(1) = a \in \dot{B}_\infty^{0,\infty}$. Note that in this example, the transpose operator ${}^t T$ is exactly the pseudo-product of a and f. The operator T is continuous on $L^2(\mathbb{R}^n)$ if and only if $a \in \mathrm{BMO}$.

We are about to come to the main definition. If T is an operator, defined in the weak sense, and if the distribution-kernel of T satisfies (2.3), then we shall write $T(x^\alpha) = 0$, modulo the polynomials of degree $\leq |\alpha| = m$, if, for every function $g \in \mathcal{D}(\mathbb{R}^n)$, satisfying $\int x^\beta g(x)\, dx = 0$, for every multi-index β of height $|\beta| \leq m$, we have $\int {}^t T(g) x^\alpha\, dx = 0$. We note that, outside the support of $g(x)$, we have ${}^t Tg(x) = \int S(y,x) g(y)\, dy = O(|x|^{-n-m-1})$. This can be seen by observing that $g = \sum_{|\gamma|=m+1} \partial^\gamma g_\gamma$, where $\partial^\gamma = (\partial/\partial x_1)^{\gamma_1} \cdots (\partial/\partial x_n)^{\gamma_n}$ and $g_\gamma \in \mathcal{D}(\mathbb{R}^n)$.

Definition 1. *We say that a Calderón-Zygmund operator T belongs to the class $\mathcal{A}_\infty$ if the following three conditions are satisfied.*

(2.9) *The kernel $K(x,y)$ corresponding to T is infinitely differentiable off the diagonal and satisfies*
$$|\partial_x^\alpha \partial_y^\beta K(x,y)| \leq C(\alpha, \beta)|x-y|^{-n-|\alpha|-|\beta|} \qquad \text{for } \alpha, \beta \in \mathbb{N}^n.$$

(2.10) *For all $\alpha \in \mathbb{N}^n$, $T(x^\alpha) = 0$, modulo the polynomials of degree $\leq |\alpha|$.*

(2.11) *For all $\alpha \in \mathbb{N}^n$, ${}^t T(x^\alpha) = 0$, modulo the polynomials of degree $\leq |\alpha|$.*

The following theorem provides a characterization of $\mathcal{A}_\infty$.

Theorem 1. *An operator T belongs to $\mathcal{A}_\infty$ if and only if one of the following conditions is satisfied.*

(2.12) *For every $\gamma > 0$, T belongs to the algebra $Op\mathcal{M}_\gamma$.*

(2.13) *The symbols of T and ^tT satisfy condition (2.2).*

The significance of this characterization is that it relates a description in terms of kernels to one in terms of symbols and to a characterization of the corresponding matrices.

Corollary. *The vector space $\mathcal{A}_\infty$ is an algebra of operators.*

Indeed, $Op\mathcal{M}_\gamma$ is an algebra for each $\gamma > 0$.

Let us turn to the proof of Theorem 1.

Essentially, this proof continues our examination of the classes A_γ which we started in Chapter 8. That is, estimates for the entries of the matrix corresponding to $T \in \mathcal{A}_\infty$ are found by repeating, word for word, the calculations of Chapter 8, but using (2.10) or (2.11) and applying Taylor's formula of order m to expand the function f, of Lemma 3 in Chapter 8, about x_0. The details are left to the reader.

Once we know that T belongs to $Op\mathcal{M}_\gamma$, for all $\gamma > 0$, it is easy to work out the symbol of T and to check (2.2). The calculation does not involve any difficulties and is left to the reader.

Lastly, we suppose that the symbol of T satisfies (2.2). Then Lemma 2 tells us that the kernel K of T satisfies (2.3). We get $T(x^\alpha) = 0$, modulo the polynomials of degree $\leq |\alpha|$, using the technique of approximating T by T_j, which we have already described. Repeating this argument for the transpose of T concludes the proof of Theorem 1.

A result similar to Theorem 1 was obtained by G. Bourdaud ([21]). Instead of (2.2), we use symbols $\sigma(x,\xi) \in C^\infty(\mathbb{R}^n \times \mathbb{R}^n)$ satisfying

(2.14) $$|\partial_\xi^\alpha \partial_x^\beta \sigma(x,\xi)| \leq C(\alpha,\beta)(1+|\xi|)^{|\beta|-|\alpha|}.$$

Using the terminology due to L. Hörmander, σ belongs to the class of symbols $S_{1,1}^0(\mathbb{R}^n \times \mathbb{R}^n)$.

Then we write $T \in \mathcal{B}_\infty$ if the symbols of both T and its transpose satisfy (2.14).

The operators $T \in \mathcal{B}_\infty$ are characterized by estimates of wavelet coefficients. But, to do this, we need to use the orthonormal basis consisting of ψ_λ, $\lambda \in \Lambda_j$, $j \in \mathbb{N}$ and ϕ_k, defined by $\phi_k(x) = \phi(x - k)$, $k \in \mathbb{Z}^n$. We note that Λ is the disjoint union of $\Gamma_0 = \mathbb{Z}^n$ and the Λ_j, $j \in \mathbb{N}$ and that, by abuse of language, we can write ϕ_λ instead of ϕ_k, for $\lambda \in \Gamma_0$.

Associated with $\lambda \in \Lambda_j$, $j \in \mathbb{N}$, are the dyadic cubes $Q(\lambda)$ defined by $2^j - k \in [0,1)^n$, where $\lambda = 2^{-j}(k + \varepsilon/2)$, $k \in \mathbb{Z}^n$, $\varepsilon \in E = \{0,1\}^n \setminus$

$\{(0,\ldots,0)\}$. On the other hand, if $\lambda \in \Lambda_0$, the corresponding cube is simply $\lambda + [0,1)^n$.

Lastly, we introduce a distance on the collection of all dyadic cubes of side ≤ 1 which are involved. If Q and R are two of these cubes, we let $\lambda(Q, R)$ denote the greatest lower bound of the numbers $\lambda \geq 1$ such that λQ contains R and λR contains Q (where λQ has the same centre as Q, but the diameter of λQ is λ times the diameter of Q). Then $d(Q, R) = \log_2 \lambda(Q, R)$ is a distance on our set of dyadic cubes. From this, we obtain a metric on Λ by writing $d(\lambda, \lambda') = d(Q(\lambda), Q(\lambda')) + d(\varepsilon, \varepsilon')$, where $d(\varepsilon, \varepsilon') = 0$ or 1, depending on whether $\varepsilon = \varepsilon'$ or $\varepsilon \neq \varepsilon'$ ($\varepsilon, \varepsilon' \in E$).

The characterization of operators $T \in \mathcal{B}_\infty$ is then given·by the following theorem.

Theorem 2. *Let* $T : \mathcal{S}(\mathbb{R}^n) \to \mathcal{S}'(\mathbb{R}^n)$ *be an operator which is weakly defined. A necessary and sufficient condition for* T *to belong to* $\mathcal{B}_\infty$ *is that the matrix*

$$\alpha(\lambda, \lambda') = \langle T(\psi_\lambda), \psi_{\lambda'} \rangle, \qquad (\lambda, \lambda') \in \Lambda^2$$

(with ψ_λ *replaced by* ϕ_λ *if* $\lambda \in \Gamma_0$ *and the same for* $\psi_{\lambda'}$*), satisfies the following condition: for* $N \geq 1$ *there is a constant* $C(N)$ *such that*

$$(2.15) \qquad |\alpha(\lambda, \lambda')| \leq C(N)e^{-Nd(\lambda, \lambda')}.$$

It is worth remarking that, if we had forgotten to replace ψ_λ by ϕ_λ when $\lambda \in \Gamma_0$, (2.15) would have characterized $\mathcal{A}_\infty$, which is the homogeneous version of $\mathcal{B}_\infty$. In the homogeneous version, the big cubes play a rôle which is as important as that of the small cubes. The distribution-kernels $S(x, y)$ of the operators $T \in \mathcal{B}_\infty$ decrease rapidly at infinity, as do the derivatives with respect to x and y, and the big cubes are not involved in the analysis of these operators.

The proof of Theorem 2 is similar to that of Theorem 1, so we shall omit it.

In order to get an idea of the algebra $\mathcal{B}_\infty$ we give some examples. We start with that given by the class of symbols $S_{1,\delta}^0$, $0 \leq \delta < 1$. Recall that $\sigma(x, \xi) \in S_{1,\delta}^0$, if $|\partial_\xi^\alpha \partial_x^\beta \sigma(x, \xi)| \leq C(\alpha, \beta)(1 + |\xi|)^{\delta|\beta|-|\alpha|}$. We verify that $Op\, S_{1,\delta}^0$ is a self-adjoint algebra of continuous operators on $L^2(\mathbb{R}^n)$ and that $Op\, S_{1,\delta}^0$ is contained in $\mathcal{B}_\infty$.

Another example of "natural" operators belonging to $\mathcal{B}_\infty$ is provided by Bony's paradifferential operators, to which we shall return in Chapter 16. We consider the symbols $\sigma(x, \xi)$, belonging to $C^\infty(\mathbb{R}^n \times \mathbb{R}^n)$, such that

$$|\partial_\xi^\alpha \sigma(x, \xi)| \leq C(\alpha)(1 + |\xi|)^{-|\alpha|}$$

and having the following property: for every $\xi \in \mathbb{R}^n$, the Fourier transform of $\sigma(x, \xi)$, regarded as a function of x, is a function or distribution whose support is contained in the ball $|\cdot| \leq |\xi|$. Then the corresponding operators $T = \sigma(x, D)$ belong to $\mathcal{B}_\infty$.

To see this, we first apply Bernstein's lemma to $\sigma(x, \xi)$, regarded as a function of x, for each fixed ξ. We get

$$|\partial_x^\beta \partial_\xi^\alpha \sigma(x, \xi)| \leq C(\alpha)(1 + |\xi|)^{-|\alpha|}|\xi|^{|\beta|}.$$

Thus $\sigma(x, \xi) \in S_{1,1}^0$ and the distribution-kernel $S(x, y)$ of $T = \sigma(x, D)$, restricted to $y \neq x$, satisfies $|\partial_x^\alpha \partial_y^\beta S(x, y)| \leq C(\alpha, \beta)|x - y|^{-n-|\alpha|-|\beta|}$.

Further, $T(x^\alpha) = 0$, modulo the polynomials of degree $\leq |\alpha|$: this can be demonstrated by adapting the proof of Lemma 3. What is more, ${}^tT(x^\alpha) = 0$, modulo the polynomials of degree $\leq |\alpha|$, because of the particular hypothesis about the symbol $\sigma(x, \xi)$. We see this by approximating $\sigma(x, \xi)$ by $\sigma_m(x, \xi) = \sigma((1 - 1/m)x, \xi)$, where $m \geq 2$, which gives, in turn, an approximation of the corresponding operator T by operators T_m, which satisfy the same hypotheses as T, uniformly in m. Then ${}^tT_m(x^\alpha) = 0$ and the corresponding result for ${}^tT$ is obtained by passing to the limit. For example, let us compute ${}^tT_m(1)$. We apply Proposition 1. This leads us to consider $\int e^{ix \cdot \xi} \sigma_m(x, \xi)\, dx$. But this integral is zero for all ξ, because of the hypothesis about the Fourier transform of the symbol.

We have just explained why T belongs to $\mathcal{A}_\infty$. The fact that T belongs to $\mathcal{B}_\infty$ is a consequence of the C^∞ regularity of the symbol, in that the estimate

$$|\partial_x^\alpha \partial_y^\beta S(x, y)| \leq C(\alpha, \beta)|x - y|^{-n-|\alpha|-|\beta|}$$

can be improved to

$$|\partial_x^\alpha \partial_y^\beta S(x, y)| \leq C(N, \alpha, \beta)|x - y|^{-N} \qquad \text{for all } N \geq 1,$$

as long as $|x - y| \geq 1$.

Having given some motivation, with these examples, for studying the algebra $\mathcal{B}_\infty$, we return to the fundamental problem of the *symbolic calculus* in the algebra of operators $\mathcal{B}_\infty$. According to A. Calderón, a symbolic calculus for a Banach algebra B is a continuous homomorphism $\chi : B \to C$, where C is a Banach algebra which is "simpler" than B and where χ has the property that $b \in B$ is invertible in B if and only if $c = \chi(b)$ is invertible in C. The Banach algebras B and C may be non-commutative.

In our case, the algebra B is $\mathcal{B}_\infty$ and C is $\mathcal{L}(L^2(\mathbb{R}^n), L^2(\mathbb{R}^n))$, the algebra of all continuous operators on $L^2(\mathbb{R}^n)$. The homomorphism χ is the continuous injection of B into C. We ask whether this injection is a symbolic calculus for $\mathcal{B}_\infty$. This comes down to knowing whether an operator $T \in \mathcal{B}_\infty$, which is invertible as a continuous operator on $L^2(\mathbb{R}^n)$, has as its inverse an operator $T^{-1} \in \mathcal{B}_\infty$. We shall give a decidedly negative answer to this question.

We remark that every operator $L \in \mathcal{B}_\infty$ is a Calderón-Zygmund operator and is thus bounded on $L^p(\mathbb{R}^n)$, for $1 < p < \infty$.

Theorem 3. *For every $p > 0$, there exists an operator $T \in \mathcal{B}_\infty$ which is an isomorphism of $L^2(\mathbb{R}^n)$ with itself, while the inverse T^{-1} is not bounded on $L^p(\mathbb{R}^n)$.*

This result was first discovered by P. Tchamitchian. We give a different example, due to P.G. Lemarié.

Recall that $\Lambda_j = 2^{-j-1}\mathbb{Z}^n \setminus 2^{-j}\mathbb{Z}^n$ and that Λ can be considered as the disjoint union of $\Gamma_0 = \mathbb{Z}^n$ and the Λ_j, $j \in \mathbb{N}$.

We define a mapping $\theta : \Lambda \to \Lambda$ by $\theta(k) = k$ if $k \in \Gamma_0 = \mathbb{Z}^n$ and $\theta(k2^{-j} + \varepsilon 2^{-j-1}) = k2^{-j} + \varepsilon 2^{-j-2}$ if $\lambda = k2^{-j} + \varepsilon 2^{-j-1} \in \Lambda_j$, $\varepsilon \in E$. We note that $\theta(\lambda) \in \Lambda_{j+1}$ and that θ defines an injective mapping of Λ into Λ.

Let $U : L^2(\mathbb{R}^n) \to L^2(\mathbb{R}^n)$ be the partial isometry corresponding to θ, which is defined by $U\phi_k = \phi_k$, $\phi_k(x) = \phi(x - k)$, $k \in \mathbb{Z}^n$, and $U\psi_\lambda = U\psi_{\theta(\lambda)}$. Since the distance from λ to $\theta(\lambda)$ is bounded as λ runs through Λ, the operator U belongs to $\mathcal{B}_\infty$.

If z is a complex number with $|z| < 1$, then $T = 1 - zU$ is invertible on $L^2(\mathbb{R}^n)$. We shall show that, for $p \geq p_0(|z|)$, the operator T^{-1} cannot be bounded on L^p. Thus T^{-1} cannot be a Calderón-Zygmund operator and, a fortiori, T^{-1} cannot belong to $\mathcal{B}_\infty$.

We have $T^{-1}(f) = \sum_0^\infty z^k T^k(f)$ and, if $f = \psi_\lambda$, this gives $T^{-1}(\psi_\lambda) = \sum_0^\infty z^k \psi_{\theta^k(\lambda)}$. It is then a straightforward matter, using Theorem 1 of *Wavelets and Operators*, Chapter 6, to calculate the L^p norm of this function and to verify that, if $|z|2^\delta \geq 1$, where $\delta = n/2 - n/p$, then $T^{-1} \notin L^p$.

When $1 < p < 2$, there similarly exists an operator $T \in \mathcal{A}_\infty$ which is an isomorphism on $L^2(\mathbb{R}^n)$ but whose inverse T^{-1} is not bounded on L^p.

To see this, we essentially repeat the preceding argument, by defining an operator U by $U(\psi_\lambda) = 0$, if $\lambda \notin \theta(\Lambda)$, and $U(\psi_\lambda) = \psi_{\theta^{-1}(\lambda)}$ otherwise. Then, for $|z| < 1$, $T = 1 - zU$ is invertible on $L^2(\mathbb{R}^n)$ and the same calculation as above gives a proof that T^{-1} is not bounded on L^p when $|z| \geq 2^{n(1/2-1/p)}$.

The algebras $\mathcal{A}_\infty$ and $\mathcal{B}_\infty$ may seem pathological. In fact, $\mathcal{A}_\infty$ is necessarily involved when we verify that the wavelets ψ_λ, $\lambda \in \Lambda$, arising from the Littlewood-Paley multiresolution approximation, form an unconditional basis of a classical space B of functions or distributions (such as the Besov spaces). For such a verification, it is essential to consider all continuous operators T on $L^2(\mathbb{R}^n)$ that are diagonalizable, with respect to the orthonormal basis ψ_λ, and to show that these operators are bounded on the space B. But these operators belong to $\mathcal{A}_\infty$, which is why we carry out a systematic examination of the continuity of the operators of $\mathcal{A}_\infty$ on various spaces of functions or distributions. (We shall do this in the next chapter.)

Further, if ψ_λ and $\tilde{\psi}_\lambda$, $\lambda \in \Lambda$, are unconditional wavelet bases belonging to the Schwartz class $\mathcal{S}(\mathbb{R}^n)$, the unitary operator $U : L^2 \to L^2$, defined by $U(\psi_\lambda) = \tilde{\psi}_\lambda$, $\lambda \in \Lambda$, belongs to $\mathcal{A}_\infty$. The algebra $\mathcal{A}_\infty$ is thus unavoidable, once we become interested in orthonormal bases of wavelets.

The following result is an application of the above remarks.

Proposition 2. *There exist three functions, ψ_1, ψ_2 and ψ_3 in $\mathcal{S}(\mathbb{R}^2)$ such that the wavelets*

$$2^j \psi_1(2^j x - k), \quad 2^j \psi_2(2^j x - k), \quad 2^j \psi_3(2^j x - k), \quad x \in \mathbb{R}^2, \, k \in \mathbb{Z}^2, \, j \in \mathbb{Z}$$

form an orthonormal basis of $L^2(\mathbb{R}^2)$, but such that there is no r-regular multiresolution approximation ($r \geq 1$) from which these wavelets can be obtained.

A recent result of P.G. Lemarié-Rieusset, completed by P. Auscher, shows that there is no such counter-example in dimension 1. More precisely, let $\psi(x)$ be a function in the Schwartz class $\mathcal{S}(\mathbb{R})$ such that $2^{j/2}\psi(2^j x - k)$, $j, k \in \mathbb{Z}$, is an orthonormal basis of $L^2(\mathbb{R})$. Then there must necessarily be a function $\phi(x) \in \mathcal{S}(\mathbb{R})$ such that the $\phi(x-k)$, $k \in \mathbb{Z}$, together with the $2^{j/2}\psi(2^j x - k)$, $j \geq 0$, $k \in \mathbb{Z}$, form an orthonormal basis of $L^2(\mathbb{R})$. In other words, in the context of the Schwartz class, in dimension 1, every orthonormal wavelet basis arises from a multiresolution approximation.

For the proof of Proposition 2, we start with the orthonormal basis of wavelets ψ_λ, $\lambda \in \Lambda$, in dimension 2 constructed by the tensor product method from the Littlewood-Paley multiresolution approximation. We shall apply the unitary operator whose symbol is $\zeta/|\zeta|$, where $\zeta = \xi + i\eta$, in other words, $U = R_1 + iR_2$, where R_1 and R_2 are the Riesz transforms. Then the example of Proposition 2 is given by the wavelets $\tilde{\psi}_\lambda = U(\psi_\lambda)$.

To see this, let V_j, $j \in \mathbb{Z}$, denote the multiresolution approximation from which the ψ_λ are constructed. If there is a multiresolution approximation giving rise to the $\tilde{\psi}_\lambda$, then that can only be $\tilde{V}_j = U(V_j)$. We shall show that $\tilde{V}_j$, $j \in \mathbb{Z}$, is not r-regular if $r \geq 1$.

We argue by contradiction. Suppose that there exists $h \in \tilde{V}_0$, such that $h \in L^2 \cap L^1$ and that $h(x - k)$, $k \in \mathbb{Z}^2$, form an orthonormal basis of $\tilde{V}_0$. Then we have $\hat{h} = \hat{\phi} \chi \zeta / |\zeta|$, where $\chi(\xi, \eta)$ is 2π-periodic in each variable, because $\tilde{V}_0 = U(V_0)$ and because of the characterization of $\mathcal{F}V_0$. Since the sequence of functions $h(x - k)$, $k \in \mathbb{Z}^2$, is orthonormal, it follows that $\sum |\hat{h}(\xi + 2k\pi)|^2 = 1$. But the same condition is satisfied by $\hat{\phi}$, so $|\chi(\xi, \eta)| = 1$, almost everywhere. Now, in the construction of the Littlewood-Paley multiresolution approximation, $\hat{\phi}(\xi, \eta) > 0$, if $-\pi \leq \xi, \eta \leq \pi$. Since $h \in L^1(\mathbb{R}^2)$, $\hat{h}$ is continuous on the square $S_0 = [-\pi, \pi]^2$ and the same is true for the product $\chi(\xi, \eta)(\xi + i\eta)/\sqrt{\xi^2 + \eta^2}$.

We now form the one-parameter family Γ_ε, $0 < \varepsilon \leq 1$, of contours $\varepsilon \partial S_0$, where ∂S_0 is the oriented boundary of S_0. We calculate the winding numbers about 0 of the image curves $\chi(\varepsilon \partial S_0)$. When $\varepsilon > 0$ is small enough, $\chi(\xi, \eta) = c(\xi - i\eta)/\sqrt{\xi^2 + \eta^2} + o(1)$, where c is a constant of modulus 1. Thus the winding number of $\chi(\varepsilon \partial S_0)$ is -1 for small enough ε. On the other hand, if $\varepsilon = 1$, we use the periodicity of χ to see that opposite sides of the square are traversed in opposite directions. It is thus clear that the winding number of $\chi(\partial S_0)$ about 0 is 0. On $S_0 \setminus \{0\}$, the function χ is continuous and of modulus 1. We now need only note that, for all values of $\varepsilon > 0$, $0 \notin \varepsilon \partial S_0$, so that $\chi(\varepsilon \partial S_0)$ is contained in $|z| = 1$. We have obtained a contradiction.

3 Commutators and Calderón's improved pseudo-differential calculus

The pseudo-differential calculus is like that mythological bird, the phoenix, which is reborn from its own ashes. Its first birth was at the end of the 1930s, the founding fathers being Giraud ([119]) and Marcinkiewicz ([183]). The second birth took place at the end of the 1950s and clearly benefited from the theory of distributions, developed by Schwartz during the 1940s.

The third birth, or renaissance, is the one to claim our interest. In order to deal with linear partial differential equations having coefficients which are only slightly regular and, above all, in order to approach the problem of the regularity of solutions of non-linear partial differential equations, Calderón decided to make the pseudo-differential calculus include the operators A of pointwise multiplication by functions $a(x)$ which are only slightly regular with respect to x (the precise meaning is given below). Of course, Calderón wanted to keep what had been gained during the previous decades: the classical pseudo-differential operators $T \in \mathcal{O}p\, S_{1,0}^m$.

The difficulty of any pseudo-differential calculus lies in that two commutative algebras of operators confront each other: the algebra X of operators of pointwise multiplication by functions $a(x)$ of a given regularity (infinite differentiability, in the usual case) and the algebra Y of differential operators with constant coefficients and of those which can be algebraically formed from them (that is, the convolution operators $T \in \mathcal{O}p\, S_{1,0}^m$).

Their confrontation is expressed by the fact that the operators $S \in X$ do not commute with the operators $T \in Y$. That lack of commutativity means we must calculate the commutators $[S, T]$, where $S \in X$ and $T \in Y$. The simplest example comes from Leibniz's rule, which gives the commutator $[A, D_j]$, where A is pointwise multiplication by the C^1 (or Lipschitz) function $a(x)$ and $D_j = \partial/\partial x_j$. Then $A_j = [A, D_j]$ is the operator of pointwise multiplication by $a_j(x) = -\partial a/\partial x_j$, which belongs to $L^\infty(\mathbb{R}^n)$.

Calderón tried to extend Leibniz's rule to the case $[A, T]$, where A is the preceding operator and $T \in \mathcal{O}p\, S_{1,0}^1$. In 1965 ([37]) he showed that this commutator was always bounded on $L^2(\mathbb{R}^n)$ when $a(x)$ is a Lipschitz function and, in a way, this does extend Leibniz's rule.

Calderón's theorem can now be obtained very simply, using the $T(1)$ theorem. We shall present this proof and, following Calderón, we shall use it to show that there exist algebras of pseudo-differential operators which have minimal reguarity with respect to x.

Theorem 4. *Let A be the operator of pointwise multiplication by a Lipschitz function $a(x)$. Then, for every classical pseudo-differential operator $T \in \mathcal{O}p\, S_{1,0}^1$ of order 1, the commutator $[A, T]$ is a Calderón-Zygmund operator.*

Conversely, if $[A, T]$ is bounded on $L^2(\mathbb{R}^n)$, for $T = \partial/\partial x_j$, $1 \le j \le n$, then $\partial a/\partial x_j \in L^\infty(\mathbb{R}^n)$ and $a(x)$ is thus a Lipschitz function.

To prove Theorem 4, we find it convenient to make two simplifications. On the one hand, we may suppose that $\sigma(x, 0) = 0$, by replacing $\sigma(x, \xi)$ by $\sigma(x, \xi) - \sigma(x, 0)$, if necessary. In terms of the operators, this means that we are altering $T = \sigma(x, D)$ by an operator of pointwise multiplication by a function of x and this does not affect the commutator $[A, T]$.

The other simplification is to replace T by $\sum_1^n T_j D_j$, where $D_j = \partial/\partial x_j$. This amounts to writing $\sigma(x, \xi) = \xi_1 \sigma_1(x, \xi) + \cdots + \xi_n \sigma_n(x, \xi)$, where $\sigma_j(x, \xi) \in S_{1,0}^0$. This can be done, because $\sigma(x, 0) = 0$, which enables us to use classical results on ideals of differentiable functions (in fact, we need to return to the proofs in order to get the necessary estimates on the $\sigma_j(x, \xi)$).

Let us first set out the formal structure of the proof. The following considerations will be rigorously justified later. Let $\Omega \subset \mathbb{R}^n \times \mathbb{R}^n$ be the open set $y \neq x$. The restriction to Ω of the distribution-kernel $S(x, y)$ of T is an infinitely differentiable function such that

$$|\partial_x^\alpha \partial_y^\beta S(x, y)| \leq C(\alpha, \beta)|x - y|^{-n-1-|\alpha|-|\beta|} .$$

As a consequence, the restriction $K(x, y)$ to Ω of the distribution-kernel of $[A, T]$ is $(a(x) - a(y))S(x, y)$, and satisfies

$$|K(x, y)| \leq C\|\nabla a\|_\infty |x - y|^{-n} ,$$

$$|(\partial/\partial x_j)K(x, y)| \leq C\|\nabla a\|_\infty |x - y|^{-n-1},$$

for $1 \leq j \leq n$, and

$$|(\partial/\partial y_j)K(x, y)| \leq C\|\nabla a\|_\infty |x - y|^{-n-1},$$

for $1 \leq j \leq n$.

To prove that $[A, T]$ is bounded on $L^2(\mathbb{R}^n)$, we apply the $T(1)$ theorem.

We start by establishing the weak continuity property. We need to show that $|\langle [A, T]u, v \rangle| \leq C\|\nabla a\|_\infty R^n$, for every ball $B \subset \mathbb{R}^n$, centre x_0 and radius R, and every pair of C^1 functions u, v, whose supports lie in B and which are "adapted" to B in the sense that $\|u\|_\infty \leq 1$, $\|\nabla u\|_\infty \leq R^{-1}$, $\|v\|_\infty \leq 1$ and $\|\nabla v\|_\infty \leq R^{-1}$.

To obtain this estimate of $|\langle [A, T]u, v \rangle|$, we replace $a(x)$ by $a(x) - a(x_0)$, which leaves $[A, T]$ unaltered. We can then find bounds for $\|AT(u)\|_{L^2(B)}$ and $\|TA(u)\|_2$ and the Cauchy-Schwarz inequality then gives the desired estimate.

In fact

$$\|AT(u)\|_{L^2(B)} \leq \|a\|_{L^\infty(B)}\|T(u)\|_2 \leq R\|\nabla a\|_\infty \sum_1^n \|T_j \partial_j u\|_2$$

$$\leq CR\|\nabla a\|_\infty \sum_1^n \|\partial_j u\|_2 \leq c'R^{n/2}\|\nabla a\|_\infty .$$

Similarly,

$$\|TA(u)\|_2 \leq C \sum_1^n \|\partial(au)/\partial x_j\|_2 \leq C'\|\nabla a\|_\infty \|u\|_2 + CR\|\nabla a\|_\infty \|\nabla u\|_2$$

$$\leq C''R^{n/2}\|\nabla a\|_\infty .$$

We then write $[A, T](1) = a(x)T(1) - T(a) = -T(a) = -\sum_1^n T_j(D_j a)$. The last function is in BMO, because the T_j are Calderón-Zygmund operators and thus continuous mappings from L^∞ to BMO.

To show the recasting of $[A, T](1)$ above is not merely formal, it is enough to approximate T by operators T_m whose symbols are truncated at the origin and at infinity (with respect to the variable ξ). This technique was described in the proofs of Lemmas 2 and 3 and we let that suffice.

The transpose of $[A, T]$ is, of course, $-[A, {}^tT]$, so the computation is similar to that which we have just described.

Following Calderón, we describe a new algebra of symbols.

Definition 2. *The symbols $\sigma(x, \xi)$ of Calderón's improved pseudo-differential calculus are the continuous functions on $\mathbb{R}^n \times \mathbb{R}^n \setminus \{0\}$ which satisfy the following conditions:*

(3.1) $\sigma(x, \lambda\xi) = \sigma(x, \xi)$ *for all $\lambda > 0$, $x \in \mathbb{R}^n$ and $\xi \neq 0$;*

(3.2) $|\partial_\xi^\alpha \sigma(x, \xi)| \leq C_\alpha$ *if $|\xi| = 1$;*

(3.3) $|(\partial/\partial x_j)\partial_\xi^\alpha \sigma(x, \xi)| \leq C_\alpha'$ *if $|\xi| = 1$ and $1 \leq j \leq n$.*

Definition 3. *A continuous linear operator $S : L^2(\mathbb{R}^n) \to L^2(\mathbb{R}^n)$ is regularizing of order 1 if the $2n$ operators $D_j S$ and $S D_j$ (where $D_j = \partial/\partial x_j$) are also continuous on $L^2(\mathbb{R}^n)$.*

We come to the statement of Calderón's theorem.

Theorem 5. *Let $\mathcal{C}$ be the collection of all operators T which can be written $T = T_1 + T_2$, where $T_1 = \sigma(x, D)$ with a symbol $\sigma(x, \xi)$ satisfying (3.1), (3.2) and (3.3) and where T_2 is regularizing of order 1. Then $\mathcal{C}$ is an algebra of operators. The set $\mathcal{I}$ of regularizing operators is an ideal in $\mathcal{C}$ and $\mathcal{C}/\mathcal{I}$ is a commutative Banach algebra, isomorphic to the algebra of symbols satisfying (3.1), (3.2) and (3.3).*

In other words, if the symbols $\sigma(x, \xi)$ and $\tau(x, \xi)$ of $\sigma(x, D)$ and $\tau(x, D)$ satisfy (3.1), (3.2) and (3.3), then the operator $\tau(x, D)\sigma(x, D)$ can be written as $T_1 + T_2$, where the symbol of T_1 is the product $\sigma(x, \xi)\tau(x, \xi)$ and T_2 is regularizing, of order 1.

We need the following lemma for the proof of Calderón's theorem.

Lemma 5. *Suppose that $\sigma(x, \xi) \in L^\infty(\mathbb{R}^n \times \mathbb{R}^n)$ satisfies the conditions*

(3.4) $\sigma(x, \lambda\xi) = \sigma(x, \xi)$,   *for all $\xi \in \mathbb{R}^n \setminus \{0\}$ and all $\lambda > 0$,*

(3.5) $|\partial_\xi^\alpha \sigma(x, \xi)| \leq C(\alpha)$,   *if $|\xi| = 1$, $x \in \mathbb{R}^n$, and $\alpha \in \mathbb{N}^n$.*

Then the operator $\sigma(x, D)$ is bounded on $L^2(\mathbb{R}^n)$.

To see this, we follow Calderón and Zygmund, by performing a spherical harmonic expansion of $\sigma(x, \xi)$ on the sphere $|\xi| = 1$, for fixed $x \in \mathbb{R}^n$.

By the regularity with respect to ξ, this gives a norm-convergent sequence

$$\sigma(x,\xi) = \sum_0^\infty m_k(x)h_k(\xi) \qquad \text{with } \sum_0^\infty \|m_k\|_\infty \|h_k\|_\infty < \infty.$$

We extend $h_k(\xi)$ as a homogeneous function of degree 0, which is the symbol of an operater we denote by H_k, and we write M_k for the operator of pointwise multiplication by $m_k(x)$. This gives

$$\sigma(x,D) = \sum_0^\infty M_k H_k$$

and the sequence of operators is convergent in norm.

Returning to the proof of the theorem, let us show that $\mathcal{I}$ is a two-sided ideal in $\mathcal{C}$. The only difficulty is to show that if $T_1 \in \mathcal{I}$ and if the symbol of $T_2 = \sigma(x,D)$ satisfies (3.1), (3.2) and (3.3) then $T_1 T_2 D_j$ is bounded on $L^2(\mathbb{R}^n)$ (writing D_j for $\partial/\partial x_j$.) We see this by working out the commutator $[D_j, T_2] = T_3$. We have $T_3 = \tau(x,D)$, where $\tau(x,\xi) = (\partial/\partial x_j)\sigma(x,\xi)$. Lemma 5 shows that T_3 is continuous on $L^2(\mathbb{R}^n)$. So $T_1 T_2 D_j = (T_1 D_j)T_2 - T_1 T_2$ and every term is continuous on $L^2(\mathbb{R}^n)$.

Now for the rest of the theorem. We systematically use the expansion as spherical harmonic of all the symbols $\sigma(x,\xi)$ satisfying (3.1), (3.2) and (3.3) and the theorem is a consequence of the following lemma.

Lemma 6. *Let A be the operator of pointwise multiplication by a Lipschitz function $a(x)$ and let H be a convolution operator whose symbol $\sigma(\xi)$ satisfies $\sigma(\lambda\xi) = \sigma(\xi)$, $\xi \neq 0$, $\lambda > 0$ and $|\partial^\alpha \sigma(\xi)| \leq C_\alpha$ when $|\xi| = 1$. Then the commutator $[A, H]$ is regularizing of order 1.*

Since $\mathcal{I}$ is a two-sided ideal, such a result allows us to interchange the positions of the operators $A_1, A_2, \ldots$ of type A and of the operators $H_1, H_2, \ldots$ of type H in such a way as to reduce every product $A_1 H_1 A_2 H_2 \ldots$ to the canonical form AH, where $A = A_1 A_2 \ldots$, $H = H_1 H_2 \ldots$ and the symbol of AH is $a(x)\sigma(\xi)$. Calderón's theorem then follows.

To prove Lemma 6, we write, for example, $D_j[A, H] = A_j H + [A, T_j]$, where A_j is the operator of pointwise multiplication by $\partial a/\partial x_j$ and where $T_j = D_j H$. We then apply the commutator theorem (Theorem 4).

4 The pseudo-differential version of Leibniz's rule

We consider integers $m \geq 1$ and $s \in [0, m]$, together with a convolution operator T defined by a symbol satisfying the homogeneous estimates $|\partial_\xi^\alpha \tau(\xi)| \leq C_\alpha |\xi|^{s-|\alpha|}$.

Let B be the operator of pointwise multiplication by a function $b(x)$ satisfying $\|\partial_x^\beta b(x)\|_\infty \leq C < \infty$, for $0 \leq |\beta| \leq m$.

We intend to obtain the pseudo-differential equivalent of Leibniz's rule giving the derivative of the product. In fact, we have the following result.

Theorem 6. *With the above notation,*

$$(4.1) \qquad TB = \sum_{|\alpha| \leq m} B_\alpha T_\alpha + R_m \,,$$

where B_α is the operation of multiplication by $\partial_x^\alpha b(x)$, T_α is a convolution operator whose symbol is $(-i)^{|\alpha|}(\alpha!)^{-1}\partial_\xi^\alpha \tau(\xi)$, and where R_m is regularizing of order $m - s$, in the sense that

$$(4.2) \qquad \|R_m \partial^\gamma f\|_2 \leq C \sum_{|\beta| \leq m} \|\partial^\beta b\|_\infty \|f\|_2 \,, \qquad \text{for } |\gamma| = m - s.$$

At first sight, (4.1) seems to be the same as the usual formula for calculating the symbol of a product of two differential operators. The new aspect is that the hypotheses about the regularity of the function b are minimal.

Theorem 6 is due to Calderón and leads to a refinement of the pseudo-differential calculus described by Theorem 5. Once again, we shall use the $T(1)$ theorem to prove this result. For our greater convenience, we replace the operator T defined by the symbol τ by approximations whose symbols are truncations of τ which are zero in neighbourhoods of the origin and infinity. By abuse of notation, we shall use T to denote these approximations. Passing to the limit poses no problems, as long as we obtain estimates which are uniform.

If $K(x,y)$ denotes the distribution-kernel of T, then that of T_α is $(\alpha!)^{-1}(y - x)^\alpha K(x,y)$ and the distribution-kernel of R_m is thus

$$R_m(x,y) = \left(b(y) - \sum_{|\alpha| \leq m} \frac{1}{\alpha!}(y - x)^\alpha \partial_x^\alpha b(x) \right) K(x,y) \,.$$

Lastly, the kernel of $R_m \partial^\gamma$ is $(-1)^{|\gamma|}\partial_y^\gamma R_m(x,y)$.

We show, first of all, how to reduce the problem to the case $s = 0$. To pass from s to $s - 1$, we split T up into $T_1 D_1 + \cdots + T_n D_n$, where the symbols $\tau_j(\xi)$ of T_j are of order $s - 1$. For example, we can put $\tau_j(\xi) = -i\xi_j|\xi|^{-2}\tau(\xi)$. Then

$$TB = T_1 D_1 B + \cdots + T_n D_n B$$

$$= T_1 B_1 + \cdots + T_n B_n + T_1 B D_1 + \cdots + T_n B D_n \,.$$

If the theorem has been proved for the pairs $(s-1, m-1)$ and $(s-1, m)$, we can decompose $T_j B_j$ and $T_j B$ using (4.1) to obtain $2m$ error terms $R_m^{(j)}$ and $S_m^{(j)}$ which are regularizing of order $m - s$ and $m - s + 1$. Thus,

$R_m = R_m^{(1)} + \cdots + R_m^{(n)} + S_m^{(1)}D_1 + \cdots + S_m^{(n)}D_n$, which is regularizing of order $m - s$.

We intend to apply the $T(1)$ theorem to the operator $R_m\partial^\gamma$ when $|\gamma| = m$ and $s = 0$. But, to do that, we need to use a kernel having a certain regularity with respect to x and y. Now, $R_m(x, y)$ contains terms involving $\partial_x^\alpha b(x)$ with $|\alpha| = m$: this function belongs to $L^\infty(\mathbb{R}^n)$ and these terms have to be treated separately.

So we consider the kernels $\partial_x^\alpha b(x)\partial_y^\gamma\{(y - x)^\alpha K(x, y)\}$, where $|\alpha| = |\gamma| = m$. The corresponding operators are bounded on $L^2(\mathbb{R}^n)$, because we can ignore $\partial_x^\alpha b(x) \in L^\infty(\mathbb{R}^n)$ and because $\partial_y^\gamma\{(y - x)^\alpha K(x, y)\}$ is a convolution operator whose symbol is $\xi^\gamma \partial_\xi^\alpha \tau(\xi)$. By hypothesis, $\xi^\gamma \partial_\xi^\alpha \tau(\xi) \in L^\infty(\mathbb{R}^n)$.

Once these special cases have been dealt with, we apply Leibniz's rule to compute $\partial_y^\gamma R_m(x, y)$. The derivatives are applied to $K(x, y)$ and to $b(y) - \sum_{|\alpha| \le m-1}(\alpha!)^{-1}(y - x)^\alpha \partial_x^\alpha b(x)$. In the latter case, this amounts to replacing $b(y)$ by $\partial_y^\beta b(y)$ and m by $m - |\beta|$. Arguing by induction on the integer m, we need only consider the case where all the derivatives are applied to $K(x, y)$, and we put

$$Z_m(x, y) = \left\{ b(y) - \sum_{|\alpha| \le m-1}(\alpha!)^{-1}(y - x)^\alpha \partial_x^\alpha b(x) \right\} \partial_y^\gamma K(x, y).$$

Let Z_m denote the corresponding operator. We now apply the $T(1)$ theorem to it. The estimates of the size and regularity of the kernel follow immediately, because $|\partial_x^\alpha \partial_y^\beta K(x, y)| \le C_{\alpha,\beta}|x - y|^{-n-|\alpha|-|\beta|}$, for all $\alpha, \beta \in \mathbb{N}^n$.

All that remains to do is to verify the weak continuity property and, finally, to compute $Z_m(1)$ and $Z_m^*(1)$.

If u is a C^1 function of compact support, we can compute $Z_m(u)$ by integrating by parts with respect to the variable y. That is, $\partial_y^\gamma K(x, y)$ becomes $\partial_y^{\gamma'} K(x, y)$, where $|\gamma'| = |\gamma| - 1$ and we differentiate either $b(y) - \sum_{|\alpha| \le m-1}(\alpha!)^{-1}(y - x)^\alpha \partial_x^\alpha b(x)$, or $u(y)$, with respect to y. In the first case, we get an operator of type Z_{m-1} which is bounded, by the induction hypothesis. In the second, we find we have a kernel $Y_m(x, y)$ whose singularity is integrable: $|Y_m(x, y)| \le C(b)|x - y|^{-n+1}$. We need give no further details.

Working out $Z_m(1)$ gives $Z_m(1) = (-1)^{|\gamma|}T(\partial^\gamma b)$, using the same ideas. This function belongs to BMO because T is a Calderón-Zygmund operator and $\partial^\gamma b \in L^\infty$.

To finish, we deal with ${}^tZ_m(1) = \int Z_m(x, y)\,dx$, which we regard as a function of y. We use the fact that $K(x, y)$ is a convolution kernel to write $\partial_y^\gamma K(x, y) = (-1)^{|\gamma|}\partial_x^\gamma K(x, y)$. This allows us to integrate by parts with respect to x to find ${}^tZ_m(1)$. More precisely, if $\alpha = (\alpha_1, \ldots, \alpha_n)$ and $\gamma = (\gamma_1, \ldots, \gamma_n)$, where $\gamma_j \geq 1$, we put $\bar{\alpha} = (\alpha_1, \ldots, \alpha_j + 1, \ldots, \alpha_n)$ and $\gamma' = (\gamma_1, \ldots, \gamma_j - 1, \ldots, \gamma_n)$. After integration by parts, we get

$$
{}^tZ_m(1) = (-1)^{|\gamma|} \sum_{|\alpha|=m-1} \frac{1}{\alpha!} \int (y-x)^\alpha \partial_x^{\gamma'} K(x,y) \partial_x^{\bar{\alpha}} b(x)\, dx.
$$

We have used the identity

$$
\frac{\partial}{\partial x_j} \left\{ b(y) - \sum_{|\alpha| \leq m-1} \frac{1}{\alpha!}(y-x)^\alpha \partial_x^\alpha b(x) \right\} = - \sum_{|\alpha|=m-1} \frac{1}{\alpha!}(y-x)^\alpha \partial_x^{\bar{\alpha}} b(x).
$$

We conclude by observing that $\partial^{\bar{\alpha}} b \in L^\infty$ and that $(y-x)^\alpha \partial_x^{\gamma'} K(x,y)$ is the kernel of a Calderón-Zygmund operator, which is bounded as an operator from $L^\infty(\mathbb{R}^n)$ to $\mathrm{BMO}(\mathbb{R}^n)$.

5 Higher order commutators

Let $H : L^2(\mathbb{R}) \to L^2(\mathbb{R})$ be the Hilbert transform. Its symbol is $-i\operatorname{sgn}\xi$ and it can also be defined as convolution by the distribution $\mathrm{PV}(\pi x)^{-1}$.

Put $D = -i(d/dx)$ and consider the operators $D^k H$. The corresponding symbols are $-i\xi^k \operatorname{sgn}\xi$, which obviously satisfy the inequality $|(d/d\xi)^m \tau(\xi)| \leq C_m |\xi|^{k-m}$ if $\xi \neq 0$. In fact $C_m = k(k-1)\cdots(k-m+1)$.

From 1966, Calderón suggested studying iterated commutators. To define them, we start with k Lipschitz functions $a_1(x), \ldots, a_k(x)$ and let $A_1, \ldots, A_k$ denote the operators of pointwise multiplication by those functions. We then form the operator Γ_k, defined by

$$
(5.1) \qquad k!\Gamma_k = [A_1, [A_2, \ldots, [A_k, D^k H] \ldots]].
$$

The definition of the Γ_k raises problems when $k \geq 2$. For example, $\Gamma_2 = A_1 A_2 T_2 - A_1 T_2 A_2 - A_2 T_2 A_1 + T_2 A_1 A_2$, where we have used T_2 to denote $D^2 H$. It is not obvious that the terms $A_1 T_2 A_2$ and $A_2 T_2 A_1$ make sense as operators from $\mathcal{D}$ to $\mathcal{D}'$. In fact, it is not a good idea to expand Γ_2 in this way and we shall use a different method to give a meaning to Γ_2.

Below, we shall verify that the distribution-kernel of Γ_k is

$$
\frac{i^k}{\pi} \mathrm{PV} \frac{(a_1(x) - a_1(y)) \cdots (a_k(x) - a_k(y))}{(x-y)^{k+1}}.
$$

At a certain formal level, we can say that $(i^k/\pi)\,\mathrm{PV}(x-y)^{-k-1}$ is the distribution-kernel of $T_k = (k!)^{-1} D^k H$ and that, each time we make an operator commute with A_j, we multiply its kernel by $a_j(x) - a_j(y)$.

If all the functions $a_j(x)$ are equal and real-valued, then, up to the coefficient $-1/2$, the generating series $\sum_0^\infty \delta^k \Gamma_k$, for $\delta > 0$, is the operator defined by the Cauchy kernel $(2\pi)^{-1}(z(x) - z(y))^{-1}$, where $z(x) = x - i\delta a(x)$, $x \in \mathbb{R}$. The curve Γ, whose parametric representation is $z(x)$, $x \in \mathbb{R}$, is thus the graph of the Lipschitz function $-\delta a(x)$. The problem has become that of the continuity of the Cauchy kernel $(2\pi)^{-1} \operatorname{PV} \int_\Gamma f(w)(z - w)^{-1} dw$ on $L^2(\Gamma \, ds)$, where ds is the arclength measure on the Lipschitz graph Γ. This continuity problem will be resolved at the end of this chapter but a further, systematic, account will be given in Chapter 12.

Calderón exploited this relationship, between higher order commutators and complex analysis, to the utmost ([39]).

The relationship with complex analysis disappears entirely once we work in $\mathbb{R}^n$ and replace each $D^k H$ by a convolution operator T_k whose symbol $\tau(\xi)$, $\xi \neq 0$, satisfies the following estimates of "homogeneous type":

$$(5.2) \qquad |\partial^\alpha \tau(\xi)| \le C(\alpha)|\xi|^{k-|\alpha|} \qquad \text{where } \xi \neq 0 \text{ and } \alpha \in \mathbb{N}^n.$$

We again let $A_1, \ldots, A_k$ be the operators of pointwise multiplication by Lipschitz functions $a_1(x), \ldots, a_k(x)$. With this notation, we have

Theorem 8. *The iterated commutators* $\Gamma_k = [A_1, [A_2, \ldots, [A_k, T_k] \ldots]]$ *are all Calderón-Zygmund operators and the norms* $\|\Gamma_k\|$ *of* $\Gamma_k : L^2 \to L^2$ *satisfy*

$$(5.3) \qquad \|\Gamma_k\| \le C(k, n, \tau)\|\nabla a_1\|_\infty \ldots \|\nabla a_k\|_\infty,$$

where the constant $C(k, n, \tau)$ *depends, in fact, only on the constants* $C(\alpha)$ *of (5.2).*

The structure of this statement allows us to use a standard technique of approximation to T_k in order to prove the theorem. To do this, we replace the symbol $\tau(\xi)$ by the symbols $\tau_m(\xi) = \chi(m^{-1}\xi)(1 - \chi(m\xi))\tau(\xi)$, where $\chi \in \mathcal{D}(\mathbb{R}^n)$ equals 1 in a neighbourhood of the origin. The symbols τ_m satisfy (5.2) uniformly in m, which will let us pass to the limit, once the theorem has been proved under the supplementary hypothesis that $\tau(\xi) \in \mathcal{D}(\mathbb{R}^n)$ and that $\tau(\xi)$ is zero in a neighbourhood of the origin.

The kernel $K(x, y)$ of T_k is a convolution kernel. Qualitatively, $K(x, y)$ is an infinitely differentiable function which is $O(|y - x|^{-N})$ for every $N \ge 1$, as $|y - x|$ tends to infinity. Quantitatively, we have

$$(5.4) \qquad |\partial_x^\alpha \partial_y^\beta K(x, y)| \le C(\alpha, \beta)|x - y|^{-n-k-|\alpha|-|\beta|}.$$

So the kernel $L(x, y)$ of Γ_k is $(a_1(x) - a_1(y)) \cdots (a_k(x) - a_k(y))K(x, y)$ and it is clear that $L(x, y)$ satisfies the Calderón-Zygmund estimates.

To prove that Γ_k is continuous on $L^2(\mathbb{R}^n)$, we use induction on k and apply the $T(1)$ theorem.

We first show that Γ_k is weakly continuous. To that end, we calculate $\Gamma_k(f)$, where f is a C^1 function of compact support. The calculation is done by writing $T_k = S_k^{(1)} D_1 + \cdots + S_k^{(n)} D_n$, where the symbols of the $S_k^{(j)}$ satisfy (5.2), with $k-1$ instead of k, and where D_j denotes $(\partial/\partial x_j)$. Then, writing $S_k^{(j)}(x-y)$ for the kernels of the $S_k^{(j)}$, we get

$$\Gamma_k(f) = -\sum_1^n \int (a_1(x)-a_1(y)) \cdots (a_k(x)-a_k(y)) \frac{\partial}{\partial y_j} S_k^{(j)}(x-y) f(y) \, dy \, .$$

We can then integrate by parts, which gives two kinds of term. In the first kind, we differentiate one of the $a_j(x) - a_j(y)$, $j = 1, \ldots, k$. This leads to terms of the form

$$\Gamma_{k-1}^{(1)} \left(\frac{\partial a_1}{\partial y_j} f \right) \, , \qquad \Gamma_{k-1}^{(2)} \left(\frac{\partial a_2}{\partial y_j} f \right) \, , \ldots \, ,$$

whose L^2 norms can be estimated by the induction hypothesis. The second kind of term is that in which f is differentiated. The kernel we then use is $(a_1(x) - a_1(y)) \cdots (a_k(x) - a_k(y)) S_k^{(j)}(x - y)$. It is bounded in modulus by $C \|\nabla a_1\|_\infty \cdots \|\nabla a_k\|_\infty |x - y|^{-n+1}$, and the estimate is trivial, because f is a C^1 function of compact support.

This way of organizing the argument gives a precise formulation of the general definition of $\Gamma_k(f)$ when f is a C^1 function of compact support. At the same time, it gives the weak continuity of Γ_k.

Exactly the same kind of argument gives $\Gamma_k(1)$. If we go through the steps above, we get $\Gamma_k(1)$ in the form of a linear combination of terms $\Gamma_{k-1}^{(j)}(D_j a_l)$, where $D_j = (\partial/\partial x_j)$. By the induction hypothesis, these terms are in BMO.

Finally, ${}^t\Gamma_k$ has exactly the same structure as Γ_k, except that T_k is replaced by ${}^t T_k$. So ${}^t\Gamma_k(1) \in$ BMO.

6 Takafumi Murai's proof that the Cauchy kernel is L^2 continuous

We return to dimension 1 and to the "historical" Calderón commutators. We begin with a Lipschitz function $A : \mathbb{R} \to \mathbb{C}$, with which we associate the distributions $\mathrm{PV}\big((A(x) - A(y))^k/(x - y)^{k+1}\big)$ belonging to $\mathcal{S}'(\mathbb{R}^2)$. So we arrive at the operators $\Gamma_k : \mathcal{S}(\mathbb{R}) \to \mathcal{S}'(\mathbb{R})$ whose distribution-kernels are the distributions we have just described. Thus

$$\langle \Gamma_k(f), g \rangle = \lim_{\varepsilon \downarrow 0} \int \int_{|x-y| \geq \varepsilon} \frac{(A(x) - A(y))^k}{(x - y)^{k+1}} g(x) f(y) \, dy \, dx \, ,$$

whenever f and g are C^1 functions of compact support.

The L^2 continuity of Γ_k is a consequence of the results of the previous section. From this, we go on to deduce that, for every $f \in L^2(\mathbb{R})$ and almost all $x \in \mathbb{R}$,

$$\lim_{\varepsilon \downarrow 0} \int_{|x-y| \geq \varepsilon} \frac{(A(x) - A(y))^k}{(x-y)^{k+1}} f(y) \, dy$$

exists.

By using the general results of Chapter 7 (Cotlar's inequality) we see that it is enough to prove that the limit exists when f is a C^1 function of compact support. In that case, we can integrate by parts, observing that $(\partial/\partial y)(x-y)^{-k} = k(x-y)^{-k-1}$, to get, as in the previous section, two kinds of term. One kind involves the operator Γ_{k-1} and is dealt with using the induction hypothesis and the fact that $A'f \in L^2$. The other kind of term comes from differentiating f and leads to an absolutely convergent integral. The integrals of these terms tend to 0 almost everywhere because

$$A'(x) = \lim_{\varepsilon \downarrow 0} \frac{A(x + \varepsilon) - A(x - \varepsilon)}{2\varepsilon}$$

for almost all $x \in \mathbb{R}$.

So, for every $f \in L^2(\mathbb{R})$, every $k \in \mathbb{N}$ and almost all $x \in \mathbb{R}^n$,

$$\Gamma_k f(x) = \lim_{\varepsilon \downarrow 0} \int_{|x-y| \geq \varepsilon} \frac{(A(x) - A(y))^k}{(x-y)^{k+1}} f(y) \, dy \,.$$

We make one final observation: about the growth of the norms $\|\Gamma_k\|$ of the operators $\Gamma_k : L^2(\mathbb{R}) \to L^2(\mathbb{R})$. Using the obvious estimates on the size and regularity of the kernel $(A(x) - A(y))^k/(x-y)^{k+1}$, we get $\|\Gamma_k\| \leq \pi C^k \|A'\|_\infty^k$. This method does not give a value for the constant, which comes from two multiplicative factors. The first is due to the presence of a constant C_0, which we do not know how to evaluate with any precision, in the statement

(6.1) $$\|T\| \leq C_0 \|T(1)\|_{\mathrm{BMO}} + C_1 \,,$$

where T is defined by an antisymmetric kernel $K(x, y)$ satisfying

$$|K(x,y)| \leq \frac{1}{|x-y|} \quad \text{and} \quad \left| \frac{\partial K(x,y)}{\partial x} \right| \leq \frac{1}{(x-y)^2} \,.$$

The other factor appears when we apply the fact that every Calderón-Zygmund operator is continuous as a mapping from L^∞ to BMO. In fact, both problems are related to a third, which is the lack of a norm canonically associated with the (L^∞, BMO) continuity of Calderón-Zygmund operators.

However, we can remark that the generating series $\sum_0^\infty \delta^k T_k$ converges in operator norm if $\delta > 0$ is small enough. This lets us prove the continuity of the Cauchy kernel on Lipschitz curves of slope less than δ. This gives the result obtained by A. Calderón in 1974 ([38]).

Now this "local" result leads to the following global one. *If $A : \mathbb{R} \to \mathbb{R}$ is a Lipschitz function, then the kernel $(x - y + i(A(x) - A(y)))^{-1}$ defines a bounded operator on $L^2(\mathbb{R})$.*

The proof we give here is based on a fundamental idea due to Guy David ([1973]), which consists of considering the set $\mathcal{E}_k$ of all operators corresponding to Lipschitz functions A satisfying $\|A'\|_\infty \le (3/2)^k$. and of using the the "rising sun lemma" to prove, by induction on k, that the operators $T \in \mathcal{E}_k$ are continuous.

David's proof is a little more complicated than the one we are going to describe, which is due to Takafumi Murai. Murai considers the operator T_A, whose distribution-kernel is PV $E_A(x, y)$, where

$$E_A(x, y) = \frac{1}{x - y} e^{i \frac{A(x) - A(y)}{(x - y)}} .$$

This operator has the following remarkable properties:

(6.2) $$T_B = e^{i\lambda} T_A ,$$

where λ is a real constant and $B(x) = \lambda x + A(x)$;

(6.3) $$T_B = -T_A^\star \qquad \text{if } B = -A;$$

(6.4) $$\int_0^\infty T_{\lambda A} e^{-\lambda} \, d\lambda \quad \text{is the Cauchy operator}$$

defined by the kernel $(x - y - i(A(x) - A(y)))^{-1}$;

(6.5) $$\|T_A\| \le \pi e^{C\|A'\|_\infty} ,$$

for a certain constant C.

The first three properties are obvious and we get the fourth by expanding T_A as $\sum_0^\infty i^k (k!)^{-1} \Gamma_k$ and using $\|\Gamma_k\| \le \pi C^k \|A'\|_\infty^k$.

It will be obvious that the Cauchy kernel is continuous (without any restriction on $\|A'\|_\infty$), if we can replace the exponential growth in (6.5) by a slow rate of growth. This slow rate, with respect to λ, will be balanced by the term $e^{-\lambda}$ in (6.4) and the integral there will converge in operator norm.

To find an upper bound for $\|T_A\|_\infty$, we use the $T(1)$ theorem and get

(6.6) $$\|T_A\| \le C_0 \|T_A(1)\|_{\text{BMO}} + C_1(1 + \|A'\|_\infty) ,$$

where the BMO norm is defined by

$$\|b\|_{\text{BMO}} = \sup_I \inf_{\gamma \in \mathbb{C}} \frac{1}{|I|} \int_I |b(x) - \gamma| \, dx .$$

The second term of the right-hand side of (6.6) comes from obvious estimates of $|E_A(x, y)|$ and $|(\partial/\partial x) E_A(x, y)|$.

We intend to show that, for real-valued $A(x)$,

(6.7) $$\|T_A\| \le C(1 + \|A'\|_\infty)^5 .$$

In fact, it will be enough to prove the corresponding inequality for $\|T_A(1)\|_{\mathrm{BMO}}$ and the basic idea is to let $\mathcal{E}_k$ denote the set of operators T_A, for which $0 \leq A'(x) \leq (3/2)^k$ and then to show, by induction on k, that $\|T_A(1)\|_{\mathrm{BMO}} \leq C(1 + (3/2)^k)^5$ when $T_A \in \mathcal{E}_k$. The step from $\mathcal{E}_k$ to $\mathcal{E}_{k+1}$ is a consequence of the rising sun lemma applied to the graph of $A(x)$.

We should observe that doing this for the special case $0 \leq A'(x) \leq M$ is enough. The general case $M \leq A'(x) \leq M$ follows by using (6.2) with $B(x) = Mx + A(x)$, which gives $0 \leq B'(x) \leq 2M$.

The rising sun lemma.

Our account follows that of Zygmund ([239], page 31).

Let $I = [a, b]$ be an interval in the real line and let $A(x)$ be an increasing Lipschitz function on I, such that $0 \leq A'(x) \leq M$. Put $m = (A(b) - A(a))/(b - a)$ and let λ be a "threshold value" satisfying $0 \leq \lambda \leq m$.

We define $B_I(x)$ by

(6.8) $$B_I(x) = \lambda x + \sup_{a \leq t \leq x} (A(t) - \lambda t).$$

In other words, $B_I(x) - \lambda x$ is the smallest increasing function on I, such that $B_I(x) - \lambda x \geq A(x) - \lambda x$, for $x \in I$ (that is, $B_I(x) \geq A(x)$ on I).

Since $A(x)$ is a Lipschitz function, the same is true for $B_I(x)$. Indeed, if $a \leq x \leq x' \leq b$,

$$\sup_{a \leq t \leq x'} (A(t) - \lambda t) \leq \sup_{a \leq t \leq x} (A(t) - \lambda t) + (M - \lambda)(x' - x),$$

which gives

$$B_I(x') - B_I(x) \leq M(x' - x).$$

The definition of $B_I(x)$ gives

$$B_I(x') - B_I(x) \geq \lambda(x' - x).$$

Thus $B_I(x)$ can be defined as follows: $B_I(x)$ is the smallest Lipschitz function (or, even, the smallest absolutely continuous function) satisfying $B'(x) \geq \lambda$ almost everywhere on I and $B_I(x) \geq A(x)$ on I.

Let $E \subset [a, b]$ be the compact set $\{x : B_I(x) = A(x)\}$ and let $\Omega = [a, b] \setminus E$. It is worth noting that $a \in E$. On the other hand, b need not belong to E. This means that Ω (assuming it is non-empty) consists of open disjoint intervals (a_k, b_k) and, possibly, an interval $(a', b]$.

Having established the above notation, here is the rising sun lemma.

Lemma 7. $B_I(x) = \lambda x + c_k$ on every connected component I_k of Ω and the measure $|\Omega|$ of Ω satisfies $|\Omega| \leq (M - m)(M - \lambda)^{-1}|I|$.

To prove the rising sun lemma, we put $f(x) = A(x) - \lambda x$ and $F(x) =$

$B_I(x) - \lambda x$. Thus $F(x) = \sup_{a \leq t \leq x} f(t)$. By the definition of E, for every open contiguous interval (a_k, b_k) of E, we have $F(a_k) = f(a_k)$ and $F(b_k) = f(b_k)$. Since F is non-decreasing, $F(b_k) \geq F(a_k)$. If we can show that $F(b_k) = F(a_k)$, it will follow that $F(x)$ is constant on $[a_k, b_k]$, as required. Now, if $F(b_k) = f(b_k) > f(a_k) = F(a_k)$, choose $c_k \in (a_k, b_k)$ such that $f(c_k) > f(a_k)$ and $d_k \in (a_k, c_k]$ such that $f(d_k) = \sup\{f(x) : x \in [a_k, c_k]\}$. But this implies that $f(d_k) = F(d_k)$, so that $d_k \in E$, which contradicts the definition of (a_k, b_k).

To conclude this part of the proof, we need to consider the case where $(a', b]$ is a connected component of Ω. Then $F(b) > f(b)$ and $F(a') = f(a')$. We see that $F(b) = f(c)$, for some $c \in [a', b)$. If $F(b) > f(a')$, then $a' < c$ and $F(c) = f(c)$, contrary to the definition of Ω. Hence, $F(b) = F(a')$.

To find an upper bound for $|\Omega|$, we observe that $A(a) = B_I(a)$. Thus

$$A(b) - A(a) \leq B_I(b) - B_I(b) = \int_a^b B_I'(x)\, dx$$

$$= \int_E B_I'(x)\, dx + \int_\Omega B_I'(x)\, dx$$

$$\leq M|E| + \lambda|\Omega| = M|I| - (M - \lambda)|\Omega|.$$

We shall use this lemma to examine the behaviour of the kernel $E_A(x, y)$ when x and y belong to the same interval I. We replace the function A by the function B_I of the rising sun lemma (the sequel will show why this is a useful thing to do.) To simplify the notation, we drop the subscript I from B_I and get

(6.9) $|E_A(x, y) - E_B(x, y)| \leq M(x - y)^{-2}(\mathrm{dist}(x, E) + \mathrm{dist}(y, E))$,

because

$$|B_I(x) - A(x)| = B_I(x) - A(x) \leq M\, \mathrm{dist}(x, E).$$

To evaluate the upper bound of the norms of the Calderón-Zygmund operators T_A when $0 \leq A' \leq M$, we introduce the following estimating function:

$$\sigma(A, I) = \int_I \left| \int_I E_A(x, y)\, dy \right| dx = \|T_A(\chi_I)\|_{L^1(I)}.$$

This was first considered by Journé in a result which anticipated the $T(1)$ theorem.

We then write $\sigma(A) = \sup_I |I|^{-1} \sigma(A, I)$ and, lastly, put

(6.10) $\tau(M) = \sup\{\sigma(A) : 0 \leq A'(x) \leq M\}.$

We aim to prove the following lemmas.

Lemma 8. *There exists a constant C_0 such that*

(6.11) $\|T_A(1)\|_{\mathrm{BMO}} \leq \sigma(A) + C_0(1 + \|A'\|_\infty).$

Lemma 9. *There exists a constant C_1 such that, for all $M > 0$,*

(6.12) $$\tau(M) \leq 4\tau(2M/3) + C_1(1+M).$$

These lemmas easily give (6.7). Indeed, we start from the continuity of the operators T_A, when $\|A'\|_\infty \leq 1$. The continuity is a direct result of the $T(1)$ theorem applied to the Calderón commutators, as we have indicated.

Repeated application of the estimate (6.12) gives $\tau(M) \leq C_2(1+M)^5$, as can be seen by considering the intervals $(3/2)^k \leq M \leq (3/2)^{k+1}$.

Then Lemma 8 provides $\|T_A(1)\|_{\text{BMO}} \leq C_3(1+\|A'\|_\infty)^5$ and the $T(1)$ theorem allows us to conclude.

All that is left to do is to prove the two lemmas.

The first is easy and is based on the estimates $|E_A(x,y)| \leq |x-y|^{-1}$ and $|(\partial/\partial x)E_A(x,y)| \leq (1+M)(x-y)^{-2}$. We are trying to show that, for every interval $I \subset \mathbb{R}$, there is a constant γ_I, such that

(6.13) $$\frac{1}{|I|}\int_I |T_A(1) - \gamma_I|\, dx \leq \sigma(A,I) + C(1+M).$$

To this end, we split the function 1 into $f + u + v + w$, where f is the characteristic function of I, while u and v are the characteristic functions of the two intervals J and J' adjacent to I and of the same length. The fourth function w is the characteristic function of the complement of $3I$.

Now, $\int_I \int_J |x-y|^{-1}\, dx\, dy = (\log 2)|I|$ and the same is true for J'. Thus, $\int |T_A(u)|\, dx \leq (\log 2)|I|$ and the same holds for $\int_I |T_A(v)|\, dx$. Finally, the kernel's regularity with respect to x enables us to estimate $\int_{(3I)^c} |E_A(x,y) - E_A(x_0,y)|\, dy$, for every $x \in I$ (where x_0 denotes the centre of I).

We go on to the proof of Lemma 9, which is quite clearly the key to the method of David and Murai.

The following remark is the starting point: $\sigma(B,I) = \sigma(A,I)$ every time that $B(x) = A(x) + px + q$ on I, where $p, q \in \mathbb{R}$, or when $B(x) = -A(x)$ on I.

We intend to replace $A(x)$ by $B(x)$ on I, hoping that $A(x) = B(x)$ on a substantial subset $E \subset I$; we shall want $|E| \geq (1/4)|I|$. But we shall also require the function $B(x)$ to satisfy $M/3 \leq B'(x) \leq M$ on I. This will let us replace M by $2M/3$, by substituting $B'(x) - M/3$ for $B'(x)$, and thus "reduce the slope".

Let us give the details of this algorithm. We write $I = [a,b]$ and distinguish between the cases

$$m = \frac{A(b) - A(a)}{b-a} \geq \frac{M}{2} \quad \text{and} \quad m = \frac{A(b) - A(a)}{b-a} < \frac{M}{2}.$$

In the first case, we apply the rising sun lemma directly, with $\lambda = M/3$, and get $|\Omega| \leq (3/4)|I|$.

In the second case, we replace $A(x)$ by $\tilde{A}(x) = Mx - A(x)$. Then

$$\sigma(\tilde{A}, I) = \sigma(A, I), \qquad 0 \leq \tilde{A}'(x) \leq M \quad \text{and} \quad \frac{\tilde{A}(b) - \tilde{A}(a)}{b - a} \geq \frac{M}{2},$$

which brings us back to the first case.

Applying the rising sun lemma to A, or $\tilde{A}$, with $\lambda = M/3$, gives us a function which we call B (rather than B_I) to simplify the notation. To finish, we put $\tilde{B}(x) = B(x) - M/3$, getting $\tilde{B}'(x) \in [0, 2M/3]$. We get $\sigma(B, I) = \sigma(\tilde{B}, I) \leq \tau(2M/3)|I|$, by the definitions of τ and σ.

To compare $\sigma(A, I)$ with $\sigma(B, I)$, we intend to prove the basic inequality

$$(6.14) \qquad \sigma(A, I) \leq \sigma(B, I) + \sum_{k \geq 0} \sigma(A, I_k) + C(1 + M)|I|,$$

where the I_k are the connected components of the open set $\Omega \subset I$ produced by applying the rising sun lemma to A or $\tilde{A}$, with $\lambda = M/3$.

Once (6.14) has been established, we can conclude the proof of the lemma. Indeed, $\sigma(A, I_k) \leq \sigma(A)|I_k|$ and $\sum |I_k| \leq (3/4)|I|$ imply that

$$(6.15) \qquad \sigma(A, I) \leq \tau(2M/3)|I| + \frac{3}{4}\sigma(A)|I| + C(1 + M)|I|.$$

We divide both sides of (6.15) by $|I|$ and take the supremum, over all intervals I, of the left-hand side. This gives

$$(6.16) \qquad \sigma(A) \leq \tau(2M/3) + \frac{3}{4}\sigma(A) + C(1 + M).$$

But we know that $\sigma(A)$ is finite, because all the operators T_A are Calderón-Zygmund operators. It follows that

$$(6.17) \qquad \sigma(A) \leq 4\tau(2M/3) + 4C(1 + M)$$

and it is enough to take the supremum, over all functions A with $0 \leq A'(x) \leq M$, of the left-hand side of (6.17) to get (6.12).

We still have to prove (6.14). We know that

$$(6.18) \qquad E_A(x, y) = E_B(x, y) + R(x, y),$$

where the error term $R(x, y)$ satisfies

$$(6.19) \qquad |R(x, y)| \leq M(x - y)^{-2}(\text{dist}(x, E) + \text{dist}(y, E)).$$

We also know that, on each of the intervals I_k which make up Ω, we have $B(x) = \lambda x + c_k$ (with $\lambda = M/3$) so that, for x and y belonging to the same I_k, we have $E_B(x, y) = e^{i\lambda}(x - y)^{-1}$.

These two properties are enough to prove (6.14). Indeed,

$$\sigma(A, I) = \int_I \left| \int_I E_A(x, y) \, dy \right| dx$$

$$\leq \int_I \left| \int_I E_B(x, y) \, dy \right| dx + \int_I \left| \int_I R(x, y) \, dy \right| dx$$

$$= \sigma(B, I) + \eta.$$

To estimate the error term η, we let G denote the set of ordered pairs $(x, y) \in I \times I$ which do not belong to the union of the sets $I_k \times I_k$. In other words, x and y do not belong to the same interval I_k. Then

$$\eta \leq \int \int_G |R(x, y)| \, dx \, dy + \sum_{k \geq 0} \int_{I_k} \left| \int_{I_k} R(x, y) \, dy \right| dx.$$

The upper bound $\int \int_G |R(x, y)| \, dx \, dy \leq 4M|I|$ follows immediately from (6.19).

We are left with the terms $\sigma_k = \int_{I_k} |\int_{I_k} R(x, y) \, dy| \, dx$. To deal with them, we go back to (6.18), which gives

$$\sigma_k \leq \int_{I_k} \left| \int_{I_k} E_A(x, y) \, dy \right| dx + \int_{I_k} \left| \int_{I_k} E_B(x, y) \, dy \right| dx.$$

The first of the integrals is $\sigma(I_k)$ and the second can be calculated, by the remark after (6.19), to give $\int_{I_k} |\int_{I_k} (x - y)^{-1} \, dy| \, dx \leq \pi|I_k|$, because the Hilbert transform is unitary on $L^2(\mathbb{R}^n)$.

7 The Calderón-Zygmund method of rotations

By the method of rotations, we can construct Calderón-Zygmund operators whose actions on $L^2(\mathbb{R}^n)$ are like taking the mean (or Bochner integral), over all directions, of Calderón-Zygmund operators acting on functions of one real variable.

We are now in a position to think of the method as the prototype of the transference methods developed systematically by Coifman and Weiss ([77]).

For example, Calderón and Zygmund used the method of rotations to establish the continuity of the Riesz transforms R_j, $1 \leq j \leq n$, on every $L^p(\mathbb{R}^n)$, $1 < p < \infty$, as a corollary of the corresponding result for the Hilbert transform.

The method of rotations, as well as the other techniques described in this section, is based on the description of operators in terms of their kernels. That is, we start from a function $K(x, y)$ which, to begin with, will belong to $L^\infty(\mathbb{R}^n \times \mathbb{R}^n)$, to avoid the problems associated with singular integrals. We then define the operator T by

(7.1) $$Tf(x) = \int K(x,y)f(y)\,dy\,,$$

where f belongs, for example, to the space E of continuous functions of compact support.

We shall let S^{n-1} denote the unit sphere in $\mathbb{R}^n$ and write ω_{n-1} for its surface area.

Theorem 9. *Let $K(x,y) \in L^\infty(\mathbb{R}^n \times \mathbb{R}^n)$ define an operator T by (7.1). For each $x_0 \in \mathbb{R}^n$ and every $\nu \in S^{n-1}$, let $T_{(x_0,\nu)}$ be the operator, acting on functions of one real variable, whose distribution-kernel is*

$$k(s,t;x_0,\nu) = |s-t|^{n-1}K(x_0 + s\nu, x_0 + t\nu) \qquad \text{for } s,t \in \mathbb{R}.$$

Let p satisfy $1 \le p \le \infty$ (the extreme values are not excluded), and suppose that the operators $T_{(x_0,\nu)}$ are uniformly bounded on $L^p(\mathbb{R})$ with operator norms (on $L^p(\mathbb{R})$) not exceeding 1.

Then T is bounded on $L^p(\mathbb{R}^n)$ and the norm of T does not exceed $\omega_{n-1}/2$.

Put $g = Tf$. We integrate using polar co-ordinates, centre x, to get

$$\begin{aligned}
g(x) &= \int K(x,y)f(y)\,dy \\
&= \frac{1}{2}\int_{S^{n-1}} \left\{ \int_{-\infty}^{\infty} K(x, x+t\nu)f(x+t\nu)|t|^{n-1}\,dt \right\} d\sigma(\nu) \\
&= \frac{1}{2}\int_{S^{n-1}} g_\nu(x)\,d\sigma(\nu)\,,
\end{aligned}$$

where

$$g_\nu(x) = \int_{-\infty}^{\infty} K(x, x+t\nu)f(x+t\nu)|t|^{n-1}\,dt\,.$$

We shall show that $\|g_\nu(x)\|_{L^p(\mathbb{R}^n)} \le \|f\|_{L^p(\mathbb{R}^n)}$, which will give $\|g\|_p \le (\omega_{n-1}/2)\|f\|_p$, by convexity.

Since ν is fixed, we can choose new co-ordinate axes, if necessary, so that $\nu = (0,0,\ldots,0,1)$ and, with respect to those co-ordinates, we write $x = (x',s)$, where $x' \in \mathbb{R}^{n-1}$ and $s \in \mathbb{R}$. Then

$$\|g_\nu\|_{L^p(\mathbb{R}^n)} = \left(\int\int_{\mathbb{R}^{n-1}\times\mathbb{R}} |g_\nu(x'+s\nu)|^p\,dx'\,ds \right)^{\frac{1}{p}}.$$

Applying Fubini's theorem, we first calculate

$$\int_{\mathbb{R}} |g_\nu(x' + s\nu)|^p \, ds$$

$$= \int_{-\infty}^{\infty} \left| \int_{-\infty}^{\infty} K(x' + s\nu, x' + (s+t)\nu) f(x' + (s+t)\nu) |t|^{n-1} dt \right|^p ds$$

$$= \int_{-\infty}^{\infty} \left| \int_{-\infty}^{\infty} K(x' + s\nu, x' + t\nu) f(x' + t\nu) |t - s|^{n-1} dt \right|^p ds$$

$$\leq \int_{-\infty}^{\infty} |f(x' + t\nu)|^p \, dt \,,$$

by the hypothesis about the operators $T_{(x',\nu)}$. We then integrate over $x' \in \mathbb{R}^n$ to finish the proof.

How do we apply the method of rotations?

We start with a kernel $K(x, y)$, $x, y \in \mathbb{R}^n$, $y \neq x$, satisfying

(7.2) $\qquad\qquad K(y, x) = -K(x, y) \,,$

(7.3) $\qquad\qquad |K(x, y)| \leq C|x - y|^{-n} \,,$

(7.4) $\qquad |(\partial/\partial x_j)K(x, y)| \leq C|x - y|^{-n-1} \qquad (1 \leq j \leq n) \,.$

Then the distribution PV $K(x, y) \in \mathcal{S}'(\mathbb{R}^n \times \mathbb{R}^n)$ defines an operator $T : \mathcal{S}(\mathbb{R}^n) \to \mathcal{S}'(\mathbb{R}^n)$. We want to know whether T is bounded on $L^2(\mathbb{R}^n)$, so we look at the kernels

$$k(s, t; x_0, \nu) = |s - t|^{n-1} K(x_0 + s\nu, x_0 + t\nu) \qquad \text{for } s, t \in \mathbb{R}.$$

They satisfy the analogous inequalities to (7.3) and (7.4) in one variable, and are antisymmetric too.

Thus these kernels define operators $T_{(x_0,\nu)} : \mathcal{S}(\mathbb{R}) \to \mathcal{S}'(\mathbb{R})$.

We then suppose that, for a certain constant C,

(7.5) $\qquad\qquad \|T_{(x_0,\nu)}(f)\|_{L^2(\mathbb{R})} \leq C\|f\|_{L^2(\mathbb{R})} \,,$

uniformly with respect to $x_0 \in \mathbb{R}^n$ and $\nu \in S^{n-1}$.

Theorem 10. *With the above hypotheses, T is bounded on $L^2(\mathbb{R}^n)$ and $\|T\| \leq C'$.*

The constant C' depends only on the dimension n and the constants appearing in (7.3), (7.4) and (7.5).

To prove the theorem, we consider the truncated operators T^ε and $T^\varepsilon_{(x_0,\nu)}$ defined by the truncated kernels

$$K(x, y)\chi_{\{|x-y| \geq \varepsilon\}} \qquad \text{and} \qquad k(s, t; x_0, \nu)\chi_{\{|s-t| \geq \varepsilon\}} \,.$$

We denote the truncated kernels by K^ε and k^ε and pass from the former to the latter by the same operation as we used to obtain k from K ("the method of rotations commutes with truncations").

So, by Theorem 9, for every $\varepsilon > 0$,

(7.6) $$\|T^\varepsilon\| \le \frac{1}{2}\omega_{n-1} \sup_{x_0} \sup_\nu \|T^\varepsilon_{(x_0,\nu)}\| \,.$$

Now, Cotlar's theorem in Chapter 7 implies the existence of a constant C, such that, for all $\varepsilon > 0$,

(7.7) $$\|T^\varepsilon_{(x_0,\nu)} f\|_2 \le C\|f\|_2 \,.$$

Combining (7.6) and (7.7) gives $\|T^\varepsilon\| \le C'$. Then the antisymmetry of the kernel guarantees that $\langle T^\varepsilon f, g\rangle$ tends to $\langle Tf, g\rangle$, as $\varepsilon \to 0$, for all $f, g \in \mathcal{S}(\mathbb{R}^n)$. The uniform estimate $\|T^\varepsilon\| \le C'$ thus gives the weak convergence of the T^ε to $T \in \mathcal{L}(L^2, L^2)$.

Here is a remarkable application of this method, due to Calderón.

Theorem 11. *Let $n, m \ge 1$ be integers and suppose that $A : \mathbb{R}^n \to \mathbb{R}^m$ is a Lipschitz function. Let $F : \mathbb{R}^m \to \mathbb{R}$ be an infinitely differentiable odd function. Then the antisymmetric kernel*

(7.8) $$K(x,y) = F\left(\frac{A(x) - A(y)}{|x - y|}\right) |x - y|^{-n}$$

defines a bounded operator on $L^2(\mathbb{R}^n)$.

To see this, we apply the method of rotations, which brings us back to the case $n = 1$. Next, we observe that $|x-y|^{-1}F((A(x)-A(y))/|x-y|) = (x-y)^{-1}F((A(x)-A(y))/(x-y))$, because F is an odd function. We then take advantage of the inequality $|A(x)-A(y)| \le M|x-y|$ to replace F on the ball $|u| \le M$ of $\mathbb{R}^m$, by an infinitely differentiable periodic function, of period $4M$ in each variable. We expand this periodic function as a Fourier series to get, for $|u| \le M$,

$$F(u) = \sum_{k \in \mathbb{Z}^n} \alpha(k) e^{i\delta k \cdot u} \,,$$

where $\delta = \pi/2M$ and the $\alpha(k)$ are rapidly decreasing. The kernel $(x-y)^{-1}F((A(x)-A(y))/(x-y))$ can thus be written as $\sum \alpha(k) G_k(x, y)$, where $G_k(x,y) = (x - y)^{-1} e^{i\delta k \cdot (A(x) - A(y))/(x-y)}$. The norm of the operator G_k defined by the kernel $G_k(x, y)$ does not, therefore, exceed $C(1 + \delta|k|\|A'\|_\infty)^5$ and the series $\sum \alpha(k) G_k$ is absolutely convergent in $\mathcal{L}(L^2, L^2)$.

We finish this section with two applications of Theorem 11, drawn respectively from potential theory and from complex analysis. We shall give more details of the first example in Chapter 15.

Let $A : \mathbb{R}^n \to \mathbb{R}$ be a Lipschitz function, let $S \subset \mathbb{R}^{n+1}$ be its graph and let $d\sigma$ be the surface measure on S. We start from an electric charge density $g \in L^2(S, d\sigma)$ and let $V(x)$, $x \in \mathbb{R}^{n+1}$, be the potential created by the charge density. We suppose that $n \geq 2$. Then

$$(7.9) \qquad V(x) = c_n \int_S |x-y|^{-n+1} g(y)\, d\sigma(y)\,.$$

Now, consider the electromagnetic field arising from the potential V. Ignoring the normalization constants, we write $\mathbf{E} = -\nabla V$.

Let $\mathbf{1}$ denote the vector $(0,0,\ldots,0,1) \in \mathbb{R}^{n+1}$. It is well-known in electrostatics that $\mathbf{E}(x)$ has a discontinuity as it passes through the surface S. For $x \in S$, we set

$$(7.10) \qquad \mathbf{E}_+(x) = \lim_{\varepsilon \downarrow 0} \mathbf{E}(x+\varepsilon\mathbf{1}) \quad \text{and} \quad \mathbf{E}_-(x) = \lim_{\varepsilon \downarrow 0} \mathbf{E}(x-\varepsilon\mathbf{1})\,.$$

These limits exist, for almost all $x \in S$, as we shall show in Chapter 15. Finally, the discontinuity of the electromagnetic field is given, for almost all $x \in S$, by

$$(7.11) \qquad \mathbf{E}_+(x) - \mathbf{E}_-(x) = \gamma_n g(x)\mathbf{n}(x)\,,$$

where $\mathbf{n}(x)$ is the upwards $(\mathbf{1} \cdot \mathbf{n}(x) > 0)$ unit vector normal to the surface S.

We then have the following result.

Theorem 12. *If $\|\nabla A\|_\infty \leq M < \infty$, then, for every $g \in L^2(S, d\sigma)$, we have*

$$(7.12) \qquad \left(\int_S |\mathbf{E}_+(x)|^2\, d\sigma(x)\right)^{1/2} \leq C(M,n) \left(\int_S |g(x)|^2\, d\sigma(x)\right)^{1/2}$$

where $C(M,n)$ depends only on the upper bound M and the dimension n.

To prove this result, we use an algorithm to write $\mathbf{E}_+(x)$ in terms of $g(x)$. The algorithm is that of the Calderón-Zygmund operators and gives

$$(7.13) \ \mathbf{E}_+(x) = \frac{1}{2}\gamma_n g(x)\mathbf{n}(x) + \mathrm{PV}\, c_n \int_S (x-y)|x-y|^{-n-1} g(y)\, d\sigma(y)\,.$$

Of course, it is the second term on the right-hand side of (7.13) that is problematic. We deal with it by using the parametric representation of S given by $x = (u, A(u))$, $u \in \mathbb{R}^n$, and the vectorial kernel $(x-y)|x-y|^{-n-1}$, restricted to $L^2(S)$, becomes

$$\left(\frac{u-v}{(|u-v|^2 + (A(u)-A(v))^2)^{(n+1)/2}}, \frac{A(u)-A(v)}{(|u-v|^2 + (A(u)-A(v))^2)^{(n+1)/2}}\right)$$

and this then acts on $L^2(\mathbb{R}^n)$.

The continuity of this operator is then a consequence of Theorem 11.

Stein and Weiss discovered how the Hardy spaces of functions holomorphic in the upper half-plane could be generalized to $\mathbb{R}^n \times (0, \infty)$.

They replaced the pair $(u(z), v(z))$ of real and imaginary parts of a holomorphic function by the gradient of a harmonic function.

Identity (7.11) then appears in the guise of a generalization of Plemej's formula, which occurs in the following context.

Let Γ be a rectifiable curve, without double points, in the complex plane parametrized by the arclength s which runs through the whole real line. We suppose that $\lim_{s \to \infty} |z(s)| = \lim_{s \to -\infty} |z(s)| = \infty$. Let Ω_1 and Ω_2 be the open sets whose boundaries are Γ. We define the Hardy spaces $H^2(\Omega_1)$ and $H^2(\Omega_2)$ as the closures in $L^2(\Gamma, ds)$ of the rational functions $P(z)/Q(z)$ which are zero at infinity and whose poles belong to Ω_2 and Ω_1, respectively.

The problem is then to determine whether

(7.14) $$L^2(\Gamma, ds) = H^2(\Omega_1) + H^2(\Omega_2),$$

where the sum is direct, but not necessarily orthogonal.

We shall examine this question systematically in the course of Chapter 12. Here, we just want to remark that (7.14) is equivalent to the Cauchy kernel's defining a bounded operator on $L^2(\Gamma, ds)$, by

$$Tf(z) = \text{PV}\, \frac{1}{i\pi} \int_\Gamma \frac{1}{z - w} f(w)\, dw\,.$$

Using the parametrization given by the arclength, everything reduces to knowing whether $(z(s) - z(t))^{-1}$ defines a bounded operator on $L^2(\mathbb{R})$.

A rectifiable curve is a Lavrentiev curve if there exists a constant $\delta > 0$ such that, for all s and t, we have $|z(s) - z(t)| \geq \delta|s - t|$.

With this definition, we can state

Theorem 13. *For every Lavrentiev curve, the Cauchy kernel defines a bounded operator on $L^2(\Gamma, ds)$ and this operator is a Calderón-Zygmund operator.*

To see this, we let F be an infinitely differentiable and odd function on $\mathbb{R}^2$, which coincides with $1/z$ when $\delta \leq |z| \leq 1$. We can then write

$$\frac{1}{z(s) - z(t)} = F\left(\frac{z(s) - z(t)}{s - t}\right) \frac{1}{s - t}$$

and it is enough to apply Theorem 11.

10

Operators corresponding to singular integrals: their continuity on Hölder and Sobolev spaces

1 Introduction

The kernels $|x - y|^{-n+\lambda}$, $0 < \lambda < n$, whose Fourier transforms in the sense of distributions are $c(n, \lambda)|\xi|^{-\lambda}$, define the *fractional integration* operators. Both the kernels and their Fourier transforms are locally integrable and it is easy to prove, without further ado, that convolution with $|x|^{-n+\lambda}$ is a continuous operator from $\dot{C}^s$ to $\dot{C}^{s+\lambda}$, for all $s > 0$.

The $\dot{C}^s$ spaces are the homogeneous Hölder spaces. For convenience, we once again give their definition.

If $0 < s < 1$ then $f \in \dot{C}^s$ when f is Hölder of exponent s, that is, when f is a continuous function modulo the constant functions and

$$|f(x) - f(y)| \leq C|x - y|^s.$$

If $s = 1$, we have to replace the usual C^1 by the Zygmund class Λ_*, which is the space of continuous functions, modulo the affine functions, such that, for all $x, y \in \mathbb{R}^n$,

$$|f(x + y) + f(x - y) - 2f(x)| \leq C|y|.$$

Lastly, if $s > 1$, we write $s = m + r$, where $0 < r \leq 1$, and $f \in \dot{C}^s$ means that f is a C^m function, modulo polynomials of degree not exceeding m ($m + 1$ if $s = m + 1$), such that all the partial derivatives $\partial^\alpha f$, $|\alpha| = m$, belong to $\dot{C}^r$.

The theorem about fractional integration, indicated above, extends to operators which are not convolution operators. Consider two exponents λ and γ, with $0 < \lambda < \gamma \leq 1$, and suppose that there is a function

$K(x,y)$ which satisfies the inequality $|K(x,y)| \leq C|x-y|^{-n+\lambda}$, for all x and y, as well as the inequality

(1.1) $\qquad |K(x',y) - K(x,y)| \leq C|x'-x|^{\gamma}|x-y|^{-n+\lambda-\gamma},$

when $|x'-x| \leq |x-y|/2$.

Suppose that $s > 0$ and $s+\lambda < \gamma$. Then the operator T defined by the kernel $K(x,y)$ can be extended as a continuous linear operator on the homogeneous space $\dot{C}^s$ if and only if $T(1) = 0$, modulo the constants. This statement is proved by following through the steps of the proof for the fractional integration operators. In the latter case, the condition $T(1) = 0$ is automatically satisfied, since T is a convolution operator. We refer the reader to [217] or [239].

The purpose of this chapter is to go a bit further and replace the exponent λ above by 0. Then the kernel $K(x,y)$ is no longer integrable over y for fixed x and we use the ideas of Chapter 7, by saying that a weakly continuous linear operator $T : \mathcal{D}(\mathbb{R}^n) \to \mathcal{D}'(\mathbb{R}^n)$ corresponds to K (as opposed to being defined by K) if, for all functions $f \in \mathcal{D}(\mathbb{R}^n)$,

$$Tf(x) = \int K(x,y)f(y)\,dy,$$

whenever x is not in the support of f.

We suppose that K satisfies (1.1) with $\lambda = 0$. It becomes necessary to introduce an extra hypothesis which, in some manner, compensates for the failings of the integral $\int K(x,y)f(y)\,dy$. The hypothesis is that T is weakly continuous on $L^2(\mathbb{R}^n)$ and this is a necessary condition for the continuity of $T : \dot{C}^s \to \dot{C}^s$. For $0 < s < \gamma$, the continuity of $T : \dot{C}^s \to \dot{C}^s$ is equivalent to $T(1) = 0$, modulo the constant functions. The proof of this result is remarkably simple if we use wavelets. We then follow Lemarié ([164]) and prove "with bare hands" that T is continuous on the homogeneous Sobolev spaces $\dot{B}^s$.

Finally, we shall examine the subtler case of the inhomogeneous Hölder and Besov spaces.

2 Statement of the theorems

We recall how to define homogeneous Sobolev spaces. Let $\mathcal{S}_0(\mathbb{R}^n) \subset \mathcal{S}(\mathbb{R}^n)$ denote the subspace of functions $f \in \mathcal{S}(\mathbb{R}^n)$, all of whose moments vanish. For every $s \in \mathbb{R}$, we equip $\mathcal{S}_0(\mathbb{R}^n)$ with the pre-Hilbert structure given by

$$\langle f, g \rangle_s = \frac{1}{(2\pi)^n} \int_{\mathbb{R}^n} \hat{f}(\xi)\bar{\hat{g}}(\xi)|\xi|^{2s}\,d\xi$$

and we let $\dot{B}^s$ denote the corresponding Hilbert space. We shall give a description of $\dot{B}^s$ as a space of functions or distributions.

The case $s = 0$ is trivial, $\dot{B}^0$ being $L^2(\mathbb{R}^n)$.

We then consider the range $-n/2 < s < n/2$. If $0 < s < n/2$, we define the exponent q by $1/2 - 1/q = s/n$ and the homogeneous case of the Sobolev embedding theorem tells us that $\dot{B}^s$ is canonically embedded in $L^q(\mathbb{R}^n)$ and thus is a space of functions.

If $-n/2 < s < 0$, $\dot{B}^s(\mathbb{R}^n)$ is a space of tempered distributions and $L^p(\mathbb{R}^n) \subset B^s(\mathbb{R}^n)$ if $1/p - 1/2 = -s/n$. It can be shown that $\mathcal{S}(\mathbb{R}^n)$ is dense in $\dot{B}^s(\mathbb{R}^n)$.

The embedding $\mathcal{S}(\mathbb{R}^n) \subset \dot{B}^s(\mathbb{R}^n)$ does not hold if $s \leq -n/2$ and is replaced by $\mathcal{S}_0(\mathbb{R}^n) \subset \dot{B}^s(\mathbb{R}^n)$. Similarly, the inclusion $L^p(\mathbb{R}^n) \subset \dot{B}^s(\mathbb{R}^n)$ does not hold when $s \leq -n/2$ and is replaced by $H^p(\mathbb{R}^n) \subset \dot{B}^s(\mathbb{R}^n)$, where $1/p - 1/2 = -s/n$. The reader is referred to [75] or [217] for the main properties of the Stein and Weiss spaces $H^p(\mathbb{R}^n)$.

If $s = -t$, the spaces $\dot{B}^s$ and $\dot{B}^t$ are canonically dual because

$$\int f(x)\bar{g}(x)\,dx = \frac{1}{(2\pi)^n}\int \hat{f}(\xi)|\xi|^s \bar{\hat{g}}(\xi)|\xi|^t\,d\xi$$

and because the modulus of the integral can be estimated using the Cauchy-Schwarz inequality.

For $s \geq n/2$, the properties of the spaces $\dot{B}^s$ are "dual" to those of $\dot{B}^t$, for $t \leq -n/2$.

The embedding $H^1(\mathbb{R}^n) \subset \dot{B}^{-n/2}$ is mirrored by $\dot{B}^{n/2} \subset$ BMO. Like BMO, $\dot{B}^{n/2}$ is a space of functions modulo the constants.

When $s = n/2 + \gamma$, $\gamma > 0$, $\dot{B}^s$ is canonically embedded in the homogeneous Hölder space $\dot{C}^\gamma$. That is, like $\dot{C}^\gamma$, $\dot{B}^s$ is a space of continuous functions modulo the polynomials of degree not exceeding γ.

The homogeneity of $\dot{B}^s$ is given by $\|U_g(f)\|_{\dot{B}^s} = a^{-s}\|f\|_{\dot{B}^s}$, when $g(x) = ax + b$, $a > 0$, $b \in \mathbb{R}^n$ and $U_g f(x) = a^{-n/2} f((x - b)/a)$.

For $0 < s < 1$, the norm on $\dot{B}^s$ can be defined without recourse to the Fourier transform. For each $f \in \mathcal{S}(\mathbb{R}^n)$,

$$c(s,n)\|f\|_{\dot{B}^s} = \left(\int\int |f(x) - f(y)|^2 |x - y|^{-n-2s}\,dx\,dy\right)^{1/2}.$$

For $s = 1$, $\|f\|_{\dot{B}^1} = \|\nabla f\|_2$ and all the other $\dot{B}^s$ norms can be calculated using these identities.

We now take a continuous linear operator $T : \mathcal{D}(\mathbb{R}^n) \to \mathcal{D}'(\mathbb{R}^n)$. When $0 < \gamma \leq 1$, we write $T \in \mathcal{L}_\gamma$ if the distribution-kernel of T, restricted to the open set $\{(x,y) : y \neq x\} \subset \mathbb{R}^n \times \mathbb{R}^n$, becomes a function $K(x,y)$ such that

$$(2.1) \qquad |K(x,y)| \leq C|x - y|^{-n},$$

(2.2) $|K(x', y) - K(x, y)| \leq C|x' - x|^\gamma |x - y|^{-n-\gamma},$

if $|x' - x| \leq |x - y|/2$.

We do not require any regularity with respect to y.

For $\gamma > 1$, we write $\gamma = m + r$, where $0 < r \leq 1$, and replace (2.2) by the two following conditions:

(2.3) $|\partial_x^\alpha K(x, y)| \leq C|x - y|^{-n-\alpha}$ for $|\alpha| \leq m$;

(2.4) $|\partial_x^\alpha K(x', y) - \partial_x^\alpha K(x, y)| \leq C|x' - x|^r |x - y|^{-n-\gamma},$

for $|\alpha| = m$ and $|x' - x| \leq |x - y|/2$.

If $T \in \mathcal{L}_\gamma$, we can define $T(x^\alpha)$, for $|\alpha| \leq m$, as follows. We let $\phi(x)$ denote a function in $\mathcal{D}(\mathbb{R}^n)$ which is 1 at 0 and set $\phi_\varepsilon(x) = \phi(\varepsilon x)$. Then, for $|\alpha| \leq m$, there exist "floating" polynomials $P_{\alpha,\varepsilon}(x)$, whose degrees do not exceed $|\alpha|$, such that the difference $T(x^\alpha \phi_\varepsilon) - P_{\alpha,\varepsilon} = S_{\alpha,\varepsilon}$ converges, in the sense of distributions, to a distribution $S_\alpha \in \mathcal{S}'(\mathbb{R}^n)$, which we denote by $T(x^\alpha)$.

To see this, it is enough to show that, if $\psi \in \mathcal{D}(\mathbb{R}^n)$ and $\int x^\alpha \psi(x) \, dx = 0$, for $|\alpha| \leq m$, then $\langle T(x^\alpha \phi_\varepsilon), \psi \rangle$ tends to a limit. We write this integral as $\langle x^\alpha \phi_\varepsilon, {}^t T(\psi) \rangle$ and observe that ${}^t T(\psi)(x)$ is $O(|x|^{-n-\gamma})$ at infinity. Lebesgue's dominated convergence theorem lets us conclude the argument.

Theorem 1. *Suppose that $T \in \mathcal{L}_\gamma$ and that $0 < s < \gamma$. Let σ denote the integer part of s. Then T can be extended as a continuous linear operator on the homogeneous Hölder space $\dot{C}^s$ if and only if T is weakly continuous on $L^2(\mathbb{R}^n)$, in the sense of the theorem of David and Journé, and if $T(x^\alpha) = 0$, modulo the polynomials of degree not exceeding σ, for $|\alpha| \leq \sigma$.*

These conditions imply that, for $0 < s < \gamma$, T can be extended as a continuous linear operator on each homogeneous Sobolev space $\dot{B}^s$ and, for $1 \leq p, q \leq \infty$, on each Besov space $\dot{B}_p^{s,q}$.

Before we prove this result, let us examine some examples.

3 Examples

It may seem strange, at first sight, to use the homogeneous spaces $\dot{C}^s$ rather than the usual Hölder spaces $C^s = \dot{C}^s \cap L^\infty$.

The reason we do this is that the Hilbert transform (in dimension 1) and the Riesz transforms (in dimension n) are not bounded on C^s, whereas they are bounded on $\dot{C}^s$.

Indeed, the norm of f in C^s is

$$\|f\|_\infty + \sup_{x \neq y} \frac{|f(y) - f(x)|}{|y - x|^s},$$

if $0 < s < 1$. It therefore follows that, for $f \in \mathcal{D}(\mathbb{R}^n)$,

$$\lim_{N \to \infty} \|f(N^{-1}x)\|_{C^s} = \|f\|_\infty .$$

An operator which commutes with dilations and is continuous on C^s is therefore continuous on L^∞. But neither the Hilbert transforms nor the Riesz transforms are continuous on L^∞.

The problem is the behaviour of the kernels $K(x, y)$ of those operators, when $|x - y|$ tends to infinity. That problem disappears in our second group of examples.

We consider the symbols $\sigma(x, \xi) \in S^0_{1,1}(\mathbb{R}^n \times \mathbb{R}^n)$ which, as we recall, are subject to the conditions

(3.1) $$|\partial_\xi^\alpha \partial_x^\beta \sigma(x, \xi)| \leq C(\alpha, \beta)(1 + |\xi|)^{|\beta| - |\alpha|} .$$

It can be seen immediately that the corresponding kernel $K(x, y)$ satisfies

(3.2) $$|\partial_x^\alpha \partial_y^\beta K(x, y)| \leq C(\alpha, \beta)|x - y|^{-n - |\alpha| - |\beta|} ,$$

when $|x - y| \leq 1$, and

(3.3) $$|\partial_x^\alpha \partial_y^\beta K(x, y)| \leq C(\alpha, \beta, N)|x - y|^{-N} ,$$

for all $N \geq 1$, when $|x - y| \geq 1$ (and $\alpha, \beta \in \mathbb{N}^n$).

To apply Theorem 1, we must make sure that $T(x^\alpha) = 0$, modulo the polynomials of degree not exceeding σ. This condition can be written in the form $\partial_\xi^\alpha \sigma(x, \xi)|_{\xi=0} = P_\alpha(x)$, where $P_\alpha(x)$ is a polynomial whose degree is not greater than σ. To reduce to this special case, it is enough to write $\sigma(x, \xi)$ as $\sigma(x, \xi)\phi_0(\xi) + \sigma(x, \xi)\phi_1(\xi)$, where $\phi_1(x) = 0$ in a neighbourhood of 0 and $\phi_0 \in \mathcal{D}(\mathbb{R}^n)$. This induces a decomposition $\sigma(x, D) = T = T_0 + T_1$. The operator T_0 is smoothly regularizing and Theorem 1 applies to T_1.

In view of (3.1), we can pass from the homogeneous to the inhomogeneous spaces and confirm that $T \in \mathcal{O}p\,S^0_{1,1}$ is continuous on all inhomogeneous Besov spaces $B_p^{s,q}$, where $s > 0$ and $1 \leq p, q \leq \infty$.

The operators in question are, in general, unbounded on $L^2(\mathbb{R}^n)$.

The next example is due to Calderón. Let $a : \mathbb{R} \to \mathbb{C}$ be a Lipschitz function with $\|a'\|_\infty \leq M$. We consider the distribution-kernel $K(x, y) = \mathrm{PV}((a(x) - a(y))/(x - y)^2)$ and the operator $T : \mathcal{D}(\mathbb{R}) \to \mathcal{D}'(\mathbb{R})$ defined by this kernel. Then T is continuous on $L^2(\mathbb{R}^n)$ (Calderón, 1965), but not on the Sobolev space H^s, when $s > 0$, or on its homogeneous version $\dot{B}^s$. By the same token, T is not continuous on $\dot{C}^s$.

On the other hand, the distribution-kernel

$$K_1(x, y) = \mathrm{PV}\,\frac{a(x) - a(y) - (x - y)a'(y)}{(x - y)^2} = \frac{\partial}{\partial y}\frac{a(x) - a(y)}{x - y}$$

leads to a better operator T_1, in the sense that T_1 is bounded on the

Sobolev space H^s, for $0 \le s \le 1$ and on the homogeneous Hölder space $\dot{C}^s$, when $0 < s < 1$.

The difference between T and T_1 is that $T_1(1) = 0$.

The following example is similar. This time, we suppose that $a(x)$ is real-valued, but still Lipschitz, and we consider the curve Γ in the complex plane defined by $z(x) = x + ia(x)$. Γ is the graph of $a(x)$. We then form the Cauchy kernel $\text{PV}(z(x) - z(y))^{-1}$ which we set against $\text{PV}\, z'(y)(z(x) - z(y))^{-1}$. Let T and T_1 denote the corresponding operators. Both are continuous on $L^2(\mathbb{R}^n)$, but T_1 satisfies the further condition $T_1(1) = 0$, modulo the constants, and the operator T_1 is continuous on the Sobolev space H^s, for $0 \le s \le 1$, and on the homogeneous Hölder space $\dot{C}^s$, for $0 < s < 1$.

A similar example is given by the double-layer potential on the boundary of a Lipschitz open set.

We let $D \subset \mathbb{R}^{n+1}$ be a connected open set whose boundary ∂D is given locally by the graph of a Lipschitz function and we let $K : L^2(\partial D) \to L^2(\partial D)$ be the operator defined by the double-layer potential.

Using local co-ordinates, ∂D is represented by a graph $t = a(x)$, where $x \in \mathbb{R}^n$ and $a : \mathbb{R}^n \to \mathbb{R}$ is Lipschitz. Then the distribution-kernel $K(x, y)$ of the operator K is given by

$$K(x, y) = \text{PV} \frac{1}{\omega_n} \frac{a(x) - a(y) - (x - y) \cdot \nabla a(y)}{(|x - y|^2 + (a(x) - a(y))^2)^{(n+1)/2}},$$

where ω_n is the surface area of the unit sphere $S^n \subset \mathbb{R}^{n+1}$. It is well known that, if $a : \mathbb{R}^n \to \mathbb{R}$ is only Lipschitz, then the operator K is not compact. Fabes, Jodeit and Rivière ([104]) showed that K was compact when $a(x)$ was a C^1 function.

The continuity of the operator K on the space $L^2(\partial D)$ is an example of Calderón's programme and is a consequence, as we have shown, of the method of rotations and of the continuity of the operators related to the Cauchy kernel on Lipschitz curves.

The operator K is continuous on $H^s(\mathbb{R}^n)$, if $0 \le s \le 1$, and on the homogeneous Hölder spaces $\dot{C}^s$, if $0 < s < 1$. On the other hand, these last two properties do not hold if we replace the numerator by the apparently simpler expression $a(x) - a(y)$.

Here is a final example of an application, or illustration, of Theorem 1. Let A denote the operator of pointwise multiplication by a function $a(x)$ belonging to the inhomogeneous Hölder space C^s, $s > 0$, and let T denote a classical pseudo-differential operator of order 1 whose symbol $\tau(x, \xi)$ belongs to the Hörmander class $S^1_{1,0}$.

Let $\tilde{A}$ be the operator of pointwise multiplication by $\tilde{a}(x)$, where $\tilde{a} = T(a)$. The function $\tilde{a}$ belongs to C^{s-1}.

We observe (and can prove) the following curious phenomenon: the commutator $[T, A]$ maps C^r continuously to C^r when $0 < r \leq s - 1$, but, if $r > s - 1$, the same commutator only maps C^r into C^{s-1} and we cannot do better.

On the other hand, $R = [T, A] - \tilde{A}$ is a continuous mapping of C^r into C^r when $0 < r \leq s$ and, again, when $r > s$, R maps C^r to C^s.

However, in this case, R is not compact or regularizing. Correcting the commutator $[T, A]$ by subtracting $\tilde{A}$ is not the correction which the pseudo-differential calculus requires.

Why is this? We choose $\tau(x, \xi) = \sqrt{1 + |\xi|^2}$, or $T = (I - \Delta)^{1/2}$, and $a(x) = e^{i\alpha \cdot x}$, for some $\alpha \in \mathbb{R}^n$. Then the commutator $[T, A]$ is defined by the symbol $\{(1 + |\xi + \alpha|^2)^{1/2} - (1 + |\xi|^2)^{1/2}\}e^{i\alpha \cdot x}$. The classical way to deal with this is to consider the asymptotic expansion

$$\left\{\alpha \cdot \frac{\xi}{|\xi|} + \frac{1}{2|\xi|}\left(1 + |\xi|^2 - \left(\alpha \cdot \frac{\xi}{|\xi|}\right)^2\right) + \cdots\right\}e^{i\alpha \cdot x}.$$

This means that the correction required by the standard pseudo-differential calculus is defined by the symbol $\alpha \cdot \xi|\xi|^{-1}e^{i\alpha \cdot x}$. The corresponding operator is not $\tilde{A}$ but $A\sum_1^n \alpha_j R_j$, where the R_j are the Riesz transforms (whose symbols are $\xi_j/|\xi|$) and A is still the operator of pointwise multiplication by $a(x)$.

The example we have just given can also be interpreted by remarking that, if L is an operator defined by a symbol $\lambda(x, \xi)$ in $L^\infty(\mathbb{R}^n \times \mathbb{R}^n)$, and if the operator L satisfies conditions (2.1) and (2.2), then the continuity of L on the homogeneous Hölder space $\dot{C}^s$, $0 < s < 1$, is given by $L(1) = 0$, modulo the constant functions, that is, by $\lambda(x, 0) = c$, a constant independent of x.

To correct an operator which does not yet satisfy this condition is to subtract the operator of pointwise multiplication by the function $\lambda(x, 0)$. On the other hand, the standard procedure of pseudo-differential calculus is to expand the symbol about the point at infinity, and not about the origin . . . .

4 Continuity of T on homogeneous Hölder spaces

Let us begin by showing that, if $T \in \mathcal{L}_\gamma$ and if $T(x^\alpha) = 0$, for $|\alpha| \leq \sigma$ (the integer part of $s < \gamma$), then T is continuous on the homogeneous Hölder space $\dot{C}^s$.

We start with $f \in \dot{C}^s$. To show that $T(f)$ belongs to the selfsame space, it is enough to prove that $|\langle T(f), \psi_\lambda\rangle| \leq C2^{-j(n/2+s)}$, where ψ_λ, $\lambda \in \Lambda$, forms an orthonormal basis of wavelets which are sufficiently regular. Now $\langle T(f), \psi_\lambda \rangle = \langle f, {}^tT(\psi_\lambda) \rangle$ and $g_\lambda = {}^tT(\psi_\lambda)$ satisfies the following three conditions:

(a) g_λ is a continuous linear form on the space C_0^s of C^s functions with compact support;

(b) $g_\lambda(x) = O(|x|^{-n-\gamma})$, as $|x| \to \infty$;

(c) all the moments $\int x^\alpha g_\lambda(x) \, dx$ vanish for $|\alpha| \le \sigma$.

Then we know that g_λ defines a continuous linear functional on the homogeneous Hölder space $\dot{C}^s$ (*Wavelets and Operators*, Chapter 6, Theorem 6).

We have seen that the expression $\langle T(f), \psi_\lambda \rangle$ is well defined. That we get the estimate $|\langle T(f), \psi_\lambda \rangle| \le C2^{-j(n/2+s)}$ is a consequence of the hypotheses on T under the action of the affine group. We shall not pursue this any further, because the details are the same as those already given when finding bounds for the entries of matrices corresponding to Calderón-Zygmund operators in an orthonormal basis (Chapter 8, section 3).

To complete the proof of Theorem 1, we establish that the conditions that $T(x^\alpha) = 0$, for $|\alpha| \le \sigma$, and that T is weakly continuous, are necessary for the continuity of T on $\dot{C}^s$.

The polynomials of degree not greater than σ are formally equivalent to 0 in $\dot{C}^s$, which is a quotient space. So, when $|\alpha| \le \sigma$, T must satisfy the condition $T(x^\alpha) = 0$, modulo the polynomials of degree not greater than σ, in order to be extensible as a continuous operator on $\dot{C}^s$.

This argument is simple because we have tacitly supposed that T has already been extended to $\dot{C}^s$. If that is not the case, that is, if T is only defined on $\mathcal{D}(\mathbb{R}^n)$, with values in $\mathcal{D}'(\mathbb{R}^n)$, we argue as follows. $\mathcal{D}(\mathbb{R}^n)$ is dense in $\dot{C}^s$ with the $\sigma(\dot{C}^s, B_1^{-s,1})$ topology (defined by treating $\dot{C}^s$ as the dual of the Besov space $B_1^{-s,1}$). The functions $\psi \in \mathcal{D}(\mathbb{R}^n)$ satisfying $\int x^\alpha \psi(x) \, dx = 0$, for $|\alpha| \le s$, form a dense subset of $B_1^{-s,1}$. It is now easy to see that, if $T : \mathcal{D}(\mathbb{R}^n) \to \mathcal{D}'(\mathbb{R}^n)$ is continuous when $\mathcal{D}(\mathbb{R}^n)$ has the $\dot{C}^s$ norm and if the hypotheses of Theorem 1 are satisfied, then T can be extended to the whole of $\dot{C}^s$. We have reduced the problem to the situation where T is defined on $\dot{C}^s$.

Let us show that the weak continuity on L^2 is equally necessary. Indeed, let B be an arbitrary ball in $\mathbb{R}^n$ with centre x_0 and radius $R > 0$. Let u be a C^s function, adjusted for B, whose support lies in B. Then we know that $f = T(u)$ belongs to the homogeneous space $\dot{C}^s$. This is not enough to find a uniform bound for f on B, because f is only defined up to a polynomial of degree not greater than σ. It remains to find this "floating polynomial", that is, to find a concrete representation of f. We obtain this by letting $x_1 \in \mathbb{R}^n$ denote a point such that $|x_1 - x_0| = 2R$

and calculating the derivatives $\partial^\alpha f(x_1)$ explicitly, for $|\alpha| \le \sigma$. We can then apply the lifting formulas for functions in the homogeneous space $\dot{C}^s$ (*Wavelets and Operators*, Chapter 6, section 4). The details raise no difficulties and are left to the reader.

5 Continuity of operators in $\mathcal{L}_\gamma$ on homogeneous Sobolev spaces

To begin with, we suppose that $0 < s < \gamma \le 1$. We give two proofs of the continuity.

The first uses a weakened form, described in the following lemma, of the continuity of $T \in \mathcal{L}_\gamma$ on a homogeneous Hölder space.

Lemma 1. *Let* $T : \mathcal{D}(\mathbb{R}^n) \to \mathcal{D}'(\mathbb{R}^n)$ *be an operator belonging to* $\mathcal{L}_\gamma$ *and satisfying the hypotheses of Theorem 1. Suppose that* $u \in \dot{C}^s$ *is a function with support in the unit ball. For all* $x \in \mathbb{R}^n$ *and* $t > 0$, *put* $u_{(x,t)}(y) = u((y - x)/t)$. *Then*

$$(5.1) \qquad \|Tu_{(x,t)}\|_\infty \le C\|u\|_{\dot{C}^s}.$$

Indeed, we know that the norm of $u_{(x,t)}$ in the homogeneous space $\dot{C}^s$ is $t^{-s}\|u\|_{\dot{C}^s}$. Since T is bounded on $\dot{C}^s$, we can bound the $\dot{C}^s$ norm of $Tu_{(x,t)}$ by Ct^{-s}. To conclude, we let x_1 denote a point of $\mathbb{R}^n$ such that $|x_1 - x| = 2t$ and then successively estimate $Tu_{(x,t)}(x_1)$, using the size of the kernel of T, followed by $Tu_{(x,t)}(y)$, where $|x - y| \le 4t$, using the definition of the $\dot{C}^s$ norm.

Equipped with this lemma, continuity on $\dot{B}^s$ may be established as follows. We let $\xi \in \mathcal{D}(\mathbb{R}^n)$ be a radial function, such that $\xi(u) = 1$, if $|u| \le 2$, and put $1 = \xi(u) + \eta(u)$.

Then, if $f \in \mathcal{D}(\mathbb{R}^n)$ and $g = T(f)$,

$$g(y) - g(x) = g_1(x,y) + g_2(x,y) + g_3(x,y) + g_4(x,y),$$

where, for $y \ne x$, and taking account of the condition $T(1) = 0$,

$$g_1(x,y) = \int (K(y,u) - K(x,u))(f(u) - f(x))\,\eta\left(\frac{u - x}{|y - x|}\right) du,$$

$$g_2(x,y) = -\int K(x,u)(f(u) - f(x))\,\xi\left(\frac{u - x}{|y - x|}\right) du,$$

$$g_3(x,y) = \int K(y,u)(f(u) - f(y))\,\xi\left(\frac{u - x}{|y - x|}\right) du,$$

and, lastly,

$$g_4(x,y) = (f(y) - f(x)) \int K(y,u)\,\xi\left(\frac{u - x}{|y - x|}\right) du.$$

All these terms are estimated by brute force, without appealing any further to the "cancellations of the kernel", which were fully exploited in the proof of the lemma.

For the first, we choose an exponent α, satisfying $s < \alpha < \gamma$, and write

$$|g_1(x,y)| \leq$$

$$C \int_{|u-x| \geq 2|y-x|} |x-y|^\gamma |u-x|^{-n-\gamma} |f(u) - f(x)|\, du$$

$$= C|x-y|^\gamma \int_{|u-x| \geq 2|y-x|} |u-x|^{-n/2-\gamma+\alpha} |u-x|^{-n/2-\alpha} |f(u) - f(x)|\, du\,.$$

The Cauchy-Schwarz inequality then gives

$$|g_1(x,y)|^2 \leq C'|y-x|^{2\alpha} \int_{|u-x| \geq 2|y-x|} |u-x|^{-n-2\alpha} |f(u) - f(x)|^2\, du\,.$$

We conclude the estimate of $\int\int |g_1(x,y)|^2 |x-y|^{-n-2s}\, dx\, dy$ by integrating with respect to y, u and, finally, x to give

$$C'' \int\int |f(u) - f(x)|^2 |u-x|^{-n-2s}\, du\, dx = C'' N_s^2(f)\,.$$

The case of g_2 is similar. We suppose that the support of ξ is contained in $|u| \leq 10$ and get

$$|g_2(x,y)| \leq C \int_{|u-x| \leq 10|y-x|} |x-u|^{-n} |f(u) - f(x)|\, du\,.$$

We then introduce an exponent β, such that $0 < \sigma < \beta$, and write $|x-u|^{-n} = |x-u|^{-n/2+\beta} |x-u|^{-n/2-\beta}$. After applying the Cauchy-Schwarz inequality, we conclude as above.

$$|g_3(x,y)| \leq C \int_{|u-y| \leq 11|x-y|} |y-u|^{-n} |f(u) - f(y)|\, du$$

and we need only interchange the roles of x and y in the treatment of $g_2(x,y)$ to conclude.

Lastly,

$$|g_4(x,y)| \leq C|f(y) - f(x)|$$

by the lemma.

Thus, T is continuous on $\dot{B}^s$, because

$$C \left(\int\int |f(y) - f(x)|^2 |x-y|^{-n-2s}\, dx\, dy \right)^{1/2}$$

is an upper bound for

$$\left(\int\int |g(y) - g(x)|^2 |x-y|^{-n-2s}\, dx\, dy \right)^{1/2}\,.$$

To deal with the case $1 \leq s \leq \gamma$, we use a different proof, which also applies when $0 < s < \gamma \leq 1$. Let ψ_λ, $\lambda \in \Lambda$, denote an orthonormal basis of wavelets with compact support and regularity C^q, where $q > \gamma$. We shall find upper bounds for the moduli of the matrix entries $\tau(\lambda, \lambda') = (T(\psi_\lambda), \psi_{\lambda'})$. As usual, we put $\lambda = 2^{-j}k + 2^{-j-1}\varepsilon$ and $\lambda' = 2^{-j'}k' + 2^{-j'-1}\varepsilon'$, where $j, j' \in \mathbb{Z}$, $k, k' \in \mathbb{Z}^n$ and $\varepsilon, \varepsilon' \in E = \{0, 1\}^n \setminus \{(0, \ldots, 0)\}$.

If $j' \geq j$, we get a bound for $|\tau(\lambda, \lambda')|$ in exactly the same way as in the proof of the $T(1)$ theorem. This gives

$$(5.2) \qquad |\tau(\lambda, \lambda')| \leq C 2^{-(j'-j)(n/2+\gamma)} \left(\frac{2^{-j}}{2^{-j} + |k2^{-j} - k'2^{-j'}|} \right)^{n+\gamma}.$$

If $j' \leq j$, we use the regularity of the function $g_\lambda = T(\psi_\lambda)$ and the rate of decrease at infinity of the derivatives of g_λ. Indeed, $g_\lambda \in C^s$, if $0 < s < \gamma$, and then $|\partial^\alpha g_\lambda(x)| = O(|x|^{-n-|\alpha|})$, for $|\alpha| < \gamma$. It is easy to get explicit values for the implied constants in these properties.

Let $C2^{-j}$ denote the diameter of the support of ψ_λ. Then, for the case $|k2^{-j} - k'2^{-j'}| \geq 2C2^{-j'}$, we can integrate

$$\tau(\lambda, \lambda') = \int \psi_{\lambda'}(x) g_\lambda(x)\, dx$$

by parts to get

$$(5.3) \qquad |\tau(\lambda, \lambda')| \leq C 2^{-(j-j')n/2} \left(\frac{2^{-j'}}{2^{-j'} + |k2^{-j} - k'2^{-j'}|} \right)^{n+\gamma}.$$

If $|k2^{-j} - k'2^{-j'}| \leq 2C2^{-j'}$, we use the estimate of $|g_\lambda(x)|$ and the fact that we are integrating over the compact support of $\psi_{\lambda'}$. We get

$$(5.4) \qquad |\tau(\lambda, \lambda')| \leq C(1 + |j' - j|)2^{-(j-j')n/2}.$$

Having done this, let us return to the continuity of T on the homogeneous Sobolev space $\dot{B}^s$: it is equivalent to the continuity of the matrix τ_s on $l^2(\Lambda)$, where the entries of τ_s are given by

$$\tau_s(\lambda, \lambda') = 2^{-s(j-j')}\tau(\lambda, \lambda')$$

for $(\lambda, \lambda') \in \Lambda \times \Lambda$. This is because $\sum \alpha(\lambda)\psi_\lambda(x)$ is in $\dot{B}^s$ if and only if $\sum_{\lambda \in \Lambda} 4^{js}|\alpha(\lambda)|^2 < \infty$.

If $0 < \gamma' < \gamma - s$ and $0 < \gamma' < s$, we can immediately verify that τ_s belongs to the algebra $\mathcal{M}_{\gamma'}$, with the consequence that τ_s is bounded on $l^2(\Lambda)$.

We have completed the proof of Theorem 1.

The above proof that T is continuous on $\dot{B}^s = \dot{B}_2^{s,2}$, for $0 < s < \gamma$, extends to all the homogeneous Besov spaces $\dot{B}_p^{s,q}$, for which $0 < s < \gamma$ and $1 \leq p, q \leq \infty$. This can be seen by repeating the proof of Theorem 4 in Chapter 8 word for word.

We can go even further and establish that T is continuous on the homogeneous Sobolev spaces $\mathcal{L}^{p,s}$, $1 < p < \infty$, as long as $0 < s < \gamma$.

To do this, we construct an isomorphism $\Lambda_s : \mathcal{L}^{p,s} \to L^p(\mathbb{R}^n)$ by choosing $\Lambda_s(\psi_\lambda) = 2^{js}\psi_\lambda$ and referring to Proposition 1 of Chapter 6 of *Wavelets and Operators*. Λ_s imitates $(-\Delta)^{s/2}$.

Since the matrix τ_s belongs to $\mathcal{O}p\,\mathcal{M}_{\gamma'}$ for $0 < \gamma' < \gamma - s$, it follows that $T = \Lambda_s^{-1}\mathcal{C}\Lambda_s$, where $\mathcal{C}$ is an operator in $\mathcal{O}p\,\mathcal{M}_{\gamma'}$. In particular, $\mathcal{C}$ is a Calderón-Zygmund operator. $\mathcal{C}$ is thus bounded on $L^p(\mathbb{R}^n)$ and the same holds for T as an operator on $\mathcal{L}^{p,s}$, for $0 < s < \gamma$.

In the next section, we shall see that, if $K(x,y)$ and its derivatives with respect to x are required to decrease fast enough at infinity, then our operator T is also bounded on the inhomogeneous Sobolev spaces $W^{p,s}$, $1 < p < \infty$, when $0 < s < \gamma$.

Observe that the direct proof of the continuity of T on $\dot{B}^s$ (due to P.G. Lemarié [164]) gives another proof of the $T(1)$ theorem. Indeed, if T is weakly continuous on L^2, if $T(1) = 0$ and ${}^tT(1) = 0$, it follows that T and its adjoint T^* are continuous on $\dot{B}^s$, for $0 < s < \gamma$. Since the dual, in the sense of distributions, of $\dot{B}^s$ is $\dot{B}^{-s}$, the continuity of T^* on $\dot{B}^s$ is equivalent to that of T on $\dot{B}^{-s}$. As T is continuous on $\dot{B}^s$ and $\dot{B}^{-s}$, we deduce that T is continuous on $L^2(\mathbb{R}^n)$, by interpolation.

6 Continuity on ordinary Sobolev spaces

Let us begin by seeing that the operators described by Theorem 1 are not, in general, continuous on ordinary (inhomogeneous) Sobolev spaces.

Here is a counter-example in dimension 1. We start with a function $\psi \in \mathcal{S}(\mathbb{R}^n)$ on which we impose the following conditions: the Fourier transform $\hat{\psi}$ of ψ is real; is even; is zero outside $2/3 \leq |\xi| \leq 4/3$; and is equal to 1 at 1.

We then form the symbol

$$\sigma(x,\xi) = \sum_1^\infty e^{-i2^{-j}x}\hat{\psi}(2^j\xi)$$

and define $T = \sigma(x, D)$ to be the corresponding operator.

The kernel $K(x,y)$ of T is $K(x,y) = \sum_1^\infty e^{-i2^{-j}x}2^{-j}\psi(2^{-j}(x-y))$, which clearly satisfies the Calderón-Zygmund estimates. The singularity is moved to infinity.

The symbol $\sigma(x,\xi)$ vanishes if $|\xi| \geq 1$, so if the operator $\sigma(x, D)$ were continuous on $H^s(\mathbb{R})$, $s > 0$, it would be continuous on $L^2(\mathbb{R})$ and vice versa. But ${}^tT(1) = \sum_1^\infty(e^{-i2^{-j}x} - 1)$ and this function does not belong to BMO. On the other hand, $T(1) = 0$ modulo the constants and, for that matter, $T(x^k) = 0$ modulo polynomials of degree not exceeding k, for all $k \in \mathbb{N}$.

The selfsame operator T cannot be continuous on the inhomogeneous Hölder spaces C^s. Indeed, let $\theta \in \mathcal{D}(\mathbb{R})$ be a function which is 1 on $[-1, 1]$. For any $f \in L^\infty(\mathbb{R})$, define g by $\hat{g}(\xi) = \theta(\xi)\hat{f}(\xi)$. Then g is in C^s for all $s > 0$. If T were continuous on C^s then T would be continuous on L^∞, because $T(f) = T(g)$. But the continuity of T on L^∞ would imply its continuity on L^2 (Chapter 8, Proposition 9).

In this example, we see that the behaviour of the kernel of T as $|y - x|$ tends to infinity appears to be bound up with the continuity of T on inhomogeneous spaces. The following result gives the precise relationship.

Theorem 2. *Let T be an operator in $\mathcal{L}_\gamma$, $\gamma > 0$. Suppose, further, that the kernel $K(x, y)$ of T satisfies the following estimates for $|x - y| \geq 1$.*

$$(6.1) \qquad |\partial_x^\alpha K(x, y)| \leq C_N |x - y|^{-N}$$

for $|\alpha| \leq \gamma$ and $N \geq 1$;

$$(6.2) \qquad |\partial_x^\alpha K(x', y) - \partial_x^\alpha K(x, y)| \leq C_N |x' - x|^r |x - y|^{-N}$$

where $m \in \mathbb{N}$ and r are defined by $\gamma = m + r$ with $0 < r \leq 1$ and where $|\alpha| = m$ and $|x' - x| \leq |x - y|/2$.

If T is weakly continuous on $L^2(\mathbb{R}^n)$ and if $T(x^\alpha) = m_\alpha(x)$ belongs to the inhomogeneous Hölder space $C^\gamma(\mathbb{R}^n)$, for $|\alpha| \leq m$, then T can be extended as a continuous linear operator on the inhomogeneous spaces C^s and H^s, for $0 < s < \gamma$.

We first remark that, because of the assumptions about the decrease at infinity of the kernel of T, the mathematical objects $T(x^\alpha)$, $|\alpha| \leq m$, have their usual meaning.

An immediate result of the theorem is that the pseudo-differential operators $T \in \mathcal{O}p\, S_{1,1}^0$ (whose symbols satisfy $|\partial_\xi^\alpha \partial_x^\beta \sigma(x, \xi)| \leq C(\alpha, \beta)(1 + |\xi|)^{|\beta| - |\alpha|}$) are continuous on the inhomogeneous Hölder spaces C^s and the Sobolev spaces H^s, when $s > 0$. We may observe that these operators are not, in general, continuous on $L^2(\mathbb{R}^n)$.

We prove Theorem 2 by combining Theorem 1 with Poincaré's inequality, which is given by the following lemma.

For $s \geq 0$, put $N_s(f) = (2\pi)^{-n/2}(\int |\hat{f}(\xi)|^2 |\xi|^{2s} \, d\xi)^{1/2}$.

Lemma 2. *There is a constant $C(s, n)$, $s \geq 0$, $n \geq 1$, such that, for every $R > 0$ and every function f in the Sobolev space H^s whose support is contained in $|x| \leq R$,*

$$\|f\|_2 \leq C(s, n) R^s N_s(f).$$

In order to apply this result, we start from an observation used in

Wavelets and Operators, Chapter 2, Lemma 6. If $\theta(x)$ belongs to $\mathcal{D}(\mathbb{R}^n)$ and if $\sum_{k \in \mathbb{Z}^n} |\theta(x - k)| \geq 1$, then, for every $s \geq 0$, $(\sum_{k \in \mathbb{Z}^n} \|f\theta_k\|_{H^s}^2)^{1/2}$ and $\|f\|_{H^s}$ are equivalent norms on H^s, where we have written $\theta_k(x) = \theta(x - k)$.

Armed with Poincaré's inequality and the possibility of localizing the calculation of the H^s norm, we split the distribution-kernel K of T into $K_1 + K_2$, where $K_1(x, y) = \eta(x - y)K(x, y)$, $\eta \in \mathcal{D}(\mathbb{R}^n)$ being 1 on the unit ball. Correspondingly, we write $T = T_1 + T_2$. It is clear that T_2 is bounded on all the H^s spaces and the whole of the problem is concentrated on T_1. The functions $T_1(x^\alpha) = m_\alpha(x) - T_2(x^\alpha)$ obviously belong to C^γ. We first consider the case where $T_1(x^\alpha) = 0$, for $0 \leq |\alpha| \leq m$.

Let $\theta \in \mathcal{D}(\mathbb{R}^n)$ be a function such that $\sum \theta(x - k) = 1$. If $f \in H^s$, we write $f = \sum \theta_k f$ and $T_1(f) = \sum T_1(\theta_k f)$. The support of $T_1(\theta_k f)$ is contained in the ball $|x - k| \leq R$ and the observations above lead to

$$\|T_1(f)\|_{H^s} \leq C(\sum \|T_1(\theta_k f)\|_{H^s}^2)^{1/2} \leq C'(\sum \|T_1(\theta_k f)\|_{\dot{B}^s}^2)^{1/2}$$
$$\leq C''(\sum \|\theta_k f\|_{\dot{B}^s}^2)^{1/2} \leq C'''\|f\|_{H^s}.$$

We still have to reduce the general case to that in which $T_1(x^\alpha) = 0$ for $|\alpha| \leq m$. We do it in the manner of the $T(1)$ theorem, correcting T_1 by an operator which we know to be bounded on C^s and H^s.

So we start with a function $\omega(x) \in \mathcal{D}(\mathbb{R}^n)$ whose mean is 1, but which is such that $\int x^\alpha \omega(x) \, dx = 0$, if $1 \leq |\alpha| \leq m$.

Let L_1 be the operator with kernel $\sum_{|\alpha| \leq m} (\alpha!)^{-1} m_\alpha(x) \partial_x^\alpha \omega(x - y)$. Clearly $L_1(x^\alpha) = m_\alpha(x)$, for $|\alpha| \leq m$. On the other hand, L_1 is bounded on the inhomogeneous spaces C^s and H^s, because these spaces are invariant under pointwise multiplication by C^γ functions. Lastly, $\tilde{T}_1 = T_1 - L_1$ satisfies the hypotheses of Theorem 2 and $\tilde{T}_1(x^\alpha) = 0$, for $|\alpha| \leq m$.

This proof also extends to the case of the inhomogeneous Sobolev spaces $W^{p,s}$, where $1 < p < \infty$ and $0 < s < \gamma$. In particular, the pseudo-differential operators T whose symbols belong to $S_{1,1}^0(\mathbb{R}^n)$ are bounded on $W^{p,s}$, for every $s > 0$ and all $p \in (1, \infty)$.

7 Additional remarks

We return to the homogeneous Sobolev space $\dot{B}^s = \dot{B}_2^{s,2}$ and suppose that $0 < s < 1$. Let $\gamma > s$ and suppose that T is an operator in $\mathcal{L}_\gamma$. We could ask under what conditions T extends to a continuous linear operator on $\dot{B}^s$. By Theorem 1, this happens if $T(1) = 0$, modulo the constant functions. But this condition is not necessary. Following Stegenga ([216]), we introduce the Banach space E_s of functions β, modulo the constant functions, such that the pseudo-product $\pi(\beta, f)$ satisfies $\|\pi(\beta, f)\|_{\dot{B}^s} \leq C\|f\|_{\dot{B}^s}$. If $s = 0$, E_s coincides with BMO. We then show ([191]) that the necessary and sufficient condition for $T \in \mathcal{L}_\gamma$ to extend continuously to $\dot{B}^s$ is that T has to be weakly continuous on $L^2(\mathbb{R}^n)$ and that $T(1)$ must belong to E_s.

Here is a condensed proof of this result. If $T \in \mathcal{L}_\gamma$ and if T is weakly continuous on L^2, it is easy to show that $\beta = T(1)$ belongs to $\dot{B}^{0,\infty}_\infty$. Then the operator $R_\beta(f) = \pi(\beta, f)$ also is an element of $\mathcal{L}_\gamma$ (for each $\gamma > 0$) and is weakly continuous. The difference $S_\beta = T - R_\beta$ satisfies $S_\beta(1) = 0$ and is weakly continuous. Because of this, S_β is continuous on $\dot{B}^s$ (Theorem 1). Thus T is continuous if and only R_β is, which is what we wanted to establish.

Let us consider the special case where T is the operator of pointwise multiplication by a function $m(x)$. The kernel $K(x, y)$ corresponding to T vanishes and T is in $\mathcal{L}_\gamma$. T is continuous on $\dot{B}^s$ if and only if T is weakly continuous and $T(1) \in E_s$. But the weak continuity means that $m(x)$ is in L^∞. The pointwise multipliers of the homogeneous Sobolev space $\dot{B}^s$ are thus the functions $m(x) \in L^\infty \cap E_s$. This is the statement of Stegenga's theorem ([216]). We should remark that E_s reduces to $\{0\}$ if $s \geq n/2$. Here we have only dealt with the case $0 < s < 1$. The case $s \geq 1$ remains open.

11

The $T(b)$ theorem

1 Introduction

We have already mentioned that David and Journé's remarkable $T(1)$ theorem has a drawback. It cannot be used to show that the Cauchy kernel on a Lipschitz curve is L^2 continuous. The criterion does not work for the following reason: if $z(x) = x + ia(x)$ and $-M \le a'(x) \le M$, then the integral PV $\int_{-\infty}^{\infty}(z(x) - z(y))^{-1}\,dy$ is not a BMO($\mathbb{R}$) function that we already know. This is because, in complex analysis, it is natural to integrate with respect to the measure $dz(y)$ and not the measure dy. Indeed $\int_{-\infty}^{\infty}(z(x) - z(y))^{-1}\,dz(y) = 0$, modulo the constant functions.

This extremely simple remark leads to the following conjecture. Let $b(x) \in L^{\infty}(\mathbb{R}^n)$ be a function with $\Re\, b(x) \ge 1$ almost everywhere. Let $K(x, y)$ be an antisymmetric kernel such that $|K(x, y)| \le C_0 |x - y|^{-n}$ and $|(\partial/\partial x_j)K(x, y)| \le C_1 |x - y|^{-n-1}$, for $1 \le j \le n$. Suppose further that $\int K(x, y)b(y)\,dy$ is in BMO($\mathbb{R}^n$). Then PV $K(x, y)$ defines an operator which is continuous on $L^2(\mathbb{R}^n)$.

This would mean that the continuity of a singular integral operator could be established by just one function b, as long as $\Re\, b \ge 1$.

To avoid the circular argument of assuming that the operator under consideration is defined on arbitrary functions $b \in L^{\infty}$, when we do not even know whether the operator is a Calderón-Zygmund operator, we restrict our attention to kernels $K(x, y)$ which satisfy the extra, qualitative condition $(1 + |x - y|)^{n+1}K(x, y) \in L^{\infty}(\mathbb{R}^n \times \mathbb{R}^n)$.

In 1984, G. David, J.L. Journé and S. Semmes proved this conjecture, which was the first formulation of the $T(b)$ theorem. P. Tchamitchian later found another, much more geometrical, version in terms of wavelets adapted to the measure $b(x)\,dx$. The $T(b)$ theorem then arises as a corollary of the existence of the wavelet basis and this approach runs parallel to the argument we used to prove the $T(1)$ theorem.

2 Statement of the fundamental geometric theorem

Let V_j, $j \in \mathbb{Z}$, be a multiresolution approximation of $L^2(\mathbb{R}^n)$ which is r-regular, for some $r \geq 1$. We say that V_j is *real* if the function ϕ and the wavelets $\psi_{j,k}^\varepsilon$ corresponding to the multiresolution approximation are real-valued.

Even though it is not strictly necessary, we shall suppose that there is an exponent $\gamma > 0$ and that there is a constant $C > 0$ such that

$$(2.1) \qquad |\partial^\alpha \phi(x)| \leq Ce^{-\gamma|x|} \qquad \text{if } |\alpha| \leq r.$$

It follows that, for $|\alpha| \leq r$, we also have

$$(2.2) \qquad |\partial^\alpha \psi_{j,k}^\varepsilon(x)| \leq C' 2^{nj/2} 2^{j|\alpha|} e^{-\gamma'|2^j x - k|},$$

for some exponent $\gamma' > 0$ and a constant $C' > 0$.

These hypotheses are satisfied in many examples. (2.1) and (2.2) could be replaced by the usual hypotheses of fast decrease at infinity, but then the proofs below would become unnecessarily complicated.

We fix a function $b(x) \in L^\infty(\mathbb{R}^n)$, with $\Re e\, b(x) \geq 1$, and define the symmetric bilinear form $B : L^2(\mathbb{R}^n) \times L^2(\mathbb{R}^n) \to \mathbb{C}$, by

$$(2.3) \qquad B(f, g) = \int_{\mathbb{R}^n} f(x)g(x)b(x)\,dx.$$

We further define $\beta(f, g) = B(f, \bar{g})$.

The bilinear form B has the following important property:

$$(2.4) \qquad \Re e\, B(f, \bar{f}) \geq \|f\|_2^2,$$

which has been studied systematically by T. Kato in [151].

This property replaces the usual (strict) positivity properties of the sesquilinear form defining a Hilbert structure.

We let $\tilde{W}_j \subset V_{j+1}$ denote the linear subspace defined by $\beta(f, g) = 0$ for all $g \in V_j$. If $b(x) = 1$, this reduces to the subspace W_j which we used in Chapter 2 of *Wavelets and Operators*.

Theorem 1. $L^2(\mathbb{R}^n)$ *is the direct sum of the linear subspaces* $\tilde{W}_j$, $j \in \mathbb{Z}$. *This means, firstly, that there are constants* $C_2 \geq C_1 > 0$, *such that, for every sequence* $f_j \in \tilde{W}_j$ *satisfying* $\sum_{-\infty}^{\infty} \|f\|_2^2 < \infty$,

$$(2.5) \qquad C_1 \left(\sum_{-\infty}^{\infty} \|f_j\|_2^2 \right)^{1/2} \leq \| \sum_{-\infty}^{\infty} f_j \|_2 \leq C_2 \left(\sum_{-\infty}^{\infty} \|f_j\|_2^2 \right)^{1/2}.$$

Further, every $f \in L^2(\mathbb{R}^n)$ *has a unique expansion* $f = \sum_{-\infty}^{\infty} f_j$, *where* $\sum_{-\infty}^{\infty} \|f_j\|_2^2 < \infty$ *and* $f_j \in \tilde{W}_j$.

If $b(x) = 1$, this result is routine and we have $L^2(\mathbb{R}^n) = \bigoplus_{-\infty}^{\infty} W_j$. This identity is the starting point for the wavelet construction. In the same way, Theorem 1 allows us to construct a Riesz basis $\tilde{\psi}_\lambda$, $\lambda \in \Lambda$, of $L^2(\mathbb{R}^n)$, with the following properties: localization and regularity are the same as in the classical case, while cancellation is defined by $\tilde{\psi}_\lambda \in \tilde{W}_j$, for $\lambda \in \Lambda_j$.

Once we have constructed this basis, we get the L^2 continuity of the singular integral operators for which $T(b) = 0$ and ${}^tT(b) = 0$ (without forgetting the weak continuity) by verifying that the matrix of T, with respect to the basis $\tilde{\psi}_\lambda$, $\lambda \in \Lambda$, belongs to $\mathcal{M}_\gamma$, for a certain $\gamma > 0$.

3 Operators and accretive forms (in the abstract situation)

We have decided to assemble some results in this section which are as simple as they are general. For the convenience of the reader, we shall give the proofs.

Let H be a complex Hilbert space and let $\langle \cdot, \cdot \rangle$ be the sesquilinear form which defines the Hilbert structure of the space. A continuous linear operator $T : H \to H$ is *accretive* if, for every $x \in H$, we have $\Re \langle T(x), x \rangle \geq 0$. T is δ-*accretive*, for $\delta > 0$, if $\Re \langle T(x), x \rangle \geq \delta \|x\|^2$. This is the same as writing $T + T^* \geq \delta I$ (where I denotes the identity operator) in the sense of self-adjoint operators.

A δ-accretive operator is an isomorphism of H. Indeed

$$\delta \|x\|^2 \leq \Re \langle T(x), x \rangle \leq |\Re \langle T(x), x \rangle| \leq \|T(x)\| \|x\|.$$

So $\delta \|x\| \leq \|T(x)\|$ and, similarly, $\delta \|x\| \leq \|T^*(x)\|$. It follows that T is an isomorphism.

Let $\beta : H \times H \to \mathbb{C}$ be a bicontinuous sesquilinear form. This means: that $|\beta(x, y)| \leq C \|x\| \|y\|$; that, for each y_0, $\beta(x, y_0)$ is a continuous linear form on H; and that, for each x_0, $\overline{\beta(x_0, y)}$ is a continuous linear form on H.

We say that β is δ-*accretive* if $\Re \beta(x, x) \geq \delta \|x\|^2$ for every $x \in H$. If

β is δ-accretive, there exists an operator $T : H \to H$ which is δ-accretive and such that $\beta(x, y) = \langle T(x), y \rangle$ for all $x, y \in H$.

This remark gives us the following lemma.

Lemma 1. *For each continuous linear form $l : H \to \mathbb{C}$, there exists a unique element $a \in H$ such that $l(x) = \beta(x, a)$.*

Indeed, by the Riesz representation theorem $l(x) = \langle x, b \rangle = \langle T(x), a \rangle$, where $T^*(a) = b$. This equation has a solution, since T^* is an isomorphism.

The following result helps with Theorem 1.

Proposition 1. *Let H be a Hilbert space, let $\beta : H \times H \to \mathbb{C}$ be a δ-accretive sesquilinear form ($\delta > 0$) and let V be a closed linear subspace of H. Define $\tilde{W}$ by*

$$(3.1) \qquad \tilde{W} = \{ x \in H : \beta(x, y) = 0 \text{ for all } y \in V \} .$$

Then H is the direct sum of V and $\tilde{W}$.

Moreover, the operator norm of the oblique projection of H on V parallel to $\tilde{W}$ depends only on δ and the constant involved in the continuity of β.

We shall check that every $c \in H$ can be written uniquely as $c = a + b$, where $a \in V$ and $b \in \tilde{W}$. Then $\beta(c, v) = \beta(a, v)$ for every $v \in V$. To find $a \in V$, we apply Lemma 1 to the linear form $\overline{\beta(c, v)} = l(v)$, defined on V. (We have to replace H by V and restrict β to $V \times V$. The restricted sesquilinear form is still δ-accretive.) Lemma 1 gives us a unique $a \in V$ such that $l(v) = \overline{\beta(a, v)}$. This concludes the proof of Proposition 1.

Returning to Theorem 1, we thus get $V_{j+1} = V_j + \tilde{W}_j$ and the norms of the oblique projections $\tilde{P}_j : V_{j+1} \to V_j$, parallel to $\tilde{W}_j$, are uniformly bounded.

We proceed to a group of remarks about the symbolic calculus of accretive operators. Following T. Kato ([151]), we have

Proposition 2. *Let $T : H \to H$ be a δ-accretive operator. There exists a unique accretive operator S such that $S^2 = T$. We denote this operator by $T^{1/2}$. It is δ'-accretive for a certain $\delta' > 0$, which depends only on δ and $\|T\|$. Finally, the inverse $T^{-1/2}$ of $T^{1/2}$ is given by*

$$(3.2) \qquad T^{-1/2} = \frac{1}{\pi} \int_0^\infty (\lambda + T)^{-1} \lambda^{-1/2} \, d\lambda .$$

Indeed, the spectrum $\sigma(T)$ of T is contained in the compact subset of the complex plane defined by $\Re e \, z \geq \delta$, $|z| \leq \|T\|$. The function $z^{-1/2}$ is holomorphic in a neighbourhood of the spectrum. It follows that the right-hand side of (3.2) defines an operator whose square is T^{-1}.

To show that $T^{1/2}$ is δ'-accretive, for a certain $\delta' > 0$, we use the following lemma, whose very simple proof is left to the reader.

Lemma 2. *Let* $T : H \to H$ *be a* δ-accretive operator. *Then* T^{-1} *is* $\delta \|T\|^{-2}$ accretive.

From the right-hand side of (3.2), we get

$$\Re \langle T^{-1/2}x, x \rangle \geq \frac{\|x\|^2}{\pi} \int_0^\infty \frac{\lambda + \delta}{(\lambda + \|T\|)^2} \lambda^{-1/2} \, d\lambda = c\|x\|^2 \, .$$

Thus $T^{-1/2}$ is c-accretive and, applying the lemma, so is $T^{1/2}$.

A different method of calculating $T^{-1/2}$, which avoids the functional calculus, uses the following observation.

Lemma 3. *If* T *is* δ-accretive and if $\lambda > (2\delta)^{-1}\|T\|^2$, *then*

(3.3) $$\|T - \lambda\| < \lambda \, .$$

Indeed,

$$\|T - \lambda\| = \|(T^* - \lambda)(T - \lambda)\|^{1/2}$$

and

$$(T^* - \lambda)(T - \lambda) = T^*T - \lambda(T^* + T) + \lambda^2 \leq \|T\|^2 - 2\lambda\delta + \lambda^2 < \lambda^2 \, ,$$

where the inequalities are to be taken in the sense of the order relation on self-adjoint operators. Thus $\|T - \lambda\| < \lambda$.

Returning to the calculation of $T^{-1/2}$, we write $T = \lambda - (\lambda - T) = \lambda(I - \lambda^{-1}(\lambda - T))$ and we take $\lambda > (2\delta)^{-1}\|T\|^2$. We put $R = \lambda^{-1}(\lambda - T)$, so $\|R\| < 1$. With this notation,

$$T^{-1/2} = \lambda^{-1/2}(1 - R)^{-1/2} = \sum_0^\infty c_k R^k \qquad \text{where } c_k = O(k^{-1/2}).$$

For the uniqueness of S, we refer the reader to T. Kato's well-known book ([151]).

4 Construction of bases adapted to a bilinear form

Let H be a Hilbert space. A Riesz basis e_j, $j \in J$, of H is a total subset of H for which there are constants $C_2 \geq C_1 > 0$ such that, for any choice of coefficients α_j,

(4.1) $$C_1 \left(\sum_{j \in J} |\alpha_j|^2 \right)^{1/2} \leq \left\| \sum_{j \in J} \alpha_j e_j \right\| \leq C_2 \left(\sum_{j \in J} |\alpha_j|^2 \right)^{1/2} .$$

A matrix $M = (m(j,k))_{(j,k) \in J \times J}$ is called δ-accretive if, for every complex sequence $\xi_j \in l^2(J)$,

$$(4.2) \qquad \Re \sum_j \sum_k m(j,k)\xi_j \bar{\xi}_k \geq \delta \sum_j |\xi_j|^2 .$$

Let H be a complex Hilbert space. Let $B : H \times H \to \mathbb{C}$ be a jointly continuous bilinear symmetric form: $B(x,y) = B(y,x)$ for all $x, y \in H$.

Let e_j, $j \in J$, be a Riesz basis of H. We consider the matrix B whose coefficients are $B(e_j, e_k)$, $j, k \in J$. This matrix is bounded on $l^2(J)$. Indeed, if $\xi_j \in l^2(J)$ and $\eta_j \in l^2(J)$, we have

$$\sum \sum \xi_j \eta_k B(e_j, e_k) = B(x,y),$$

where $x = \sum \xi_j e_j$ and $y = \sum \eta_k e_k$. Thus

$$\left| \sum \sum \xi_j \eta_k B(e_j, e_k) \right| \leq C \|x\| \|y\| \leq C' \left(\sum |\xi_j|^2 \right)^{1/2} \left(\sum |\eta_k|^2 \right)^{1/2}.$$

We shall make systematic use of the following result.

Proposition 3. *Let H be a Hilbert space over the complex field and let $B : H \times H \to \mathbb{C}$ be a jointly continuous, symmetric, bilinear form. Suppose that e_j, $j \in J$, is a Riesz basis of H such that the matrix $(B(e_j, e_k))_{(j,k) \in J \times J}$ is δ-accretive, for some $\delta > 0$. Then there exists a Riesz basis f_j, $j \in J$, such that $B(f_j, f_k) = 0$ or 1, depending on whether $j \neq k$ or $j = k$.*

If, moreover, $J = \mathbb{Z}^n$ and if, for a certain exponent $\alpha > 0$ and some constant $C > 0$, we have

$$(4.3) \qquad |B(e_j, e_k)| \leq C e^{-\alpha |j-k|},$$

then there exist an exponent $\beta > 0$ and a constant $C' > 0$ such that

$$(4.4) \qquad f_j = \sum \gamma(j,k) e_k ,$$

where

$$(4.5) \qquad |\gamma(j,k)| \leq C' e^{-\beta |j-k|}.$$

In the proof of the proposition, we let B also denote the matrix $(B(e_j, e_k))_{j,k \in J}$ and we write $\gamma(j,k)$ for the entries of the accretive matrix $B^{-1/2}$. Since $B^{-1/2}$ is an isomorphism of $l^2(J)$, the vectors f_j, defined by (4.4), form a Riesz basis of H.

We show now that $B(f_j, f_k) = \delta_{jk}$. Firstly, note that ${}^t B = B$, because B is symmetric. Hence, applying (3.2), $B^{-1/2} = {}^t B^{-1/2}$. Thus $\gamma(j,k) = \gamma(k,j)$. This gives

$$B(f_j, f_k) = \sum_l \sum_m \gamma(j,l)\gamma(k,m) B(e_l, e_m)$$

$$= \sum_l \sum_m \gamma(j,l) B(e_l, e_m) \gamma(m,k) .$$

These are just the entries of the matrix $B^{-1/2} B B^{-1/2} = 1$ and this establishes the first statement of Proposition 3.

To prove the second, fix $\lambda > \|B\|^2/2\delta$. By Lemma 3, $B = \lambda(1 - R)$, where $\|R\| < 1$, which gives $B^{-1/2} = \sum_0^\infty c_k R^k$, with $c_k = O(k^{-1/2})$. The entries of the matrix R, denoted by $r(j, k)$, satisfy (4.3). The entries $r_k(j, j')$ of the matrix R^k are calculated using the identity

$$(4.6) \qquad r_k(j, j') = \sum_{j_1} \cdots \sum_{j_{k-1}} r(j, j_1) r(j_1, j_2) \cdots r(j_{k-1}, j').$$

To estimate $|r_k(j, j')|$, we split α into $\beta + \nu$, where β is chosen arbitrarily, subject to $0 < \beta < \alpha$. Using the triangle inequality on $\mathbb{Z}^n$ we then replace $e^{-\beta|j-j_1|-\cdots-\beta|j_{k-1}-j'|}$ by $e^{-\beta|j-j'|}$.

This leaves us the task of dealing with

$$e^{-\nu|j-j_1|} \ldots e^{-\nu|j_{k-1}-j'|}.$$

We replace the last term by 1 and sum successively over $j_1, \ldots, j_{k-1}$, using the obvious inequality

$$\sum e^{-\nu|k|} \le C(\nu).$$

In the end, this gives

$$|r_k(j, j')| \le C^k (C(\nu))^k e^{-\beta|j-j'|}.$$

On the other hand, if we let r denote the operator norm of R, we have

$$0 \le r < 1 \qquad \text{and} \qquad |r_k(j, j')| \le r^k.$$

As a result of these estimates, by using logarithmic convexity, we may assert that, for $k \in \mathbb{N}$ and $j, j' \in \mathbb{Z}^n$,

$$(4.7) \qquad\qquad |r_k(j, j')| \le \omega^k e^{-\alpha'|j-j'|},$$

where $0 \le \omega < 1$ and α' depend only on r, C, $C(\nu)$ and β.

Now we have got this far, we may sum the estimates and get the advertised upper bound for the moduli of the entries of the matrix $B^{-1/2}$.

It can be shown that, if the exponential decrease of $|B(e_j, e_k)|$, as $|j - k|$ tends to infinity, is replaced by rapid decrease, then the matrix entries $\gamma(j, k)$ also decrease rapidly as $|j - k|$ tends to infinity.

5 Tchamitchian's construction

We come back to the multiresolution approximation V_j of $L^2(\mathbb{R}^n)$. Recall that the functions ϕ and ψ_{jk}^ε are assumed to be *real*-valued and r-regular, where $r \ge 1$. We suppose, further, that these functions decrease exponentially, together with their derivatives of order less than or equal to r.

We now put $\phi_{jk}(x) = 2^{nj/2}\phi(2^j x - k)$, $j \in \mathbb{Z}$, $k \in \mathbb{Z}^n$, and let B_j denote the matrix $(B(\phi_{jk}, \phi_{jl}))_{(k,l)\in\mathbb{Z}^n\times\mathbb{Z}^n}$. We first remark that the matrices B_j are uniformly bounded on $l^2(\mathbb{Z}^n)$ and uniformly δ-accretive, with $\delta = 1$. Indeed, if ξ_k and η_l are square-summable sequences,

$$\sum_k \sum_l B(\phi_{jk}, \phi_{jl}) \xi_k \eta_l = B(f, g)$$

where $f = \sum \xi_k \phi_{jk}$ and $g = \sum \eta_l \phi_{jl}$. Now

$$|B(f, g)| \leq C \|f\|_2 \|g\|_2 \qquad \text{where } C = \|b\|_\infty,$$

and, thus,

$$|\sum_k \sum_l B(\phi_{jk}, \phi_{jl}) \xi_k \eta_l| \leq C (\sum |\xi_k|^2)^{1/2} (\sum |\eta_k|^2)^{1/2}.$$

Also,

$$\Re \sum_k \sum_l B(\phi_{jk}, \phi_{jl}) \xi_j \bar{\xi}_l = \Re B(f, \bar{f}) \geq \|f\|_2^2.$$

(Here, $\bar{f} = \sum \bar{\xi}_k \phi_{jk}$, because the functions ϕ_{jk} are real-valued.)

Using Proposition 3, we get a new sequence $\tilde{\phi}_{jk}$, $k \in \mathbb{Z}^n$, which is a Riesz basis of V_j. In fact

(5.1) $$C (\sum |\xi_k|^2)^{1/2} \leq \| \sum \xi_k \tilde{\phi}_{jk} \|_2 \leq C' (\sum |\xi_k|^2)^{1/2},$$

where C and C' are independent of j,

(5.2) $$B(\tilde{\phi}_{jk}, \tilde{\phi}_{jl}) = \delta_{kl},$$

and, finally,

(5.3) $$|\partial^\alpha \tilde{\phi}_{jk}(x)| \leq C' 2^{nj/2} 2^{j|\alpha|} e^{-\gamma' |2^j x - k|} \qquad \text{for } |\alpha| \leq r,$$

where $\gamma' > 0$ and $C' > 0$ depend only on n, C, and the exponent γ of (2.1).

Let W_j denote the orthogonal complement of V_j in V_{j+1}. Now $\tilde{W}_j$ is the subspace defined by $B(f, u) = 0$, for all $u \in V_j$. We observe immediately that, if $u \in V_j$, then the same is true for the conjugate function $\bar{u}$. So we can apply Proposition 1 to get $V_{j+1} = V_j \oplus \tilde{W}_j$.

Let $\tilde{P}_j : V_{j+1} \to V_j$ denote the non-orthogonal projection with kernel $\tilde{W}_j$. It is given by

$$\tilde{P}_j = \sum_k B(f, \tilde{\phi}_{jk}) \tilde{\phi}_{jk}.$$

We next write $T_j : W_j \to \tilde{W}_j$ for the inverse of the orthogonal projection of $\tilde{W}_j$ on W_j. Then, $T_j = 1 - \tilde{P}_j$ and T_j is an isomorphism of W_j onto $\tilde{W}_j$. Moreover, for a constant C independent of j, we get, for all $f \in W_j$,

(5.4) $$\|f\|_2 \leq \|T_j(f)\|_2 \leq C \|f\|_2.$$

For the time being, we shall assume the following result.

Proposition 4. *The operator T whose restriction to each W_j coincides with T_j is continuous on $L^2(\mathbb{R}^n)$.*

In the next section we shall see that this is a very easy consequence of the $T(1)$ theorem.

Now consider the "usual" wavelets ψ_{jk}^ε, $\varepsilon \in E$, obtained by the algorithm of *Wavelets and Operators*, Chapter 3. For the functions $h_{jk}^\varepsilon = T_j(\psi_{jk}^\varepsilon)$, we get

Lemma 4. *For every $j \in \mathbb{Z}$, the matrix $B(h_{jk}^\varepsilon, h_{jk'}^{\varepsilon'})$ is δ-accretive.*

Following Lemarié's method, we prove this lemma by noting that $B(h_{jk}^\varepsilon, h_{jk'}^{\varepsilon'}) = B(h_{jk}^\varepsilon, \psi_{jk'}^{\varepsilon'})$. This identity holds, because $h_{jk'}^{\varepsilon'} - \psi_{jk'}^{\varepsilon'} \in V_j$ and $B(f, u) = 0$, when $f \in \tilde{W}_j$ and $u \in V_j$. The wavelets ψ_{jk}^ε are real-valued, so $\bar{h}_{jk'}^{\varepsilon'} - \psi_{jk'}^{\varepsilon'} \in V_j$ (since also $\bar{V}_j = V_j$). So we arrive at the identity $B(h_{jk}^\varepsilon, h_{jk'}^{\varepsilon'}) = B(h_{jk}^\varepsilon, \bar{h}_{jk'}^{\varepsilon'})$ and the matrix on the right-hand side is δ-accretive, because the h_{jk}^ε form a Riesz basis of $\tilde{W}_j$.

We now apply Proposition 3, once more, to get another Riesz basis $\tilde{\psi}_\lambda$, $\lambda \in \Lambda_j$, of $\tilde{W}_j$, such that, for $\lambda, \lambda' \in \Lambda_j$, we have $B(\tilde{\psi}_\lambda, \tilde{\psi}_{\lambda'}) = 0$, if $\lambda \neq \lambda'$, and 1, if $\lambda = \lambda'$.

But, by our construction, we also have $B(\tilde{\psi}_\lambda, \tilde{\psi}_{\lambda'}) = 0$, if $\lambda \in \Lambda_j$, $\lambda' \in \Lambda_{j'}$ and $j \neq j'$. Indeed, by symmetry, we need only consider the case $j > j'$, when $\tilde{\psi}_\lambda \in \tilde{W}_j$, whereas $\tilde{\psi}_{\lambda'} \in V_j$. We conclude by applying the definition of $\tilde{W}_j$.

We now have a collection of functions $\tilde{\psi}_\lambda$, $\lambda \in \Lambda$, at our disposal, satisfying

$$(5.5) \qquad \left\| \sum_{\lambda \in \Lambda} \alpha(\lambda) \tilde{\psi}_\lambda(x) \right\| \leq C \left(\sum_{\lambda \in \Lambda} |\alpha(\lambda)|^2 \right)^{1/2},$$

$$(5.6) \qquad B(\tilde{\psi}_\lambda, \tilde{\psi}_{\lambda'}) = \delta_{\lambda\lambda'},$$

and

$$(5.7) \qquad |\partial^\alpha \tilde{\psi}_\lambda(x)| \leq C 2^{nj/2} 2^{j|\alpha|} e^{-\gamma|2^j x - k|}, \qquad \text{if } |\alpha| \leq r$$

(where $\gamma > 0$ is no longer the γ we started with).

Inequality (5.5) follows from the continuity of the operator T. In fact,

$$f = \sum_{\lambda \in \Lambda} \alpha(\lambda) \tilde{\psi}_\lambda(x) = \sum_j \sum_{\lambda \in \Lambda_j} \alpha(\lambda) \tilde{\psi}_\lambda(x) = \sum_{j \in \mathbb{Z}} f_j,$$

where $f_j \in \tilde{W}_j$. Thus $f_j = T_j(g_j)$, where $g_j \in W_j$, and $f = T(g)$, with $g = \sum_{-\infty}^\infty g_j$. Then

$$\|f\|_2 \leq C\|g\|_2 = C(\sum \|g\|_2^2)^{1/2} \leq C'(\sum \|f\|_2^2)^{1/2},$$

because of (5.4). We conclude by observing that the $\tilde{\psi}_\lambda$, $\lambda \in \Lambda_j$, form a Riesz basis of $\tilde{W}_j$ and that the constants are uniform in j.

By duality, inequalities (5.5) and (5.6) imply the converse of inequality (5.5). Indeed, consider an arbitrary finite form $g(x) = \sum_{\lambda \in \Lambda} \beta(\lambda) \tilde{\psi}_\lambda$ such that $\sum |\beta(\lambda)|^2 \leq 1$. Then $\|g\|_2 \leq C$ and, for every sum $f(x) = \sum \alpha(\lambda) \tilde{\psi}_\lambda(x)$, we have $B(f, g) = \sum \alpha(\lambda) \beta(\lambda)$. However $|B(f, g)| \leq \|f\|_2 \|g\|_2 \|b\|_\infty \leq C'\|f\|_2$, which can be summarized as

$$(5.8) \qquad \left| \sum \alpha(\lambda) \beta(\lambda) \right| \leq C'\|f\|_2 \qquad \text{as long as } \sum |\beta(\lambda)|^2 \leq 1.$$

We get $(\sum |\alpha(\lambda)|^2)^{1/2} \leq C'\|f\|_2$, as claimed, by taking the supremum of the left-hand side of (5.8).

The functions $\tilde{\psi}_\lambda$ are wavelets which are adapted to $b(x)$. For each fixed j, they form a Riesz basis of $\tilde{W}_j$ as λ runs through Λ_j. To show that the whole collection is a Riesz basis of $L^2(\mathbb{R}^n)$, we must show that the (algebraic) sum of the $\tilde{W}_j$ is dense in $L^2(\mathbb{R}^n)$.

Since the union of the V_m is dense in $L^2(\mathbb{R}^n)$, it will be enough to know how to approximate $f \in V_m$ by a sequence f_j belonging to the algebraic sum referred to above.

Now, $V_m = V_{m-1} + \tilde{W}_{m-1}$ and thus $f = f_1 + g_1$, where $f_1 \in V_{m-1}$. We iterate this decomposition to get $f = f_N + r_N$, where $f_N \in V_{m-N}$ and r_N belongs to the algebraic sum of the $\tilde{W}_j$, for $j \geq m - N$.

We shall show that $\|f_N\|_2 \to 0$ as $N \to \infty$. To begin with, we observe that the inclusion $V_j \subset V_{j+1}$ implies that $B(u, v) = 0$ if $u \in \tilde{W}_j$, $v \in V_{j'}$ and $j' \leq j$. In particular, $B(r_N, v) = 0$ if $v \in V_{m-N}$.

Thus,

$$f_N = \sum_k B(f_N, \tilde{\phi}_{(m-N)k}) \tilde{\phi}_{(m-N)k} = \sum_k B(f, \tilde{\phi}_{(m-N)k}) \tilde{\phi}_{(m-N)k} \, .$$

The kernel $K_N(x, y)$ of this last operator satisfies

$$|K_N(x, y)| \leq C 2^{(m-N)n} (1 + 2^{m-N}|x - y|)^{-n-1} \, ,$$

by (5.3). Hence $\|f_N\|_2 \to 0$ as $N \to \infty$. The algebraic direct sum of the $\tilde{W}_j$ is thus dense in $L^2(\mathbb{R}^n)$, which proves Theorem 1.

So the wavelets $\tilde{\psi}_\lambda$, $\lambda \in \Lambda$, form a Riesz basis of $L^2(\mathbb{R}^n)$.

We have, in fact, established the following theorem.

Theorem 2. *Let V_j, $j \in \mathbb{Z}$, be a multiresolution approximation of $L^2(\mathbb{R}^n)$ satisfying properties (2.1) and (2.2). Then there exist a constant C, an exponent $\eta > 0$, and a family of functions $\tilde{\psi}_\lambda$, $\lambda \in \Lambda$, with the following properties:*

$$(5.9) \qquad \tilde{\psi}_\lambda \in V_{j+1} \, , \qquad \text{when } \lambda \in \Lambda_j;$$

$$(5.10) \qquad |\partial^\alpha \tilde{\psi}_\lambda(x)| \leq C 2^{j(|\alpha|+n/2)} e^{-\eta|2^j x - k|} \, , \qquad \text{for } |\alpha| \leq r;$$

$$(5.11) \qquad \int \tilde{\psi}_\lambda(x) \tilde{\psi}_{\lambda'}(x) b(x) \, dx = \delta_{\lambda \lambda'} \, ;$$

$$(5.12) \qquad \text{the collection } \tilde{\psi}_\lambda, \lambda \in \Lambda, \text{ is a Riesz basis of } L^2(\mathbb{R}^n).$$

Corollary 1. *Every function $f \in L^2(\mathbb{R}^n)$ can be written uniquely as $f(x) = \sum_{\lambda \in \Lambda} \alpha(\lambda) \tilde{\psi}_\lambda(x)$, with $\sum_{\lambda \in \Lambda} |\alpha(\lambda)|^2 < \infty$, where $\alpha(\lambda) = \int f(x) \tilde{\psi}_\lambda(x) b(x) \, dx$, and where, for constants $C_2 \geq C_1 > 0$,*

$$(5.13) \qquad C_1 \|f\|_2 \leq \left(\sum_{\lambda \in \Lambda} |\alpha(\lambda)|^2 \right)^{1/2} \leq C_2 \|f\|_2 \, .$$

Corollary 2. *For all $\lambda \in \Lambda$,*

$$\int \tilde{\psi}_\lambda(x) b(x) \, dx = 0 \, .$$

Indeed, let $g(x) \in V_0$ be a function which is 1 at 0. Put $g_m(x) = g(2^{-m}x) \in V_{-m}$. Then, orthogonality gives $\int \tilde{\psi}_\lambda(x) b(x) g_m(x) \, dx = 0$, for m sufficiently large. It is then enough to let m tend to infinity and apply the Lebesgue dominated convergence theorem.

We still have to prove that the operator T is continuous, a result which is the cornerstone of the construction above.

6 Continuity of T

To prove that the operator T is continuous, we apply David and Journé's $T(1)$ theorem. As above, we let ψ_{jk}^ε denote the usual basis of orthonormal wavelets ($\varepsilon \in E$, $j \in \mathbb{Z}$, and $k \in \mathbb{Z}^n$). Let D_j be the orthogonal projection of $L^2(\mathbb{R}^n)$ onto W_j. This is given by

$$D_j(f) = \sum_k \sum_{\varepsilon \in E} (f, \psi_{jk}^\varepsilon) \psi_{jk}^\varepsilon \, .$$

Then

$$T = \sum_{-\infty}^{\infty} T_j D_j = \sum_{-\infty}^{\infty} (1 - \tilde{P}_j) D_j = 1 - R \, ,$$

where $R = \sum_{-\infty}^{\infty} \tilde{P}_j D_j$. The kernel $R_j(x, y)$ of $\tilde{P}_j D_j$ is

$$\sum_{\varepsilon \in E} \sum_k \sum_l B(\psi_{jk}^\varepsilon, \tilde{\phi}_{jl}) \tilde{\phi}_{jl}(x) \psi_{jk}^\varepsilon(y) \, .$$

This can be seen by taking the composition of the kernels of $\tilde{P}_j$ and D_j. Now, the coefficients $B(\psi_{jk}^\varepsilon, \tilde{\phi}_{jl})$ satisfy

$$(6.1) \qquad |B(\psi_{jk}^\varepsilon, \tilde{\phi}_{jl})| \leq C e^{-\eta|k-l|} \, ,$$

where $\eta > 0$ and $C > 0$ are constants.

As an immediate consequence,

$$(6.2) \qquad |R_j(x, y)| \leq C 2^{nj} (1 + 2^j |x - y|)^{n-1}$$

and

$$|\nabla_x R_j(x, y)| + |\nabla_y R_j(x, y)| \leq C 2^{(n+1)j} (1 + 2^j |x - y|)^{n-2} \, .$$

Further, $\int R_j(x,y)\,dy = 0$ because the wavelets ψ^ε_{jk} are of mean 0.

These properties lead to the usual estimates for the kernel $R(x,y)$ of R, as well as the weak continuity of R.

Indeed, to show that R is weakly continuous, we have to find an upper bound for $|\sum_{-\infty}^{\infty} \int\int R_j(x,y)u(y)v(x)\,dy\,dx|$, when $u(x)$ and $v(y)$ are regular functions with supports in the same ball B of radius S.

If $2^{-j} > S$, we need only a rough estimate, which we get by using the inequality $|R_j(x,y)| \leq C2^{nj}$. On the other hand, if $2^{-j} \leq S$, we observe that $\int R_j(x,y)u(y)\,dy = \int R_j(x,y)(u(y) - u(x))\,dy$. Then

$$\left| \int R_j(x,y)(u(y) - u(x))\,dy \right| \leq \|\nabla u\|_\infty \int |R_j(x,y)||x - y|\,dy$$
$$\leq C2^{-j}\|\nabla u\|_\infty \,.$$

It is now obvious how to conclude.

We get $R(1) = 0$, again because $\int R_j(x,y)\,dy = 0$. All that remains to do is to calculate ${}^tR(1) = \sum_{-\infty}^{\infty} {}^tR_j(1)$.

Let $\varepsilon(j,l) = \int \tilde{\phi}_{jl}(x)\,dx$, put $m_j(x) = \sum_l \varepsilon(j,l)\tilde{\phi}_{jl}$, set

$$S^\varepsilon_{jk}(x) = \psi^\varepsilon_{jk}(x)m_j(x)\,,$$

and, finally, write

$$\beta^\varepsilon(j,k) = B(\psi^\varepsilon_{jk}, m_j) = \int S^\varepsilon_{jk}(x)b(x)\,dx\,.$$

With this notation, ${}^tR_j(1) = \sum_k \sum_\varepsilon \beta^\varepsilon(j,k)\psi^\varepsilon_{jk}$. To conclude the proof, we need to show that the coefficients $\beta^\varepsilon(j,k)$ satisfy the necessary and sufficient condition given by Theorem 4 of *Wavelets and Operators*, Chapter 5.

This will follow immediately from the next lemma.

Lemma 5. *The functions* S^ε_{jk}, $\varepsilon \in E$, $j \in \mathbb{Z}$, *and* $k \in \mathbb{Z}^n$, *are vaguelets (in the sense of Chapter 8, Definition 3).*

If we assume this lemma, the $\beta^\varepsilon(j,k)$ are the coefficients of $b \in L^\infty \subset$ BMO, with respect to the vaguelets and thus satisfy Carleson's quadratic estimate (Theorem 4 of *Wavelets and Operators*, Chapter 5 and the remark following the proof of the theorem).

Returning to Lemma 5, we have $|\varepsilon(j,l)| \leq C2^{-nj/2}$ (because of the properties of $\tilde{\phi}_{jl}$). As a result, $\|m_j(x)\|_\infty \leq C'$ and $\|\nabla m_j(x)\|_\infty \leq C'2^j$. Now $m_j(x)$ lies in V_j (or, more accurately, $V_j(\infty)$, using the notation of *Wavelets and Operators*, Chapter 2, section 5). Since ψ^ε_{jk} is in W_j (and $L^1(\mathbb{R}^n)$), we get $\int \psi^\varepsilon_{jk}(x)m_j(x)\,dx = 0$ (because W_j is orthogonal to V_j). The functions S^ε_{jk} satisfy the same properties of regularity, localization and cancellation as the wavelets: they are thus vaguelets.

Before leaving Theorem 2, we shall deduce a noteworthy generalization of inequality (5.5).

We define *vaguelets associated with* $b(x)$ to be any collection $w_\lambda(x)$, $\lambda \in \Lambda$, of $L^2(\mathbb{R}^n)$ functions such that, for exponents $\beta > \alpha > 0$ and some constant C,

$$(6.3) \qquad |w_\lambda(x)| \leq C2^{nj/2}(1 + |2^j x - k|)^{-n-\beta},$$

$$(6.4) \qquad |w_\lambda(x') - w_\lambda(x)| \leq C2^{(n/2+\alpha)j}|x' - x|^\alpha,$$

and

$$(6.5) \qquad \int w_\lambda(x)b(x)\,dx = 0.$$

If $b(x) = 1$, this is the same as the vaguelets of Chapter 8, section 5.

An apparently stronger formulation of Theorem 2, which is actually equivalent to the original version, is the following statement.

Proposition 5. *With the hypotheses (6.3), (6.4) and (6.5), there exists a constant C such that, for every sequence of coefficients $\alpha(\lambda) \in l^2(\Lambda)$,*

$$(6.6) \qquad \left\| \sum_{\lambda \in \Lambda} \alpha(\lambda)w_\lambda(x) \right\| \leq C\left(\sum_{\lambda \in \Lambda} |\alpha(\lambda)|^2\right)^{1/2}.$$

This, entirely general, estimate implies, in particular, inequality (5.5), on which the proof of Theorem 2 is based.

To prove (6.6), it is enough to compute the matrix of the $(w_\lambda)_{\lambda \in \Lambda}$, with respect to the basis given by the $\tilde\psi_\lambda$ (as in Theorem 2) and to verify that the matrix defines a continuous operator on $l^2(\Lambda)$.

Now the entries $\omega(\lambda, \lambda')$ are given by $\omega(\lambda, \lambda') = \int w_\lambda(x)\tilde\psi_{\lambda'}(x)b(x)\,dx$ (Corollary 1 of Theorem 2). What we have to do is to verify that, for some exponent $\gamma > 0$ and some constant $C > 0$,

$$(6.7) \quad |\omega(\lambda, \lambda')| \leq C2^{-|j'-j|(n/2+\gamma)}\left(\frac{2^{-j} + 2^{-j'}}{2^{-j} + 2^{-j'} + |k2^{-j} - k'2^{-j'}|}\right)^{n+\gamma}.$$

The continuity of the matrix will then be a consequence of Schur's lemma. Now (6.7) may be obtained by a "simple" integration by parts: if $j' \geq j$, then $w_\lambda(x)$ acts as a "flat" function, whilst the product $b(x)\tilde\psi_{\lambda'}(x)$ is "oscillating and highly localized". We leave the details to the reader, as they are the same as those in the case $b(x) = 1$.

7 A special case of the $T(b)$ theorem

We now deal with a special case of the $T(b)$ theorem, a case which is sufficient to give the continuity of the Cauchy kernel on any Lipschitz curve.

Suppose that $(1 + |x - y|)^{n+1}K(x,y)$ belongs to $L^\infty(\mathbb{R}^n \times \mathbb{R}^n)$, a qualitative hypothesis that we shall refrain from using in a quantitative form.

Next, suppose that $K(y,x) = -K(x,y)$ and that

(7.1) $$|\nabla_x K(x,y)| \le C_0 |x-y|^{-n-1}.$$

Let $b \in L^\infty(\mathbb{R}^n)$ be a function such that $\Re\, b(x) \ge 1$ almost everywhere.

Theorem 3. *Suppose that, besides the antisymmetry and (7.1), we assume that $\int K(x,y)b(y)\,dy = 0$ identically, in x; then the operator T, defined by the kernel K, is bounded on $L^2(\mathbb{R}^n)$ and the norm $\|T\|$ of $T : L^2 \to L^2$ satisfies*

(7.2) $$\|T\| \le C(C_0, n, \|b\|_\infty).$$

The above estimate is a uniform bound, that is, it is independent of the L^∞ norm of $(1+|x-y|)^{n+1}K(x,y)$.

To prove Theorem 3, we follow the proof of the $T(1)$ theorem step by step.

Consider the kernel $L(x,y) = K(x,y)b(y)$ and the corresponding operator $\mathcal{L}$. Let $\tilde{\psi}_\lambda$, $\lambda \in \Lambda$, be the wavelets of Theorem 2. We prove that $\mathcal{L}$ is continuous, by estimating the size of the entries of the matrix $(\omega(\lambda,\lambda'))_{(\lambda,\lambda')\in\Lambda\times\Lambda}$ of $\mathcal{L}$ with respect to the basis $\tilde{\psi}_\lambda$, $\lambda \in \Lambda$.

This is how we find the matrix. If we expand $f \in L^2$ as $f = \sum \alpha(\lambda)\tilde{\psi}_\lambda$, then $\mathcal{L}(f) = \sum \beta(\lambda)\tilde{\psi}_\lambda$, where

$$\beta(\lambda) = \int \mathcal{L}(f)(x)\tilde{\psi}_\lambda(x)b(x)\,dx = \sum_{\lambda'\in\Lambda}\alpha(\lambda')\int \mathcal{L}(\tilde{\psi}_{\lambda'})(x)\tilde{\psi}_\lambda(x)b(x)\,dx$$

$$= \sum_{\lambda'\in\Lambda}\omega(\lambda,\lambda')\alpha(\lambda').$$

So we get

(7.3) $$\omega(\lambda,\lambda') = \int\int K(x,y)b(y)\tilde{\psi}_{\lambda'}(y)b(x)\tilde{\psi}_\lambda(x)\,dy\,dx.$$

All that remains to do is to show that $|\omega(\lambda,\lambda')|$ is bounded by

$$C2^{-|j'-j|(n/2+\gamma)}\left(\frac{2^{-j}+2^{-j'}}{2^{-j}+2^{-j'}+|k2^{-j}-k'2^{-j'}|}\right)^{n+\gamma},$$

so that we can apply Schur's lemma.

Since the problem is entirely symmetric in λ and λ', we may suppose that $j' \le j$. We shall deal with the case $j' = j$ at the end of the proof so, for now, we assume that $j' < j$. Writing $\omega(\lambda,\lambda') = \int\tilde{\psi}_{\lambda'}(y)h_\lambda(y)\,dy$, we begin by projecting h_λ orthogonally on V_j, which, in its turn, contains $\tilde{W}_j$, by the geometry of the construction of the wavelets $\tilde{\psi}_\lambda$.

This orthogonal projection g_λ is $\sum_l \theta_j(k,l)\phi_{jl}$, where

(7.4) $$\theta_j(k,l) = \int\int K(x,y)b(y)\phi_{jl}(y)b(x)\tilde{\psi}_\lambda(x)\,dy\,dx.$$

We intend to show that

(7.5) $$|\theta_j(k,l)| \le C(1 + |k - l|)^{-n-1}.$$

We shall then be able to write the condition $T(b) = 0$ in the form $\sum_l \theta_j(k,l) = 0$, and $\omega(\lambda, \lambda')$ can be regarded as the integral of a "flat" function $\tilde{\psi}_{\lambda'}$ multiplied by a highly localized, oscillating function g_λ. We can then apply Lemma 3 and conclude.

To prove (7.5), we model the situation by a double integral of the form

$$I = \int \int K(x,y)b(y)u(y)b(x)v(x)\, dy\, dx$$

where u and v are C^1 functions whose supports are contained, respectively, in $|x - x_0| \le 1$ and $|y - y_0| \le 1$, which are such that $\|\nabla u\|_\infty \le 1$, $\|\nabla v\|_\infty \le 1$ and finally, $\int v(x)b(x)\, dx = 0$.

We shall show that

(7.6) $$|I| \le C(1 + |x_0 - y_0|)^{-n-1}.$$

If $|x_0 - y_0| \le 3$, we can use the regularity of the functions u, v, and the antisymmetry of the kernel $K(x,y)b(x)b(y)$ to write

$$I = \frac{1}{2} \int \int K(x,y)b(x)b(y)(u(y)v(x) - u(x)v(y))\, dx\, dy,$$

and to get the required inequality.

If $|x_0 - y_0| > 3$, we can integrate by parts, because the kernel is regular in x and the integral of bv is zero.

The model is not accurate for two reasons. The scale is 2^{-j} and not 1 (the scale is defined by the functions ϕ_{jl} and $\tilde{\psi}_\lambda$) and, more importantly, ϕ_{jl} and $\tilde{\psi}_\lambda$ decrease rapidly at infinity rather than having compact supports.

To reduce to the case $j = 0$, we use a change of scale which does not alter the hypotheses about $K(x,y)$ and $b(x)$ in the slightest.

To reduce to the case of functions of compact support, it is enough to make the following observation.

Lemma 6. *For every integer n, there exists $R(n) > 0$ such that the following property holds. For each $\gamma > 0$, there is a constant $C'(\gamma, n)$ such that, if $f(x)$ satisfies*

$$|f(x)| \le e^{-\gamma|x|}, \qquad |\nabla f(x)| \le e^{-\gamma|x|}, \qquad \text{and} \qquad \int f(x)b(x)\, dx = 0,$$

then f can be expanded as a series $\sum_{k \in \mathbb{Z}^n} \gamma_k f_k(x)$, where the support of each f_k is contained in $|x - k| \le R(n)$, $\|f_k\|_\infty \le 1$, $\|\nabla f_k\|_\infty \le 1$, $\int f_k b(x)\, dx = 0$, and the coefficients γ_k satisfy $|\gamma_k| \le C'(\gamma, n)e^{-\gamma|k|}$.

To simplify the notation, we restrict the proof to the case $n = 1$. We start from a C^1 function $\theta \ge 0$ with support in $[-1, 1]$ such that

$\sum_{-\infty}^{\infty} \theta(x - k) = 1$. We then expand f the "wrong way" as $f(x) - \sum_{-\infty}^{\infty} f(x)\theta(x-k) = \sum_{-\infty}^{\infty} m_k(x)$. Putting $I_k = \int m_k(x)b(x)\,dx$, we get $|I_k| \leq Ce^{-\gamma|k|}$ and $\sum_{-\infty}^{\infty} I_k = 0$. This enables us to write $I_k = J_{k+1} - J_k$, with $|J_k| \leq C'e^{-\gamma|k|}$. Then we put $\omega_k = \int \theta(x - k)b(x)\,dx$, giving $\Re e\, \omega_k \geq 1$. We now can correct m_k by replacing it by $\gamma_k f_k(x) = m_k(x) - (r_{k+1}(x) - r_k(x))$, $r_k(x) = J_k \omega_k^{-1}\theta(x-k)$. This gives $\int r_k(x)b(x)\,dx = J_k$ and thus $\int f_k(x)b(x)\,dx = 0$, as claimed. The other properties required of f_k are obviously satisfied.

Lemma 6, together with (7.6), gives (7.5), which deals with the estimate of $|\omega(\lambda, \lambda')|$ when $j' \neq j$.

The case $j' = j$ is even simpler. We go back to $\omega(\lambda, \lambda')$, defined by (7.3), and must verify that $|\omega(\lambda, \lambda')| \leq C(1+|k-k'|)^{-n-\gamma}$. This is done by exactly the same method as we used for estimating $|\theta_j(k, l)|$.

Now, to finish the proof of Theorem 3, it is enough to apply Lemma 3 of Chapter 8 and then Schur's lemma. This last part of the proof is identical to that used for the $T(1)$ theorem and is thus not repeated here.

8 Application to the L^2 continuity of the Cauchy kernel

We come back to dimension 1 and start from a real-valued Lipschitz function $a(x)$. In this case, the function $b(x)$ is $1 + ia'(x)$ and we have $\|b\|_\infty \leq (1 + M^2)^{1/2}$ for $-M \leq a'(x) \leq M$ as well as $\Re e\, b(x) = 1$.

For each $\varepsilon > 0$, we consider the truncated kernel

$$C_\varepsilon(x, y) = \frac{1}{z(x) - z(y) + i\varepsilon} - \frac{1}{z(x) - z(y) + i\varepsilon^{-1}},$$

where $z(x) = x + ia(x)$. By the geometry of the graph Γ of $a(x)$, we have $C_\varepsilon(x, y) = O((x - y)^{-2})$ as $|x - y| \to \infty$. Moreover, $C_\varepsilon(x, y)$ is bounded (but both estimates depend on $\varepsilon > 0$).

By Cauchy's formula, $\int C_\varepsilon(x, y)z'(y)\,dy = 0$. The only problem with $C_\varepsilon(x, y)$ is that it is not antisymmetric. For that reason, we consider $\Gamma_\varepsilon(x, y) = C_\varepsilon(x, y) - C_\varepsilon(y, x)$, to which we can apply Theorem 2. It is not difficult to show that, if $u(x)$ is a C^1 function of compact support,

$$(8.1) \quad \lim_{\varepsilon \downarrow 0} \int_{-\infty}^{\infty} \left(\frac{1}{z(x) - z(y) + i\varepsilon} + \frac{1}{z(x) - z(y) - i\varepsilon} \right) z'(y)u(y)\,dy$$

$$= 2 \lim_{\varepsilon \downarrow 0} \int_{|y-x| \geq \varepsilon} \frac{1}{z(x) - z(y)} z'(y)u(y)\,dy.$$

All that we need do to prove this identity is to integrate by parts. The final factor to take into account is that the Cauchy kernel is bounded on $L^2(\Gamma)$, for each Lipschitz curve Γ.

We shall discuss these matters in more detail, in our next chapter.

9 The general case of the $T(b)$ theorem

Let $K(x,y)$ be a function such that $(1 + |x - y|)^{n+1} K(x,y)$ belongs to $L^\infty(\mathbb{R}^n \times \mathbb{R}^n)$. This is a qualitative hypothesis which avoids the thorny problem of how to define an operator $T : \mathcal{D} \to \mathcal{D}'$ on functions $b(x) \in L^\infty(\mathbb{R}^n)$.

We suppose that there exist an exponent $\gamma \in (0, 1]$ and a constant C_0 such that

(9.1) $|K(x,y)| \leq C_0 |x - y|^{-n}$,

(9.2) $|K(x',y) - K(x,y)| \leq C_0 |x' - x|^\gamma |x - y|^{-n-\gamma}$,

for $|x' - x| \leq |x - y|/2$, and

(9.3) $|K(x,y') - K(x,y)| \leq C_0 |y' - y|^\gamma |x - y|^{-n-\gamma}$,

if $|y' - y| \leq |x - y|/2$.

The second group of hypotheses involves a function $b(x) \in L^\infty(\mathbb{R}^n)$, satisfying $\Re e\, b(x) \geq 1$ almost everywhere.

We require the existence of a constant C_1, such that, for each ball $B \subset \mathbb{R}^n$,

(9.4) $\int_B \left| \int_B K(x,y)b(y)\,dy \right| dx \leq C_1 |B|$

as well as

(9.5) $\int_B \left| \int_B K(x,y)b(x)\,dx \right| dy \leq C_1 |B|$.

Theorem 4. *With hypotheses (9.1) to (9.5), the norm of the operator T defined by the kernel $K(x,y)$, is not greater than a constant $C_2 = C_2(C_0, C_1, \|b\|_\infty, n)$.*

The proof of Theorem 4 is a variant of that of the $T(1)$ theorem, the principal change being that standard wavelets are replaced by wavelets adapted to the function b.

First of all, we check that conditions (9.4) and (9.5) are equivalent to the familiar conditions $T(b) \in \mathrm{BMO}$ and ${}^t T(b) \in \mathrm{BMO}$, together with the property of weak cancellation, which we may write as follows:

(9.6) $\left| \iint_{B \times B} K(x,y)b(x)b(y)u(x)v(y)\,dx\,dy \right|$

$$\leq C|B|(\|u\|_\infty \|v\|_\infty + r^2 \|\nabla u\|_\infty \|\nabla v\|_\infty),$$

whenever B is a ball of volume $|B|$, radius r, and where u, v are C^1 functions with supports in B.

We start by verifying that the constant C_1 of (9.4) lets us control the BMO norm of $T(b)$.

To do this, we let B denote an arbitrary ball (centre x_0, radius $r > 0$) and we split b into $b_1 + b_2 + b_3$, where b_1 is the product of b by the characteristic function of B, b_2 is the product of b by the characteristic function of the annulus Δ defined by $r \leq |x - x_0| \leq 2r$, and b_3 is the product of b by the characteristic function of $|x - x_0| \geq 2r$.

Now, as we have often observed, if $|x - x_0| \leq r$, then

$$|Tb_3(x) - Tb_3(x_0)| \leq C|x-x_0|^\gamma \int_{|x_0-y|\geq 2r} |y-x_0|^{-n-\gamma}|b(y)| \, dy \leq C'\|b\|_\infty \,.$$

We then find a bound for $\int_B |Tb_2(x)| \, dx$ by remarking that

$$\int\int_{B\times\Delta} |x - y|^{-n} \, dx \, dy \leq C(n)r^n \,.$$

Lastly, we get a bound for $\int_B |Tb_1(x)| \, dx$ from (9.4).

To establish (9.6), we write

$$T(bv)(x) = \int K(x,y)b(y)(v(y) - v(x)\chi(y)) \, dy + v(x) \int K(x,y)\chi(y) \, dy \,,$$

where χ is the characteristic function of B. The first integral is bounded above by $C\|\nabla v\|_\infty \int_{|x-y|\leq 2r} |x - y|^{-n+1} \, dy \leq C'\|\nabla v\|_\infty$ and the second is covered by (9.4).

In the opposite direction, suppose that (9.6) is satisfied and that $T(b)$ is in $\mathrm{BMO}(\mathbb{R}^n)$. To prove (9.4) it would be enough to apply the definition of the space BMO, as long as we knew that the "floating constant" involved in the definition of BMO could be zero.

To evaluate the "floating constant", we let $2B$ (respectively $4B$) denote B "doubled" (respectively "quadrupled") and let $v \in \mathcal{D}(\mathbb{R}^n)$ be a function which equals 1 on $2B$, equals 0 outside $4B$ and satisfies $0 \leq v(x) \leq 1$ and $\|\nabla v\|_\infty \leq C/r$, where r is the radius of B. We let $w = 1 - v$ and get, as above,

$$|T(wb)(x) - T(wb)(x_0)| \leq C\|b\|_\infty \,,$$

where x_0 is the common centre of the balls B, $2B$ and $4B$. Since $T(b)$ is in BMO, there exists a constant (the "floating constant") γ_B, such that

$$T(vb)(x) = \gamma_B + r_b(x) \qquad \text{where} \qquad \int_B |r_B(x)| \, dx \leq C|B| \,.$$

We now let $u(x)$ denote a positive C^1 function with support in B which satisfies $\|\nabla u\|_\infty \leq Cr^{-n-1}$ and has mean 1. These conditions imply, in particular, that $\|u\|_\infty \leq Cr^{-n}$.

Property (9.6) now becomes

$$\left| \int u(x)b(x)T(vb)(x) \, dx \right| \leq C' \,,$$

where we have used $4B$ instead of B. Taking account of the identity $T(vb)(x) = \gamma_B + r_B(x)$, we get $|\gamma_B \int u(x)b(x)\,dx| \leq C''$. But $\Re e \int u(x)b(x)\,dx \geq 1$, so we get $|\gamma_B| \leq C''$.

After these variations on the statement of Theorem 4, we come to the proof. It follows that of the $T(1)$ theorem word for word except for the standard wavelets being replaced by the wavelets $\tilde{\psi}_\lambda$ whose cancellation is adapted to the function b.

We begin with the construction of the pseudo-products. They are adapted to the function $b \in L^\infty(\mathbb{R}^n)$ non-linearly. Starting from $\beta \in \mathrm{BMO}(\mathbb{R}^n)$, we intend to construct a continuous linear operator T on $L^2(\mathbb{R}^n)$, whose kernel, restricted to $x \neq y$, satisfies the usual estimates and which, moreover, is such that $T(b) = \beta$ and ${}^t T(b) = 0$.

To construct T, we mimic the usual pseudo-product construction. Firstly, we put $\theta_\lambda(x) = 2^{nj}\theta(2^j x - k)$, where $\theta \in \mathcal{D}(\mathbb{R}^n)$, $\int \theta(x)\,dx = 1$, where $\lambda = 2^{-j}(k + \varepsilon/2)$, $\varepsilon \in E$. If we set $\omega(\lambda) = (\int b(x)\theta_\lambda(x)\,dx)^{-1}$, then $\Re e \int b(x)\theta_\lambda(x)\,dx \geq 1$, so that $|\omega(\lambda)| \leq 1$.

The wavelets $\tilde{\psi}_\lambda$ adapted to $b(x)$ are those that we constructed earlier. We now define

$$\pi(\beta, f) = \sum_{\lambda \in \Lambda} \omega(\lambda)\langle f, \theta_\lambda \rangle \langle \beta, b\tilde{\psi}_\lambda \rangle \tilde{\psi}_\lambda\,,$$

where $\langle u, v \rangle = \int u(x)v(x)\,dx$.

We first of all check that the kernel of the operator $T(f) = \pi(\beta, f)$ satisfies the standard estimates. The integral $\int b(x)\tilde{\psi}_\lambda(x)\,dx$ vanishes and we have $|b(x)\tilde{\psi}_\lambda(x)| = O(|x|^{-n-1})$ at infinity. This is the same as saying that $b(x)\tilde{\psi}_\lambda(x)$ is a molecule of the Stein and Weiss space H^1. Note that $\tilde{\psi}_\lambda(x)$ itself is not a molecule and does not even belong to H^1. Lastly, $|\langle \beta, b\tilde{\psi}_\lambda \rangle| \leq C 2^{-nj/2}\|\beta\|_{\mathrm{BMO}}$. The rest is easy and the kernel

$$K(x, y) = \sum_{\lambda \in \Lambda} \omega(\lambda)\gamma(\lambda)2^{nj}\theta(2^j y - k)\tilde{\psi}_\lambda(x)$$

satisfies the usual estimates when $|\gamma(\lambda)| \leq C 2^{-nj/2}$.

We move to the $L^2(\mathbb{R}^n)$ continuity. Taking Theorem 2 into account, we must show that

$$(9.7) \qquad \sum_{\lambda \in \Lambda} |\langle f, \theta_\lambda \rangle|^2 |\langle \beta, b\tilde{\psi}_\lambda \rangle|^2 \leq C^2 \|\beta\|_{\mathrm{BMO}}^2 \|f\|_2^2\,.$$

The proof of (9.7) reduces to verifying the inequality

$$(9.8) \qquad \sum_{Q(\lambda) \subset Q} |\langle \beta, b\tilde{\psi}_\lambda \rangle|^2 \leq C|Q|\,,$$

for each dyadic cube Q. Once (9.8) has been established, Carleson's inequality gives (9.7).

To get (9.8), we refer to the proof of Theorem 4 in *Wavelets and Oper-*

ators, Chapter 5. The estimate (9.8) is a consequence of the "Plancherel formula"

$$(9.9) \qquad \|f\|_2^2 \cong \sum_{\lambda \in \Lambda} |\langle f, b\tilde{\psi}_\lambda \rangle|^2 \,,$$

given by Theorem 2, together with the localization of the wavelets $b\tilde{\psi}_\lambda$ and, lastly, the condition $\int b(x)\tilde{\psi}_\lambda(x)\,dx = 0$, which enables us to eliminate the floating constants which arise in the definition of the space BMO.

We have $T(b) = \beta$, because $\pi(\beta, b) = \sum_{\lambda \in \Lambda} \langle \beta, b\tilde{\psi}_\lambda \rangle \tilde{\psi}_\lambda = \beta$ (from the Corollary of Theorem 2). Also, $^tT(b) = 0$, because $\int \tilde{\psi}_\lambda(x)b(x)\,dx = 0$.

The proof of the $T(b)$ theorem then repeats that of the $T(1)$ theorem word for word. Using pseudo-products to correct T, we reduce to an operator R which satisfies conditions (9.1), (9.2), (9.3) and (9.6) and is such that $R(b) = {}^tR(b) = 0$. The L^2 continuity of such an operator is obtained by calculating the matrix coefficients $\tau(\lambda, \lambda') = \langle R(b\tilde{\psi}_\lambda), b\tilde{\psi}_{\lambda'} \rangle$ and showing that

$$(9.10) \quad |\tau(\lambda, \lambda')| \le C2^{-|j-j'|(n/2+\gamma)} \left(\frac{2^{-j} + 2^{-j'}}{2^{-j} + 2^{-j'} + |k2^{-j} - k'2^{-j'}|} \right)^{n+\gamma} .$$

Once (9.10) has been established, the L^2 continuity of R follows from Schur's lemma.

To prove (9.10), we re-use the method employed for the antisymmetric case. The details are left to the reader.

10 The space H_b^1

As before, we assume that $b \in L^\infty(\mathbb{R}^n)$ and that $\Re e\, b(x) \ge 1$ almost everywhere. The space H_b^1 which we are now going to define is just a simple copy of the usual Hardy space and, similarly, its dual BMO_b is a copy of BMO. However, these spaces have the advantage of a cancellation adapted to the "complex measure" $b(x)\,dx$ and will thus be closely related to the $T(b)$ theorem.

We write $f \in H_b^1$ if the product bf is in the Hardy space $H^1(\mathbb{R}^n)$ (in the real version of Stein and Weiss.) In other words, the functions f in H_b^1 have an atomic decomposition $f(x) = \sum \alpha_j a_j(x)$, where $\sum |\alpha_j| < \infty$ and the functions $a_j(x)$ are atoms in the following sense: there exists a ball B_j corresponding to $a_j(x)$, such that the support of $a_j(x)$ is contained in B_j, the volume $|B_j|$ of B_j satisfies $\|a_j\|_\infty \le |B_j|^{-1}$ and $\int a_j(x)b(x)\,dx = 0$.

The following result is much less obvious.

Proposition 6. *The wavelets $\tilde{\psi}_\lambda$ adapted to $b(x)$ form an unconditional basis of H_b^1.*

We shall prove a more precise result. For every $\lambda = (k + \varepsilon/2)2^{-j} \in \Lambda$, $\varepsilon \in \{0,1\}^n \setminus \{(0,0,\ldots,0)\}$, we let $Q(\lambda)$ be the dyadic cube defined by $2^j x - k \in [0,1)^n$. Then the two following properties are equivalent:

(10.1) $(\sum_{Q(\lambda) \ni x} |\alpha(\lambda)|^2 |Q(\lambda)|^{-1})^{1/2} \in L^1(dx)$;

(10.2) $\sum_{\lambda \in \Lambda} \alpha(\lambda) \tilde{\psi}_\lambda(x) \in H_b^1$.

Indeed, let ψ_λ, $\lambda \in \Lambda$, denote the usual orthonormal wavelet basis (corresponding to $b = 1$ and with regularity $r \geq 1$). We consider the operator J, defined by $J(\psi_\lambda) = \tilde{\psi}_\lambda$. The distribution-kernel of J is thus $\sum_{\lambda \in \Lambda} \tilde{\psi}_\lambda(x)\psi_\lambda(y)$ and the restriction of this distribution-kernel to $x \neq y$ satisfies the estimates which define the Calderón-Zygmund operators. The L^2 continuity of J comes from the fact that ψ_λ and $\tilde{\psi}_\lambda$, $\lambda \in \Lambda$, are unconditional bases of $L^2(\mathbb{R}^n)$. Hence J is a Calderón-Zygmund operator. We let B denote the operator given by pointwise multiplication by $b(x)$ and we consider the operator $T = BJ$. Then ${}^t T(1) = 0$ and Theorem 3 of Chapter 7 implies that T is continuous on the usual H^1 space. This means that J is continuous as an operator from H^1 to H_b^1. We now show that $J(H^1) = H_b^1$, by verifying that $J^{-1}(H_b^1) \subset H^1$.

The distribution-kernel of J^{-1} is $\sum_{\lambda \in \Lambda} \psi_\lambda(x)\tilde{\psi}_\lambda(y)b(y)$. Thus $J^{-1} = LB$, where L is a Calderón-Zygmund operator such that ${}^t L(1) = 0$. It is enough, once again, to apply Theorem 3 of Chapter 7.

Our operator J is thus an isomorphism of H^1 with H_b^1 and the equivalence of (10.1) and (10.2) follows from Theorem 1 of *Wavelets and Operators*, Chapter 5.

In the natural duality between distributions and test functions, the dual of H_b^1 is BMO_b. A function f is in BMO_b if and only if $f = bu$, where $u \in \mathrm{BMO}$. An alternative characterization is that the sequence $\langle f, \tilde{\psi}_\lambda \rangle$, $\lambda \in \Lambda$, satisfies the usual Carleson condition.

The space H_b^1 is related, in a natural way, to the Hardy spaces arising out of complex analysis. Let $\Gamma \subset \mathbb{C}$ be the graph of a Lipschitz function $a(x)$ of a real variable x. Let Ω_2 be the open set "below" Γ and let Ω_1 be the open set "above" Γ (defined by $y > a(x)$). The Hardy space $H^1(\Omega_1)$ consists of the functions $F(z)$ which are holomorphic in Ω_1 and such that

$$\sup_{\tau > 0} \int_\Gamma |F(z + i\tau)| \, ds < \infty.$$

The Hardy space $H^1(\Omega_2)$ is defined similarly. If $F \in H^1(\Omega_1)$, then

$\lim_{\tau \downarrow 0} F(z + i\tau) = F(z)$ exists in $L^1(\Gamma)$ norm and almost everywhere. Further, for any $z_0 \in \Omega_1$,

$$F(z_0) = \frac{1}{2\pi i} \int_\Gamma \frac{F(z)}{z - z_0} \, dz \,,$$

so the boundary values completely determine the function $F \in H^1(\Omega_1)$. All that is left to do is to describe the Banach space consisting of these boundary values.

To do this, we use the parametrization of Γ by the abscissa x (where $z = x + ia(x)$, $-\infty < x < \infty$) and, with $b(x) = 1 + ia'(x)$, we get $F \in H^1(\Omega_1)$ if and only if

(10.3) $$F(z(x)) = f(x) \in H_b^1(\mathbb{R})$$

and, for each $z_2 \in \Omega_2$,

(10.4) $$\int_{-\infty}^\infty \frac{f(x)}{z(x) - z_2} \, z'(x) \, dx = 0 \,.$$

In other words, we may identify $H^1(\Omega_1)$ with a closed subspace of H_b^1. The same applies to $H^1(\Omega_2)$.

Note that H_b^1 is the direct sum of the two closed subspaces $H^1(\Omega_1)$ and $H^1(\Omega_2)$. Indeed, the decomposition $f = f_1 - f_2$ of an arbitrary function $f \in H_b^1$, as the difference of $f_1 \in H^1(\Omega_1)$ and $f_2 \in H^1(\Omega_2)$, may be obtained by applying the Cauchy operator $(2\pi i)^{-1} \int_\Gamma (\zeta - z)^{-1} f(\zeta) \, d\zeta$ to f and then taking boundary values (respectively from above and from below). The continuity of this operator on H_b^1 is a result of its continuity on $L^2(\mathbb{R})$) and of Theorem 3 in Chapter 7. Indeed, the kernel $K(x, y) = z'(x)(z(x) - z(y))^{-1}$ has the necessary regularity with respect to y and satisfies $\int K(x, y) \, dx = 0$, modulo the constant functions. To pass from these formal considerations to a precise statement, it is enough, once more, to approximate the kernel K by regular kernels $K_{\varepsilon, T}$, defined by

$$K_{\varepsilon, T}(x, y) = \frac{z'(x)}{z(x) - z(y) \pm i\varepsilon} - \frac{z'(x)}{z(x) - z(y) \pm iT} \,.$$

A noteworthy subspace of H_b^1 is the space generated by the special atoms, which we denote by $\tilde{B}_1^{0,1}$. This space consists of those functions $f \in H_b^1$ of the form $\sum \alpha(\lambda) \tilde{\psi}_\lambda$, where $\sum_{\lambda \in \Lambda} |\alpha(\lambda)| 2^{-j/2} < \infty$ and the sum of the series gives the norm of f in $\tilde{B}_1^{0,1}$. If we let $\delta(z_0)$ denote the distance from Γ of a point z_0 and we consider the case where $b(x) = z'(x) = 1 + ia'(x)$, with $-M \leq a'(x) \leq M$, then the function $\delta(z_0)(z - z_0)^{-2}$, restricted to Γ, belongs to $\tilde{B}_1^{0,1}$ and the norm of $\delta(z_0)(z - z_0)^{-2}$ in $\tilde{B}_1^{0,1}$ does not exceed $C(M)$. These "special atoms" form a total set in $\tilde{B}_1^{0,1}$. More precisely, each function $f \in \tilde{B}_1^{0,1}$ is of the form

$$(10.5) \qquad f(x) = \sum_0^\infty \frac{\lambda_k \delta(z_k)}{(z(x) - z_k)^2}, \qquad \text{where } z_k \notin \Gamma \text{ and } \sum_0^\infty |\lambda_k| < \infty.$$

The proof of this atomic decomposition is a straightforward adaptation of the proof for the case $a(x) = 0$, which the reader will find in [73].

Conversely, any series as in (10.5) converges to a function in $\tilde{B}_1^{0,1}$.

Using (10.5) we can, finally, decompose $\tilde{B}_1^{0,1}$ into two Bergman spaces $B^1(\Omega_1)$ and $B^1(\Omega_2)$, where $f_j \in B^1(\Omega_j)$ if and only if f_j is holomorphic in Ω_j and $\int\int_{\Omega_j} |f_j(z)| \, dx \, dy < \infty$, for $j = 1, 2$. To get the decomposition, it is enough to define $f_1(x)$ by grouping together all the terms in (10.5) for which $z_k \in \Omega_2$.

Let $B^\infty(\Omega_1)$ denote the Bloch space consisting of the functions which are holomorphic on Ω_1 and such that $\sup_{z \in \Omega_1} \delta(z)|f'(z)|$ is finite.

Consider the bilinear form $B(f, g) = \int_\Gamma f(z)g(z) \, dz$ defined on the ambient Hilbert space $L^2(\Gamma, ds)$. This form corresponds to the special case $b(x) = 1 + ia'(x)$ of the general definition in section 4 of this chapter. Then the dual of the Hardy space $H^2(\Omega_1)$ is precisely $H^2(\Omega_2)$, under the duality given by the bilinear form B. Indeed, $H^2(\Omega_1)$ is a closed linear subspace of $L^2(\Gamma, ds)$ and, for an element λ of the dual space, the Hahn-Banach theorem gives the representation $\lambda(f) = \int_\Gamma f(z)h(s) \, ds$, where $h \in L^2(\Gamma, ds)$. We now put $h(s)/z'(s) = g_1 + g_2$, where $g_1 \in H^2(\Omega_1)$ and $g_2 \in H^2(\Omega_2)$. We can decompose $h(s)/z'(s)$ in this way because of the $L^2(\Gamma)$ continuity of the operator defined by the Cauchy kernel. But $f(z)$ and $g_1(z)$ are in $H^2(\Omega_1)$, so $\int_\Gamma f(z)g_1(z) \, dz = 0$. We are left with $\lambda(f) = \int_\Gamma f(z)g_2(z) \, dz$, as claimed. The converse is obvious.

A similar argument would show that the dual of $H^p(\Omega_1)$ was $H^q(\Omega_2)$, where $1/p + 1/q = 1$, when $1 < p < \infty$.

Finally, the dual of $H^1(\Omega_1)$ is BMO(Ω_2) while $B^\infty(\Omega_2)$ is the dual of $B^1(\Omega_1)$, the duality still being defined by $\int_\Gamma fg \, dz$. The verification of the latter assertion is amusing. The special atoms $\delta(w)(z(x) - w)^{-2}$, $w \in \Omega_2$, generate $B^1(\Omega_1)$ and, to check that a holomorphic function $g : \Omega_2 \to \mathbb{C}$ belongs to the dual of $B^1(\Omega_1)$, it is enough to be sure that

$$(10.6) \qquad \delta(w) \left| \int_\Gamma \frac{g(z)}{(z - w)^2} \, dz \right| \le C.$$

But $\int_\Gamma g(z)(z - w)^{-2} \, dz = -2\pi i g'(w)$, so $\sup_{w \in \Omega_2} \delta(w)|g'(w)| \le C/2\pi$, by (10.6). The fact that we can restrict our considerations to functions which are holomorphic in Ω_2 is a consequence of the duality of $H^2(\Omega_1)$ and $H^2(\Omega_2)$.

It is as if the functions $(z - w)^{-2}$, $w \in \Omega_2$, were wavelets in $H^2(\Omega_1)$ and thus, by duality with $H^2(\Omega_2)$ and other spaces of holomorphic functions on Ω_2, led to an *analysis* of those spaces, classifying them by simple conditions. Of course, the roles of Ω_1 and Ω_2 are symmetric. At the same time, these same functions $(z - w)^{-2}$, $w \in \Omega_2$, are the "atoms", by the approriate linear combinations of which we can *represent* those holomorphic functions on Ω_1.

To end this chain of ideas, $\tilde{B}_\infty^{0,\infty}$ is the natural dual of $\tilde{B}_1^{0,1}$. A distribution S belongs to $\tilde{B}_\infty^{0,\infty}$ if and only if $|\langle S, \tilde{\psi}_\lambda \rangle| \leq C2^{-j/2}$, $\lambda \in \Lambda_j$, $j \in \mathbb{Z}$. In fact, $\tilde{B}_\infty^{0,\infty}$ is defined modulo the function $b(x)$. Then, when $b(x) = 1 + ia'(x)$, every distribution $S \in \tilde{B}_\infty^{0,\infty}$ may be written uniquely as $S = bS_1 + bS_2$, where $S_1 \in B^\infty(\Omega_1)$ and $S_2 \in B^\infty(\Omega_2)$.

11 The general statement of the $T(b)$ theorem

Let $T : \mathcal{D}(\mathbb{R}^n) \to \mathcal{D}'(\mathbb{R}^n)$ be a continuous linear operator. We suppose, firstly, that T is weakly continuous, which is a necessary condition for L^2 continuity. We suppose, further, that the restriction to $x \neq y$ of the distribution-kernel $S(x, y)$ of T is a function $L(x, y)$, satisfying $|L(x, y)| \leq C_0 |x - y|^{-n}$.

Then T has a natural extension to the space of Hölder functions of exponent $\eta > 0$ and compact support. If f and g are two such functions, then $\langle T(f), g \rangle$ is well-defined.

Next, we make use of the function b to state the extra conditions which will be necessary and sufficient for T to be continuous on L^2. We assume that $b \in L^\infty(\mathbb{R}^n)$ and $\Re e\, b(x) \geq 1$. We suppose, moreover, that $L(x, y) = b(x)K(x, y)b(y)$, where $K(x, y)$ satisfies (9.1), (9.2), and (9.3).

We can then state a necessary and sufficient condition for the L^2 continuity of T. This condition involves, once again, $T(1)$ and $^tT(1)$. We must define these as mathematical objects. Letting $\tilde{\psi}_\lambda$, $\lambda \in \Lambda$, denote the collection of the wavelets of Theorem 2 (of regularity $r \geq 1$), we want to make sense of the expression $\langle T(1), \tilde{\psi}_\lambda \rangle$.

To do this, we start with the most academic case of $\langle T(1), u \rangle$, where u is a Lipschitz function of compact support, satisfying $\int u(x)b(x)\,dx = 0$. Then $^tT(u) = \sigma$ is a distribution, but, outside the support of u, we have

$$\sigma(x_0) = \int b(x)K(x, x_0)b(x_0)u(x)\,dx = O(|x_0|)^{-n-\gamma},$$

since we can use the kernel's regularity with respect to x and the fact that $\int u(x)b(x)\,dx = 0$. As a result, the integral $\langle \sigma, 1 \rangle$ converges.

To pass from the special case (compact support) to the general (Lipschitz functions u of exponential decrease and such that $\int u(x)b(x)\,dx = 0$) we use Lemma 6.

The wavelet coefficients $\beta(\lambda) = \langle T(1), \tilde{\psi}_\lambda \rangle$ are thus well-defined and, if these coefficients satisfy Carleson's condition $\sum_{Q(\lambda) \subset Q} |\beta(\lambda)|^2 \leq C_1 |Q|$, we write $T(1) \in \mathrm{BMO}_b$. The same remark holds for the coefficients $\langle {}^t T(1), \tilde{\psi}_\lambda \rangle$.

We have arrived at the statement of the $T(b)$ theorem.

Theorem 5. *With the above notation, a necessary and sufficient condition for T to extend as a continuous linear operator on $L^2(\mathbb{R}^n)$ is that T is weakly continuous, $T(1) \in \mathrm{BMO}_b$ and ${}^t T(1) \in \mathrm{BMO}_b$.*

The proof of Theorem 5 is a simple variant of that of Theorem 4 and we shall not detain the reader with the details.

12 An application to complex analysis

As an application of Theorem 5, let us return to the example of the Cauchy kernel operating on the space $L^2(\Gamma, ds)$, where Γ is a Lipschitz curve. It is no longer necessary to approximate by integrable kernels. Instead, we proceed as follows.

As above, we let $a(x)$ denote a real-valued Lipschitz function of a real variable x. We then have $-M \leq a'(x) \leq M$. We then form the kernel $K(x,y) = z'(x)z'(y)/(z(x) - z(y))$. There exists a constant C_0 such that $|K(x,y)| \leq C_0 |x-y|^{-1}$ and, because the kernel K is also antisymmetric, we can define the distribution $S = \mathrm{PV}\, K$. This distribution belongs to $\mathcal{S}'(\mathbb{R}^2)$ and is the distribution-kernel of a continuous linear operator $T : \mathcal{D}(\mathbb{R}) \to \mathcal{D}'(\mathbb{R})$. To establish that T is continuous on $L^2(\mathbb{R})$, it is enough to verify that $T(1) = 0$, modulo z'.

This identity depends, essentially, on Cauchy's formula. We know that $T(1) \in \tilde{B}_\infty^{0,\infty}$. In other words, $T(1)$ is a continuous linear functional on $\tilde{B}_1^{0,1}$ which we can evaluate by calculating $\langle T(1), a \rangle$, where $a(x) = \delta(z_0)(z(x) - z_0)^{-2}$, $z_0 \notin \Gamma$. Such evaluations are enough, because the set of such atoms is a total subset of $\tilde{B}_1^{0,1}$. Finally,

$$\langle T(1), a \rangle = \lim_{\varepsilon \downarrow 0} \iint_{|\Re e\, z - \Re e\, w| \geq \varepsilon} (z - z_0)^{-2} (z - w)^{-1} \, dz \, dw = 0,$$

as can be seen by integrating first with respect to z and then with respect to w.

13 Algebras of operators associated with the $T(b)$ theorem

We suppose, once more, that $b \in L^\infty(\mathbb{R}^n)$ and that $\Re e\, b(x) \geq 1$. Let T be an operator which satisfies the hypotheses of Theorem 5. Can

T be extended as a continuous linear operator on H_b^1? The expected response is that this will be the case if T is continuous on $L^2(\mathbb{R}^n)$ and if ${}^tT(1) = 0$. But these two conditions lead to a continuous linear operator with domain H_b^1 and range in H^1 (the standard space of Stein and Weiss). To get the conclusion we want, we transform the problem. We set $T = BL$, where B is the operator of pointwise multiplication by $b(x)$.

The distribution-kernel of the operator L, restricted to $x \neq y$, is thus of the form $K(x,y)b(y)$, where $K(x,y)$, as usual, satisfies (9.1), (9.2), and (9.3).

From the continuity of the operator T on the space $L^2(\mathbb{R}^n)$ and from the condition ${}^tL(b) = 0$ (modulo the function b), it follows that L is continuous on the space H_b^1. This can be shown by writing $T = MB$ and applying Theorem 3 of Chapter 7 to M.

This same operator L is continuous on the usual BMO space if $L(1) = 0$, modulo the constant functions, as is shown by the corollary to Theorem 3 of Chapter 7.

Let $\mathcal{L}$ be the collection of operators L of the above type which are continuous on H_b^1 and BMO. We shall verify that $\mathcal{L}$ is a subalgebra of $\mathcal{L}(L^2, L^2)$. To verify this, we make use of the method of Chapter 7. We let $(\tau(\lambda, \lambda'))_{(\lambda,\lambda')\in\Lambda\times\Lambda}$ denote the matrix of L with respect to the wavelet basis ψ_λ, $\lambda \in \Lambda$, of Theorem 2. If $0 < \beta < \gamma$ (γ specifies the regularity with respect to x and y of the kernel $K(x,y)$ corresponding to L), then $\tau(\lambda, \lambda')$ belongs to the algebra $\mathcal{M}_\beta$. Conversely, if this matrix belongs to $\mathcal{M}_\beta$, then the distribution-kernel $S(x, y)$ of L is

$$\sum_\lambda \sum_{\lambda'} \tau(\lambda, \lambda')\tilde{\psi}_\lambda(x)\tilde{\psi}_{\lambda'}(y)b(y)$$

and it is easy (by repeating calculations for the case $b = 1$, word for word) to see that $L \in \mathcal{L}$ (and that $K(x,y) = \sum_\lambda \sum_{\lambda'} \tau(\lambda, \lambda')\tilde{\psi}_{\lambda'}$ satisfies (9.1), (9.2), and (9.3) with $\beta = \gamma$).

Just as in the case $b = 1$, we can introduce the Banach algebras $A(b, \gamma)$, $0 < \gamma < 1$, consisting of the operators L whose matrices with respect to the basis ψ_λ, $\lambda \in \Lambda$, belong to $\mathcal{M}_\gamma$. The definition does not depend on the choice of basis. However, in the general case, the algebras $A(b, \gamma)$ are not self-adjoint. This is what prevents us using Cotlar's lemma in the proof of Theorem 5. In fact, if $T \in A(b, \gamma)$, then T is continuous on H_b^1 and BMO. As a result, the transpose of T is continuous on H^1 and BMO$_b$. This makes it necessary to conjugate the transpose by the operator B defined by pointwise multiplication by $b(x)$, in order to return to an operator belonging to $A(b, \gamma)$.

The algebra $\mathcal{L}$ can equally well be interpreted by the use of vaguelets.

Indeed, $L \in \mathcal{L}$ if and only if, for every family w_λ, $\lambda \in \Lambda$, satisfying (6.3), (6.4), and (6.5), the functions $\tilde{w}_\lambda = L(w_\lambda)$ satisfy the same estimates (with, possibly, different values of the exponents α and β). This way of looking at things has the advantage of showing that the L^2 continuity of the operators $L \in \mathcal{L}$ follows from Proposition 5. Indeed, we take $f \in L^2$ and decompose it as $\sum_{\lambda \in \Lambda} \alpha(\lambda)\tilde{\psi}_\lambda$, with $(\sum_{\lambda \in \Lambda} |\alpha(\lambda)|^2)^{1/2} \leq C\|f\|_2$; then $L(f) = \sum_{\lambda \in \Lambda} \alpha(\lambda)\tilde{w}_\lambda$, where $\tilde{w}_\lambda = L(\tilde{\psi}_\lambda)$. To conclude, it is enough to apply Proposition 5.

14 Extensions to the case of vector-valued functions

We shall need the Hilbert space analogue of Theorem 4 when we study Hardy spaces corresponding to a Lipschitz graph (Theorem 5 of Chapter 12).

For this, we have to introduce the space $\mathrm{BMO}(\mathbb{R}^n, H)$, where H is a Hilbert space. A function $f : \mathbb{R}^n \to H$ belongs to $\mathrm{BMO}(\mathbb{R}^n, H)$ if $\|f(x)\|_H$ is locally square-integrable and if there is a constant C, such that, for each ball $B \subset \mathbb{R}^n$, there exists a vector $\gamma(B) \in H$ with

$$\left(\frac{1}{|B|} \int_B \|f(x) - \gamma(B)\|_H^2 \, dx \right)^{1/2} \leq C.$$

If $f(x)$ belongs to $\mathrm{BMO}(\mathbb{R}^n, H)$ and if $\tilde{\psi}_\lambda$, $\lambda \in \Lambda$, are the wavelets of Theorem 2, then the wavelet coefficients $\alpha(\lambda) = \int f(x)b(x)\tilde{\psi}_\lambda(x) \, dx$ are in H and satisfy Carleson's condition

$$(14.1) \qquad \sum_{Q(\lambda) \subset Q} \|\alpha(\lambda)\|_H^2 \leq C|Q|.$$

Armed with this remark, we repeat the proof of Theorem 4 and obtain the theorem in its vector-valued version, which we now describe.

We start with a Hilbert space H and a kernel $K : \mathbb{R}^n \times \mathbb{R}^n \to H$ such that $(1 + |x - y|)^{n+1} K(x, y) \in L^\infty(\mathbb{R}^n \times \mathbb{R}^n, H)$. We suppose that, for a certain exponent $\gamma \in (0, 1]$ and for a certain constant C_0, we have

$$\|K(x, y)\|_H \leq C_0 |x - y|^{-n},$$

$$\|K(x', y) - K(x, y)\|_H \leq C_0 |x' - x|^\gamma |x - y|^{-n-\gamma},$$

for $|x' - x| \leq |x - y|/2$, and, similarly,

$$\|K(x, y') - K(x, y)\|_H \leq C_0 |y' - y|^\gamma |x - y|^{-n-\gamma},$$

for $|y' - y| \leq |x - y|/2$.

We then suppose that there is a constant C_1 such that, for every ball $B \subset \mathbb{R}^n$,

$$(14.2) \qquad \int_B \left\| \int_B K(x, y)b(y) \, dy \right\|_H dx \leq C_1 |B|.$$

as well as

(14.3)
$$\int_B \left\| \int_B K(x,y)b(x)\,dx \right\|_H dy \le C_1 |B| \,.$$

Then the operator J, defined by $Tf(x) = \int K(x,y)f(y)\,dy$, where the scalar-valued function $f(x)$ is in $L^2(\mathbb{R}^n)$, is a continuous operator from $L^2(\mathbb{R}^n)$ to $L^2(\mathbb{R}^n, H)$ and

(14.4)
$$\|T\| \le C_2 = C_2(C_0, C_1, n, \|b\|_\infty) \,.$$

The proof is just a rewrite of the scalar case.

15 Replacing the complex field by a Clifford algebra

Let $a : \mathbb{R}^n \to \mathbb{R}^n$ be a Lipschitz function. We consider the kernel

$$K(x,y) = \frac{a(x) - a(y) - (x_1 - y_1)\frac{\partial a}{\partial y_1}(y) - \cdots - (x_n - y_n)\frac{\partial a}{\partial y_n}(y)}{[|x - y|^2 + (a(x) - a(y))^2]^{(n+1)/2}} \,.$$

The question that Calderón asked was whether there existed a continuous linear operator $T : L^2(\mathbb{R}^n, dx) \to L^2(\mathbb{R}^n, dx)$ which could be defined by $T(f) = \mathrm{PV} \int K(x,y)f(y)\,dy$, for $f \in L^2(\mathbb{R}^n)$.

The case $n = 2$ is given directly by Theorem 2, since, in this case, $K(x,y)$ is the imaginary part of the Cauchy kernel $(z(x) - z(y))^{-1}\,dz(y)$ associated with the Lipschitz curve $z(x) = x + ia(x)$.

By the results of Chapter 9, we know that T is, indeed, continuous on $L^2(\mathbb{R}^n)$. We shall see in a moment that $\mathrm{PV} \int K(x,y)f(y)\,dy$ exists almost everywhere, for $f \in L^2(\mathbb{R}^n)$.

What we intend to do is to describe a generalization of the $T(b)$ theorem which applies directly to the double-layer potential without involving the method of rotations.

First of all, let us recall the definition of the Clifford Algebra A_m. This algebra is a vector space of dimension 2^m over $\mathbb{R}$. A basis of A_m is consists of vectors e_S, where S is an arbitrary subset of the set $\{1, 2, \ldots, m\}$. For the time being, this is just a notation which lets us assign a subscript to 2^m vectors. We now show how multiplication is defined on this basis.

If $S = \emptyset$, e_S is the unit element of A_m. We want A_m to be an associative algebra generated by $e_1, \ldots, e_m$, where we have writen e_1 instead of $e_{\{1\}}$, etc. The relations in A_m are generated by $e_j^2 = -1$, $1 \le j \le m$, and $e_j e_k = -e_k e_j$, if $j \ne k$. This leads to the rule $e_S e_T = \pm e_R$, where R is the symmetric difference of the arbitrary subsets S and T of $\{1, \ldots, m\}$. It is easy to determine the choice of sign.

We embed the vector space $\mathbb{R}^{m+1}$ in A_m by $(x_0, x_1, \ldots, x_m) \mapsto x = x_0 + x_1 e_1 + \cdots + x_m e_m$. Such elements are called *Clifford numbers*. We then put $\bar{x} = x_0 - x_1 e_1 - \cdots - x_m e_m$ and $|x|^2 = x\bar{x} = x_0^2 + x_1^2 + \cdots + x_m^2$. As a consequence, each non-zero Clifford number is invertible in the Clifford algebra and its inverse is also a Clifford number.

To apply this to the theory of singular integrals, we take a function $b \in L_A^\infty(\mathbb{R}^n)$, whose values are Clifford numbers of $A = A_m$. Then $b(x) = b_0(x) + b_1(x)e_1 + \cdots + b_m(x)e_m$. We suppose that $b_0(x) \geq 1$. If $m = 1$, then A_1 is just the field of complex numbers and the condition reduces to $\Re e\, b(x) \geq 1$.

The operators which we consider are defined by A-valued distribution-kernels $S(x, y)$. They act on A-valued functions in $L_A^2(\mathbb{R}^n)$. This action arises from the asymmetry of the product $S(x, y)f(y)$ calculated in the sense of the multiplication on A.

The norm of $x = \sum_S \alpha_S e_S$ is $|x| = (\sum_S |\alpha_S|^2)^{1/2}$. We suppose that $S(x, y)$, restricted to $x \neq y$, satisfies

$$S(x, y) = b(x)K(x, y)b(y)$$

where

(15.1) $|K(x, y)|_A \leq C_0 |x - y|^{-n}$

(15.2) $|K(x', y) - K(x, y)|_A \leq C_1 |x' - x|^\gamma |x - y|^{-n-\gamma}$,

for $|x' - x| \leq |x - y|/2$, and, similarly,

(15.3) $|K(x, y') - K(x, y)|_A \leq C_1 |y' - y|^\gamma |x - y|^{-n-\gamma}$,

for $|y' - y| \leq |x - y|/2$.

The L^2 continuity of an operator T whose distribution-kernel $S(x, y)$ satisfies the above conditions is equivalent to the following set of conditions:

(15.4) T is weakly continuous on $L_A^2(\mathbb{R}^n)$;

(15.5) $T(1) \in \mathrm{BMO}_b$;

(15.6) $^t T(1) \in \mathrm{BMO}_b$.

The proof of this theorem may be obtained, as P. Auscher obtains it in [3], by adapting the construction of the wavelets $\tilde{\psi}_\lambda$ to the case of the Clifford algebras. We then get two unconditional bases $\tilde{\psi}_\lambda$ and $\tilde{\psi}_\lambda^\#$ of $L_A^2(\mathbb{R}^n)$. The first satisfies $\tilde{\psi}_\lambda \in V_{j+1}$ and $\int \tilde{\psi}_\lambda(x)f(x)b(x)\,dx = 0$, for $\lambda \in \Lambda_j$ and $f \in V_j$ (all the functions are A-valued). The second satisfies the transposed condition $\int b(x)f(x)\tilde{\psi}_\lambda^\#(x)\,dx = 0$, for $\lambda \in \Lambda_j$ and $f \in V_j$ (still A-valued.) Finally, $\int \tilde{\psi}_\lambda(x)b(x)\tilde{\psi}_{\lambda'}^\#(x)\,dx = 0$, whenever $\lambda \neq \lambda'$ and 1, when $\lambda = \lambda'$. Once these bases have been constructed, the proof of the scalar case can be adapted without difficulty.

To apply these considerations to the problem of the double-layer potential, we consider the kernel $L(x, y)$ which is defined on $\mathbb{R}^n \times \mathbb{R}^n$ when $x \neq y$ by

$$L(x, y) = \frac{(x_1 - y_1)e_1 + \cdots + (x_n - y_n)e_n - (a(x) - a(y))}{[|x - y|^2 + (a(x) - a(y))^2]^{(n+1)/2}},$$

where $a : \mathbb{R}^n \to \mathbb{R}$ is a Lipschitz function and $L(x, y)$ takes its values in the vector space of Clifford numbers of A_n.

We consider the function

$$b(x) = 1 - \frac{\partial a}{\partial x_1} e_1 - \cdots - \frac{\partial a}{\partial x_n} e_n$$

and form the antisymmetric kernel

$$K(x, y) = b(x)L(x, y)b(y).$$

This kernel defines a distribution PV $K(x, y) \in \mathcal{S}'(\mathbb{R}^n \times \mathbb{R}^n)$ having values in the algebra A_n and we let T denote the operator whose distribution-kernel is PV $K(x, y)$. Property (15.4) is clearly satisfied and the L^2 continuity of T will be a consequence of the identity $T(1) = 0$, which we shall now establish.

To do this, we use the following identity (which we shall need in Chapter 15):

$$\text{PV} \int L(x, y)b(y)f(y)\,dy$$

(15.7)
$$= -\int \left\{ \frac{1}{(n-1)} [|x - y|^2 + (a(x) - a(y))^2]^{(n-1)/2} \right.$$

$$\left. + \sum_1^n \frac{y_j - x_j}{|y - x|^n} e_j \lambda \left(\frac{a(y) - a(x)}{|y - x|} \right) \right\} \sum_1^n \frac{\partial f}{\partial y_j} e_j \, dy,$$

where λ is the odd function whose derivative is $(1 + t^2)^{-(n+1)/2}$.

Identity (15.7) is an immediate consequence of Green's formula and may be written in the more condensed form

$$T = (R_0 + R_1 e_1 + \cdots + R_n e_n)(e_1 \partial_1 + \cdots + e_n \partial_n)$$

where the R_j are locally integrable kernels. More precisely, $|R_j(x, y)| \leq C|x-y|^{-n+1}$, $|(\partial/\partial x_k)R_j(x, y)| \leq C|x-y|^{-n}$, $|(\partial/\partial y_k)R_j(x, y)| \leq C|x-y|^{-n}$, and $|(\partial^2/\partial x_k \partial y_l)R_j(x, y)| \leq C|x - y|^{-n-1}$. It follows that $T(1) = 0$.

16 Further remarks

After they established the $T(b)$ theorem, David, Journé, and Semmes looked for the most general possible condition on a function $b \in L^\infty(\mathbb{R}^n)$ which would lead to the L^2 continuity of *all* the singular integral operators satisfying the conditions of the $T(b)$ theorem.

First of all, they showed that the condition $\Re e\, b(x) \geq \delta > 0$ could be weakened to $\inf_Q |Q|^{-1}|\int_Q b(x)\, dx| \geq \delta > 0$, the lower bound being taken over all cubes Q in $\mathbb{R}^n$. This new sufficient condition has the advantage of applying to the study of the Cauchy kernel on a Lavrentiev curve whose parametric representation by arc length is denoted by $z(t)$, $-\infty < t < \infty$. Then $Q = [s,t]$ and we do have $(t-s)^{-1}|z(t) - z(s)| \geq \delta > 0$ with $b(t) = z'(t)$.

Finally, David, Journé, and Semmes discovered the necessary and sufficient condition on the function $b(x) \in L^\infty(\mathbb{R}^n)$ for the L^2 continuity of all the singular integral operators satisfying the hypotheses of the $T(b)$ theorem. This condition is that there exist constants $\gamma > 0$ and $\delta > 0$ such that each cube $Q \subset \mathbb{R}^n$ has a subcube Q_1 of volume $|Q_1| \geq \gamma|Q|$ for which the inequality $|Q_1|^{-1}|\int_{Q_1} b(x)\, dx| \geq \delta$ holds ([97], [98]).

Tchamitchian's construction has been the starting point for very active lines of research, for the most part as yet unpublished. R. Coifman, P. Jones, and S. Semmes modify the Haar system h_I, $I \in J$, in the simplest and naivest possible way, to construct $h_I^\#(x)$ satisfying $\int h_I^\#(x) b(x)\, dx = 0$. They decide to put $h_I^\#(x) = \alpha_I$ on the left half of I and $h_I^\#(x) = \beta_I$ on the right half. The support of $h_I^\#$ is I. By a judicious choice of these constants, $h_I^\#$, $I \in J$, is a Riesz basis of $L^2(\mathbb{R})$ and can replace Tchamitchian's basis in the proof of the $T(b)$ theorem ([64]).

12

Generalized Hardy spaces

1 Introduction

Let Γ be a rectifiable Jordan curve in the complex plane, passing through the point at infinity. Let $z(s)$ denote the arc-length parametrization of Γ. Then $\lim_{s \to \pm\infty} |z(s)| = \infty$. We shall, in fact, suppose that, whenever z_0 lies off Γ, then, for any $1 < p < \infty$, we have $\int_\Gamma |z(s) - z_0|^{-p} \, ds < \infty$. It will be convenient to suppose that Γ is oriented. The Jordan curve theorem tells us that the complement of Γ in $\mathbb{C}$ splits into two connected components Ω_1 and Ω_2.

What we shall study is the generalized Hardy space $H^p(\Omega_1)$, which consists of the functions $F(z)$ which are holomorphic in Ω_1 and have, in a very precise sense, a trace $\theta(F)$ on Γ which belongs to $L^p(\Gamma, ds)$. The space $H^p(\Omega_1)$ can be identified with a closed subspace of $L^p(\Omega_1)$, which we shall also, by abuse of language, call $H^p(\Omega_1)$. We recover the holomorphic function $F : \Omega_1 \to \mathbb{C}$ from its trace $f = \theta(F)$ by Cauchy's formula

$$(1.1) \qquad F(z) = \frac{1}{2\pi i} \int_\Gamma \frac{f(\zeta)}{\zeta - z} \, d\zeta \,,$$

when Ω_1 is the domain "to the left of" the oriented curve Γ.

In fact, matters are not quite as simple as we have suggested and there are three possible definitions of $H^p(\Omega_1)$. The definitions coincide when Ω_1 is a Smirnov domain; it is known that this condition does not depend on the exponent $p \in (1, \infty)$.

Calderón's problem is to know whether, for $1 < p < \infty$, the space

$L^p(\Gamma)$ is the direct sum of the spaces $H^p(\Omega_1)$ and $H^p(\Omega_2)$, when these spaces are realized as subspaces of $L^p(\Gamma)$ by means of their respective trace operators θ_1 and θ_2.

Thanks to a theorem of David ([93]), we now can characterize the curves Γ for which Calderón's problem has a positive response. The curves in question are those which are regular in the sense of Ahlfors: they are rectifiable curves such that, for a certain constant $C > 1$, for all $z_0 \in \mathbb{C}$, and for all $R > 0$, the measure of the set $E(z_0, R)$ of those $s \in \mathbb{R}$ for which $|z(s) - z_0| \leq R$ is not greater than CR. This condition requires Γ not to have "too many zigzags" and is violated, for example, by the rectifiable graph $|x|^\alpha \sin(1/x)$, when $1 < \alpha < 2$.

The object of this chapter is to prove David's theorem. There are now two ways to achieve this. One uses the resources of real analysis and has already been alluded to in Chapter 9. The other is based on complex analysis and gives the simplest proof of the continuity of the Cauchy kernel on Lipschitz curves.

David's proof is a geometric variant of the methods introduced by D. Burkholder and R. Gundy under the name of "good λ inequalities". This proof needs a starting point: the continuity of the Cauchy kernel on Lipschitz curves.

We shall therefore start with the special case of Hardy spaces corresponding to a Lipschitz domain. Calderón's theorem, either in the form of the identity $L^2(\Gamma) = H^2(\Omega_1) + H^2(\Omega_2)$, or as the continuity of the projection of $L^2(\Gamma)$ onto $H^2(\Omega_1)$, will get three different proofs (among them the so-called "shortest proof" of P. Jones and S. Semmes). We shall then follow David and prove his theorem by real-variable methods. On the way, we shall show how the algebras of operators defined by the $T(b)$ theorem are related to algebras associated with singular holomorphic kernels $K(z, w)$ acting on $L^2(\Gamma)$.

2 The Lipschitz case

Once and for all, $\Gamma \subset \mathbb{C}$ is the graph of a Lipschitz function $a : \mathbb{R} \to \mathbb{R}$. We have $ds = \left(1 + a'^2(x)\right)^{1/2} dx$ and, if $-M \leq a'(x) \leq M$, it follows that $dx \leq ds \leq (1 + M^2)^{1/2} dx$. The space $L^2(\Gamma, ds)$ may thus be identified with $L^2(\mathbb{R}, dx)$.

We let Ω_1 denote the open set given by $y > a(x)$ and let Ω_2 be that given by $y < a(x)$.

The definition of the Hardy spaces relies on the idea of parallel curves. We observe that, for every $\tau > 0$, $\Gamma + i\tau$ lies in Ω_1, which allows us to make the following definition.

Definition 1. *Let $F(z)$ be a function which is holomorphic on Ω_1. We say that $F(z) \in H^p(\Omega_1)$, for some $0 < p < \infty$, if*

$$\sup_{\tau > 0} \left(\int_{\Gamma + i\tau} |F(z)|^p \, ds \right)^{1/p} = \|F\|_{H^p(\Omega_1)} < \infty \,.$$

We could equally well have written $\sup_{\tau > 0} (\int_\Gamma |F(z + i\tau)|^p \, ds)^{1/p} < \infty$.

For the time being, we shall restrict attention to the case $1 < p < \infty$ and shall prove the following theorem.

Theorem 1. *Let $1 < p < \infty$. If $F \in H^p(\Omega_1)$, then the functions $F_\tau \in L^p(\Gamma)$, defined by $F_\tau(z) = F(z + i\tau)$, where $z \in \Gamma$ and $\tau > 0$, converge, almost everywhere and in L^p norm, to a function, which we denote by $\theta(F)$ and call the trace of F on Γ.*

The trace operator $\theta : H^p(\Omega_1) \to L^p(\Gamma)$ is an isomorphism of $H^p(\Omega_1)$ with a closed subspace of $L^p(\Gamma)$, which, by abuse of notation, we again denote by $H^p(\Omega_1)$.

For each function $f \in L^p(\Gamma, ds)$ the following conditions are equivalent:

(2.1) *there exists a sequence $R_j(z)$, $j \geq 1$, of rational functions which vanish at infinity, are holomorphic in a neighbourhood of $\overline{\Omega}_1$, and converge to f in $L^p(\Gamma, ds)$ norm;*

(2.2) $$\int_\Gamma \frac{f(\zeta)}{\zeta - z} \, d\zeta = 0 \qquad \text{for all } z \in \Omega_2;$$

(2.3) $$f \in H^p(\Omega_1) \,.$$

Finally, if the equivalent properties (2.1), (2.2), or (2.3) are satisfied, we have, for all $z \in \Omega_1$,

(2.4) $$F(z) = \frac{1}{2\pi i} \int_\Gamma \frac{f(\zeta)}{\zeta - z} \, d\zeta \qquad \text{where } f = \theta(F) \,.$$

The meaning of the equivalence of (2.1) and (2.2) is that Ω_1 is a Smirnov domain. It is not true that (2.2) implies (2.1) when Γ is an arbitrary rectifiable Jordan curve, as Lavrentiev has shown ([155]).

The deep meaning of Theorem 1 is that it is a form of the "maximum principle" for functions $F(z) \in H^p(\Omega_1)$. What happens is as follows.

If we know, a priori, that $F(z)$ belongs to $H^p(\Omega_1)$, without being in a position to calculate the $H^p(\Omega_1)$ norm of $F(z)$, then we may use the estimate given by $(\int_\Gamma |f(z)|^p \, ds)^{1/p}$. If, on the other hand, we lack that a priori information (namely, that $F \in H^p(\Omega_1)$), this procedure may not work. The classical counter-example is $\Omega_1 = \{x + iy : y > 0\}$ with $F(z) = (z + i)^{-1} e^{-iz}$, whose boundary values are in $L^2(\mathbb{R})$ but which does not belong to $H^2(\Omega_1)$.

Much of the line of argument we take in this section is due to the above difficulty.

The proof of Theorem 1 relies on an approximate version of (2.4), given by the following lemma. (We shall write Ω for Ω_1 and Γ_τ for $\Gamma + i\tau$, to simplify the notation.)

Lemma 1. *Let $F(z)$ be in $H^p(\Omega)$. Then, for each $\tau > 0$ and every $z_0 = x_0 + ia(x_0) + i\sigma \in \Omega$,*

$$(2.5) \qquad F(z_0) = \frac{1}{2\pi i} \int_{\Gamma_\tau} \frac{F(\zeta)}{\zeta - z_0} \, d\zeta \qquad \text{if } \sigma > \tau$$

and

$$(2.6) \qquad 0 = \frac{1}{2\pi i} \int_{\Gamma_\tau} \frac{F(\zeta)}{\zeta - z_0} \, d\zeta \qquad \text{if } \sigma < \tau.$$

To establish these two identities, we reduce to the standard Cauchy formula, replacing Γ_τ by the oriented boundary of the rectangle given by $-R \le x \le R$ and $a(x) + \tau \le y \le a(x) + \tau'$. Then we let R and τ' tend to infinity.

The integral along the top boundary can be bounded directly by

$$\int_\Gamma \frac{|F(z + i\tau')|}{|z + i\tau' - z_0|} \, ds \le \left(\int_\Gamma |F(z + i\tau')|^p \, ds \right)^{1/p} \varepsilon(\tau'),$$

where, for $1/p + 1/q = 1$,

$$\varepsilon(\tau') = \left(\int_\Gamma |z + i\tau' - z_0|^{-q} \, ds \right)^{1/q},$$

which tends to 0 as τ' tends to infinity.

The integrals on the vertical boundaries are dealt with by a technique described in Titchmarsh's treatise ([229]), which we now describe. We make the vertical boundaries "vibrate", by letting R run through the interval $[M, M+1]$. We then take the means of the Cauchy formulas we have used. This lets us replace the integrals by double integrals which are bounded above, when we take the definition of $H^p(\Omega)$ into account. The reader is referred to [229].

The proof of (2.6) is identical and is left to the reader.

The second ingredient, which we shall use systematically, is the following observation.

Lemma 2. *Let $K_\tau(z, w)$, with $z, w \in \Gamma$, be a function satisfying*

$$|K_\tau(z, w)| \le C \frac{\tau}{|z - w|^2 + \tau^2}$$

and

$$\int_\Gamma K_\tau(z, w) \, dw = 1.$$

Then, for every $f \in L^p(\Gamma)$,

(2.7)
$$\left| \int_\Gamma K_\tau(z,w) f(w) \, dw \right| \le C' f^*(z) \,,$$

where $f^*(z)$ is the Hardy-Littlewood maximal function of f. Further,

(2.8)
$$\lim_{\tau \downarrow 0} \int_\Gamma K_\tau(z,w) f(w) \, dw = f(z) \,,$$

both in $L^p(\Gamma)$ and almost everywhere on Γ.

The proof is similar to that of Lemma 5 in Chapter 7.
This lemma will be applied to the kernel

$$K_\tau(z,w) = \frac{1}{2\pi i} \left(\frac{1}{z - w - i\tau} - \frac{1}{z - w + i\tau} \right) \,.$$

It is easy to see that this kernel satisfies the conditions of the lemma.
We return to our examination of the Hardy spaces.
We consider the sequence $F(z + i/m)$, where $m \ge 1$ and $z \in \Gamma$. Since this sequence is bounded in $L^p(\Gamma)$, we can extract a subsequence which converges to $f \in L^p(\Gamma)$ in the $\sigma(L^p, L^q)$ topology, where $1/p + 1/q = 1$.
We put $\tau = m_j^{-1}$ in (2.5) and (2.6). Since $(\zeta - z_0)^{-1}$ lies in $L^q(\Gamma)$, when $z_0 \notin \Gamma$, we can pass to the limit to get

$$F(z_0) = \frac{1}{2\pi i} \int_\Gamma \frac{f(\zeta)}{\zeta - z_0} \, d\zeta \,, \qquad \text{when } z_0 \in \Omega,$$

and

$$0 = \frac{1}{2\pi i} \int_\Gamma \frac{f(\zeta)}{\zeta - z_0} \, d\zeta \,, \qquad \text{when } z_0 \notin \bar{\Omega} \,.$$

We then put $z_0 = x_0 + ia(x_0) \pm i\tau$, with $\tau > 0$, and subtract the second identity from the first to get

(2.9)
$$F(z + i\tau) = \int_\Gamma K_\tau(z,w) f(w) \, dw \,, \qquad z \in \Gamma$$

(where $z = x_0 + ia(x_0)$, and we have written w instead of ζ). It is now enough to apply Lemma 2 to get the existence of the trace operator $\theta : H^p(\Omega) \to L^p(\Omega)$. We thus have $f = \theta(F)$.

Identity (2.9), together with the Hardy-Littlewood theorem on the maximal function, gives the estimate

(2.10)
$$\| \sup_{\tau > 0} |F(z + i\tau)| \|_{L^p(\Gamma)} \le C(M, p) \|\theta(F)\|_{L^p(\Gamma)} \,,$$

for all $F \in H^p(\Omega)$.

A fortiori, the norm of F in $H^p(\Omega)$ is equivalent to the L^p norm of the trace $\theta(F)$. In other words, $H^p(\Omega)$ is equivalent to a closed subspace of $L^p(\Gamma)$.

Before continuing, we have a natural break, in the form of the next lemma.

Lemma 3. *If $F \in H^p(\Omega)$ and $G \in H^q(\Omega)$, where $1/p + 1/q = 1$ and $1 < p < \infty$, then*

$$(2.11) \qquad\qquad \int_\Gamma F(z)G(z)\, dz = 0.$$

By abuse of language we have written, and shall continue to write, F instead of $\theta(F)$, when it is clear that the corresponding integral is taken on Γ.

To prove (2.11), we calculate the integral as a limit of the corresponding integrals on Γ_τ, for $\tau > 0$. So it is enough to show that $\int_{\Gamma_\tau} F(z)G(z)\, dz = 0$. To do this, we use the same change of contour as in Lemma 1 to show that the value of this integral is independent of τ. Finally, we note that $\lim_{\tau \to \infty}(\int_{\Gamma_\tau} |F(z)|^p\, ds)^{1/p} = 0$, indeed, if $z \in \Gamma$, $|F(z + i\tau)| \leq CF^*(z) \in L^p(\Gamma)$, which allows us to apply Lebesgue's dominated convergence theorem.

Thus $\lim_{\tau \to \infty} \int_{\Gamma_\tau} F(z)G(z) = 0$, which concludes the proof of the lemma.

We return to the proof of Theorem 1, letting $\mathcal{R}(\Omega)$ denote the algebra of the restrictions to Ω of rational functions which vanish at infinity and whose poles lie outside $\overline{\Omega}$.

To characterize the closure of $\mathcal{R}(\Omega)$ in $L^p(\Gamma)$, we use the Hahn-Banach theorem. Let $g \in L^q(\Gamma)$ be a function such that $\int_\Gamma g(z)f(z)\, ds = 0$, for every $f \in \mathcal{R}(\Omega)$. Since $dz = z'(s)\, ds$ and $|z'(s)| = 1$, we may replace $g(z)$ by $h(z) = g(z)/z'(s)$. Then $\int_\Gamma h(z)f(z)\, dz = 0$, for $f \in \mathcal{R}(\Omega)$.

In particular, $\int_\Gamma h(z)(z - z_0)^{-1}\, dz = 0$, when $z_0 \notin \overline{\Omega}$. We then put

$$H(z) = \frac{1}{2\pi i} \int_\Gamma \frac{h(w)}{w - z}\, dw = \frac{1}{2\pi i} \int_\Gamma \left(\frac{1}{w - z} - \frac{1}{w - z^\star} \right) h(w)\, dw\,,$$

where $z = u + ia(u) + iv \in \Omega$ and $z^\star = u + ia(u) - iv \notin \overline{\Omega}$. Lemma 2 applies again and

$$\sup\{|H(z)| : \Re e\, z = u, z \in \Omega\} \leq Ch^*(u + ia(u)) \in L^q(\Gamma)\,.$$

A fortiori, $h(z)$ lies in $H^q(\Omega)$.

We have not finished the proof yet, because we still must show that $\int_\Gamma h(z)f(z)\, dz = 0$, for $f \in H^p(\Omega)$. But that is done by Lemma 3. Our proof has also furnished the required characterization of $H^p(\Omega)$.

It would be pleasant to be able to know that a function belongs to $H^p(\Omega)$ as a result of conditions couched purely in terms of the trace of the function on the boundary of Ω.

A well-known counter-example is given by $F(z) = e^{-iz}/(z + i)$, when $\Gamma = \mathbb{R}$. This function is holomorphic on $\Im m\, z > 0$, continuous on $\Im m\, z \geq 0$, and has a trace (in the obvious sense) on the real axis. What is more, the trace is the limit of the functions $F(x + i\varepsilon)$, for $\varepsilon > 0$. But

$F(z)$ does not lie in $H^2(\Omega)$, because $F(x+i\tau) = e^{\tau}e^{-ix}/(x+i\tau+i)$: the norm of this function is $\sqrt{\pi}e^{\tau}/\sqrt{(\tau+1)}$, which tends to infinity with τ. However, we have the following result.

Lemma 4. *Let $F(z)$ be a function which is holomorphic and bounded on Ω. Then F has a trace on $\Gamma = \partial\Omega$: $\lim_{\tau\downarrow 0} F(z+i\tau)$ exists for almost all $z \in \Gamma$. If the trace belongs to $L^p(\Gamma)$, then $F(z)$ lies in $H^p(\Omega)$.*

To see this, we first replace $F(z)$ by $F_\varepsilon(z) = F(z)/(1 - i\varepsilon z)$, which does lie in $H^p(\Omega)$, since F is bounded on Ω. Hence the trace of $F_\varepsilon(z)$ exists, which gives the first part of the lemma.

Theorem 1 gives

$$(2.12) \quad \left(\int_{\Gamma_\tau}|F_\varepsilon(z)|^p\,ds\right)^{1/p} \leq C\left(\int_{\Gamma}|F_\varepsilon(z)|^p\,ds\right)^{1/p} \leq C\left(\int_{\Gamma}|F(z)|^p\,ds\right)^{1/p}.$$

We just have to let ε tend to 0 to obtain the desired conclusion.

For our final definition of the H^p spaces, we shall use a conformal representation. This definition has the advantage of generalizing to the case of domains with rectifiable boundary.

3 Hardy spaces and conformal representations

Let Ω be a bounded, open, connected, and simply-connected subset of the complex plane. Then there exists a conformal representation $\Phi : D \to \Omega$, where D is the open unit disk. Furthermore, if Ω is a Jordan domain (that is, Ω is the bounded connected component of the comlement of a closed Jordan curve Γ), then Φ extends to a homeomorphism of $\bar{D}$ onto $\bar{\Omega}$. If $|z_0| = 1$ and $z_1 \in \Gamma$, we can impose the condition $\Phi(z_0) = z_1$ on the choice of Φ.

We can take the composition of Φ with two bilinear transformations to get a conformal representation of the open upper half-plane $P = \{z \in \mathbb{C} : \Im m\, z > 0\}$ onto an open set Ω whose boundary is an oriented Jordan curve Γ passing through the point at infinity. We shall also use Φ to denote this conformal representation, which extends to a homeomorphism of $\mathbb{R}$ onto Γ. This homeomorphism is increasing if Ω lies "to the left" of the oriented curve Γ.

Let $a : \mathbb{R} \to \mathbb{R}$ be a Lipschitz function (so $\|a'\|_\infty \leq M < \infty$). Let Γ be the graph of a, oriented from left to right. $\Omega = \{(x,y) : y > a(x)\}$ is the open set "above" Γ (that is, to the left of the oriented curve Γ).

Let Σ be the sector of the complex plane defined by $x \geq 0$ and $|y| \leq Mx$.

The next theorem, due to Calderón and improved by C. Kenig, tells us that, for each $y > 0$, the curves $\Phi(x + iy)$, $x \in \mathbb{R}$, are graphs of Lipschitz functions of slope not greater than M.

More precisely and keeping the notation above, we have the following result.

Theorem 2. *We can choose a branch of* $\log \Phi'(z)$ *such that*
$$|\Im m \, \log \Phi'(z)| \leq \theta_0 = \tan^{-1} M$$
for all $z \in P$.

To prove Theorem 2, we start with the case where Γ is a polygonal line ending with two half-lines whose slopes do not exceed M. Then the conformal representation Φ is provided by the Schwarz-Christoffel formulas ([211]). We can find N real numbers $c_1 < c_2 < \cdots < c_N$ (N is the number of vertices of Γ), a real number γ, and N real exponents γ_j, $1 \leq j \leq N$, such that $\Phi'(z) = e^{i\gamma} \prod_1^N (z - c_j)^{\gamma_j}$. We choose z^γ so that is holomorphic in P with $1^\gamma = 1$. The "angles" γ and γ_j are then related to the slopes of the polygonal line Γ: indeed, if $c_j < x < c_{j+1}$, we have $\arg \Phi'(x) = \gamma + \pi(\gamma_{j+1} + \cdots + \gamma_N)$; if $x < c_1$, we have $\arg \Phi'(x) = \gamma + \pi(\gamma_1 + \cdots + \gamma_N)$; and if $x > c_N$, we have $\arg \Phi'(x) = \gamma$. Thus
$$|\gamma_1 + \cdots + \gamma_N| \leq 2\theta_0/\pi = r < 1.$$

As a consequence, $|\Phi'(z)| = O(|z|^r)$ at infinity and, if $z = u + iv$, $v > 0$, we have

$$(3.1) \qquad \Phi'(z) = \frac{1}{\pi} \int_{-\infty}^\infty \frac{v}{(x - u)^2 + v^2} \Phi'(x) \, dx \, .$$

The complex numbers $\Phi'(x)$ lie in the sector Σ (by construction). Since Σ is a convex sector and $\Phi'(z)$ is a convex linear combination of the $\Phi'(x)$, the number $\Phi'(z)$ also lies in Σ.

We arrive at the conclusion of Theorem 2 by composing $\log$ (defined on a conical neighbourhood (excluding 0) of Σ) with $\Phi'(z)$.

We can pass to the general case by approximating Γ by a sequence Γ_j of polygonal lines such that the open sets Ω_j above Γ_j increase to Ω. To achieve this, it is enough that the Γ_j are graphs of piecewise affine functions $a_j(x)$ which decrease to $a(x)$.

To construct the $a_j(x)$, we consider the subdivision formed by the points $x = k2^{-j}$, where $k \in \mathbb{Z}$ and $|k| \leq j2^j$. We put $y(k, j) = a(k2^{-j}) + 2M2^{-j}$ and we let Γ_j denote the polygonal line whose nodes are $(k2^{-j}, y(k, j))$ and whose ends are formed by half-lines of slope M and $-M$, respectively.

It follows immediately that the functions $a_j(x)$ form a sequence which decreases, uniformly on each compact set, to $a(x)$.

We then establish that the conformal representations Φ_j converge to

Φ, by returning to the case where P is replaced by the open unit disk D and Ω becomes a bounded open set. We use the following remark (which is a consequence of Schwarz's lemma).

Lemma 5. *Let D be the open unit disk, let Ω be a bounded open simply-connected set and let Ω_j be an increasing set of simply-connected open sets such that $\Omega = \bigcup_{j \geq 0} \Omega_j$.*

Fix $z_0 \in \Omega_0$ and let $\Phi_j : D \to \Omega_j$ be the conformal representation normalized by $\Phi_j(0) = z_0$ and $\Phi'_j(0) > 0$.

Then the sequence $\Phi'_j(0)$ is increasing and the functions $\Phi_j(z)$ converge uniformly on each compact subset of D to the conformal representation $\Phi : D \to \Omega$, normalized by $\Phi(0) = z_0$ and $\Phi'(0) > 0$.

Coming back to the case of P and an unbounded Ω, we have $\Phi'_j(z) \in \Sigma$, so $\Phi'(z) \in \Sigma$ and the theorem has been proved.

Theorem 2 gives $\log \Phi'(z) = u(z) + iv(z)$, where $\sup_{z \in P} |v(z)| \leq \theta_0 < \pi/2$. The function $v(z)$ is harmonic on P and $v(i(1-z)/(1+z))$ is harmonic on the unit disk $|z| < 1$. We now use the following lemma.

Lemma 6. *Let $f(z) = u(z) + iv(z)$ be a function which is holomorphic on $|z| < 1$. Suppose that $\sup_{|z|<1} |v(z)| < \infty$. Then*

$$f(z) = \frac{i}{2\pi} \int_0^{2\pi} \frac{e^{i\theta} + z}{e^{i\theta} - z} v(e^{i\theta}) \, d\theta + C \,,$$

where C is a real constant.

The proof is simple. We know that a bounded harmonic function is the Poisson integral of its boundary values. So we get

$$v(z) = \frac{1}{2\pi} \int_0^{2\pi} \frac{1 - |z|^2}{|e^{i\theta} - z|^2} v(e^{i\theta}) \, d\theta \,.$$

We then put

$$g(z) = \frac{i}{2\pi} \int_0^{2\pi} \frac{e^{i\theta} + z}{e^{i\theta} - z} v(e^{i\theta}) \, d\theta \,.$$

By the construction, $\Im m \, g(z) = v(z)$. On the other hand, $g(z)$ is clearly holomorphic in $|z| < 1$. The function $g(z) - f(z)$ is holomorphic in $|z| < 1$ and real-valued. It is therefore constant.

In the case of the half-plane, $v(x)$ exists, satisfies $\|v\|_{L^\infty(\mathbb{R})} \leq \theta_0 < \pi/2$ and, for $z = x + iy$,

$$(3.2) \qquad v(z) = \frac{1}{\pi} \int_{-\infty}^{\infty} \frac{y}{(x-t)^2 + y^2} v(t) \, dt \,.$$

Finally, the lemma gives

$$(3.3) \qquad \log \Phi'(z) = \frac{1}{\pi} \int_{-\infty}^{\infty} \left(\frac{1}{t - z} - \frac{t}{t^2 + 1} \right) v(t) \, dt + C \,,$$

where C is a real constant.

Conversely, let us begin with a real-valued function $v \in L^\infty(\mathbb{R})$, satisfying $\|v\|_\infty < \pi/2$, and form

$$F(z) = \frac{1}{\pi} \int_{-\infty}^{\infty} \left(\frac{1}{t-z} - \frac{t}{t^2+1} \right) v(t) \, dt.$$

The problem is to know whether we can obtain a conformal representation $\Phi : P \to \Omega$, by defining Φ by $\Phi'(z) = e^{F(z)}$, and whether Ω is the open set above a Lipschitz graph Γ. To deal with this problem, we use Rouché's theorem, in the following form. Suppose that a function $U(z)$ is holomorphic in the open disk $|z| < 1$ and that $U(e^{i\theta})$, $0 \leq \theta \leq 2\pi$, is a parametric representation of a closed Jordan curve Γ. Then $U(z)$ is a conformal representation of D on the interior Ω of Γ.

Rouché's theorem can obviously be generalized to the case where D is replaced by the upper half-plane P. There we have to make the following assumptions: $\Phi(x)$, $-\infty < x < \infty$, must be the parametric representation of a Jordan curve Γ and we must have $\lim_{|z| \to \infty} |\Phi(z)| = \infty$.

Returning to our problem, we shall take the limit of a sequence of compactly supported step functions v_j, $j \in \mathbb{N}$, which tend to v and satisfy $\|v_j\|_\infty \leq \|v\|_\infty$.

We start with the case of the v_j, omitting the index j, but making sure that our estimates are uniform in j.

On each interval $[a_k, a_{k+1})$ on which $v(t) = \theta_k$, $|\theta_k| < \pi/2$, we also have $\log \Phi'_k(t) = u(t) + i\theta_k$, and the curve Γ, of which $\Phi(t)$ is a parametric representation, is a Lipschitz graph in the form of a polygonal line with a finite number of sides and beginning and ending with a half-line. Rouché's theorem applies and Φ is the conformal representation given by the Schwarz-Christoffel formulas ([211]). In particular,

$$(3.4) \qquad \Phi'(z) = \frac{1}{\pi} \int_{-\infty}^{\infty} \frac{y}{(x-t)^2 + y^2} \Phi'(t) \, dt.$$

Since the sector Σ is convex,

$$(3.5) \qquad \Phi'(z) \in \Sigma \qquad \text{for all } z \in P.$$

Following Calderón's argument ([38]), we shall show that $\omega(t) = |\Phi'(t)|$ satisfies Muckenhoupt's A_2 condition, with constants independent of j.

We first observe that $G(z) = 1/\Phi'(z)$ satisfies

$$(3.6) \qquad G(z) = \frac{1}{\pi} \int_{-\infty}^{\infty} \frac{y}{(x-t)^2 + y^2} G(t) \, dt.$$

In fact, the change from Φ' to $1/\Phi'$ corresponds to that of v to $-v$ in

(3.3). Thus $G(z) \in \Sigma$ and

$$\left(\frac{1}{\pi} \int_{-\infty}^{\infty} \frac{y}{(x-t)^2+y^2} |\Phi'(t)| \, dt\right) \left(\frac{1}{\pi} \int_{-\infty}^{\infty} \frac{y}{(x-t)^2+y^2} |\Phi'(t)|^{-1} \, dt\right)$$
$$\leq (1+M^2)\Re\, \Phi'(z)\Re\,(1/\Phi'(z))$$
$$\leq (1+M^2)^2 |\Phi'(z)||\Phi'(z)|^{-1}$$
$$= (1+M^2)^2 .$$

We then consider the interval $I = [x - y, x + y]$ and observe that $y/((x-t)^2+y^2) \geq (2y)^{-1}$ on I. As a result, we deduce that the product of the means of $|\Phi'|$ and $|\Phi'|^{-1}$ on I is uniformly bounded. So $\omega(t) = |\Phi'(t)|$ belongs to Muckenhoupt's class A_2 (Chapter 7, section 8): the constant which appears in the definition of the class only depends on M. In other words

$$(3.7) \qquad \sup_I \left(\frac{1}{|I|} \int_I \omega(t) \, dt\right) \left(\frac{1}{|I|} \int_I \frac{1}{\omega(t)} \, dt\right) \leq C(M) < \infty .$$

As a consequence (Chapter 7, section 8), there exist an exponent $\delta = \delta(M) > 0$, $\delta \leq 1$, and a constant $C'(M)$ such that

$$(3.8) \qquad \frac{\omega(E)}{\omega(I)} \leq C' \left(\frac{|E|}{|I|}\right)^\delta$$

(where $\omega(E) = \int_E \omega(t) \, dt$ etc.).

Writing (3.8) with ω replaced by $1/\omega$ and taking account of the condition

$$\left(\int_E \omega(t) \, dt\right) \left(\int_E \frac{1}{\omega} \, dt\right) \geq |E|^2 ,$$

we get

$$(3.9) \qquad \frac{\omega(E)}{\omega(I)} \geq C' \left(\frac{|E|}{|I|}\right)^{2-\delta} .$$

Applying (3.8) and (3.9), with $E = [0,1]$, $I = [0,t]$ (where $t \geq 1$), or $E = [-1,0]$, $I = [t,0]$ (where $t \leq -1$), we get

$$(3.10) \qquad C_1(M)|t|^\delta \leq |\Phi(t)| \leq C_2(M)|t|^{2-\delta} \qquad \text{if } \Phi(0) = 0.$$

Similarly, if $0 < h < 1$,

$$(3.11) \qquad |\Phi(t+h) - \Phi(t)| \geq C_3(M)h^{2-\delta}(1+|t|)^{-2+2\delta},$$

where $C_1(M) > 0$, $C_2(M) > 0$, and $C_3(M) > 0$ depend only on M. Finally, if $0 < h < 1$, we get

$$(3.12) \qquad |\Phi(t+h) - \Phi(t)| \leq C_4(M)h^\delta(1+|t|)^{2-2\delta}.$$

To arrive at these estimates, we need

$$\left|\int_a^b \Phi'(t) \, dt\right| \geq \int_a^b \Re\, \Phi'(t) \, dt \geq \sqrt{1+M^2} \int_a^b |\Phi'(t)| \, dt$$

and, having made this remark, the inequalities follow immediately.

We are now equipped to pass to the limit. We begin with a real, measurable function $v(t)$ such that $\|v\|_\infty = \theta_0 < \pi/2$. With it, we associate a sequence v_j of step functions with compact supports, such that $\|v_j\|_\infty \le \theta_0$ and $v_j(t) \to v(t)$ almost everywhere. We then construct $\Phi_j : P \to \mathbb{C}$, by requiring

$$\log \Phi_j'(z) = \frac{1}{\pi} \int_{-\infty}^{\infty} \frac{1}{t-z} v_j(t)\, dt + C_j\,, \qquad C_j \in \mathbb{R},\ z \in P,$$

$$\Re \log \Phi_j'(i) = 0, \qquad \text{and} \qquad \Phi_j(0) = 0\,.$$

The functions Φ_j are univalent in P and they satisfy the estimates (3.10), (3.11), and (3.12), uniformly in j.

To conclude, we apply the following lemma (after two appropriate bilinear transformations).

Lemma 7. *Let F_j, $j \in \mathbb{N}$, be a sequence of univalent functions on $|z| < 1$, which are continuous on $|z| \le 1$ and converge uniformly on $|z| \le 1$ to F. We suppose that $F_j(e^{i\theta})$ and $F(e^{i\theta})$, $0 \le \theta \le 2\pi$, are closed Jordan curves, which we denote by Γ_j and Γ. Then F is univalent and $F(D)$ is the inside of Γ.*

This lemma is a variant of Rouché's theorem.

To use it, we first consider the auxiliary functions $(i + \Phi_j(z))^{-1}$. Since $\Gamma_j = \Phi_j(\mathbb{R})$ is a polygonal line through 0, with slope not exceeding M, we have $|i + \Phi_j(t)| \ge c(M)(1 + |\Phi_j(t)|) \ge c'(M)(1 + |t|)^\delta$ and, thus, $|(i + \Phi_j(t))^{-1}| \le (c'(M))^{-1}(1 + |t|)^{-\delta}$. Moreover, the functions $\Phi_j(t)$ are equicontinuous: by passing to a subsequence, if necessary, we may suppose that the functions $\Phi_j(t)$ converge uniformly on every compact set. As a consequence, the $(i + \Phi_j(t))^{-1}$ converge uniformly on the whole real line. The maximum modulus principle now shows that the functions $(i + \Phi_j(z))^{-1}$ converge uniformly on P. To get to the situation in the lemma, it is enough to put $z = i(1 - \zeta)/(1 + \zeta)$, $\zeta \in D$.

We have proved the following theorem.

Theorem 3. *If $v(t) \in L^\infty(\mathbb{R})$ is a real-valued function with norm $\|v\|_\infty = \theta_0 < \pi/2$, then the holomorphic function Φ, defined by*

$$\log \Phi'(z) = \frac{1}{\pi} \int_{-\infty}^{\infty} \left(\frac{1}{t-z} - \frac{t}{1+t^2} \right) v(t)\, dt\,,$$

gives a conformal representation of the half-plane P onto the open set Ω above a Lipschitz graph Γ whose slope nowhere exceeds $\tan \theta_0$.

On the way, we have established the following properties. The boundary values $\log \Phi'(t)$, $-\infty < t < \infty$, exist almost everywhere and $|\Phi'(t)|$ is a weight function belonging to Muckenhoupt's class A_2. The Lipschitz graph Γ is represented parametrically by $z = \Phi(t)$, $t \in \mathbb{R}$, and we thus have $ds = |\Phi'(t)|\,dt$. The mapping $t \mapsto s$ is an increasing homeomorphism which preserves sets of measure zero.

Let us return to the Hardy spaces associated with Lipschitz open sets. We suppose that $a(x)$ is a real-valued Lipschitz function of one real variable. We have $\|a'\|_\infty \le M < \infty$ and let Ω denote the open set above the graph Γ of $a(x)$. Let $\Phi : P \to \Omega$ be a conformal representation which extends to an increasing homeomorphism of $\mathbb{R}$ onto Γ.

Let $F(z) = P(z)/Q(z)$ be a rational function, vanishing at infinity, whose poles do not lie in $\overline{\Omega}$. Then $G = (F \circ \Phi)\Phi'^{1/p}$ belongs to the Hardy space $H^p(P)$. To see this, we must compute $\sup_{y>0} \int_{-\infty}^\infty |G(x+iy)|^p\,dx$.

We use the change of variable $z(t) = \Phi(t+iy)$, which leads to the notation $\Gamma_y \subset \Omega$ for the curve defined by this parametric representation. We know that Γ_y is a graph whose slope is nowhere greater than M. As a consequence, $\int_{\Gamma_y} |P(z)|^p |Q(z)|^{-p}\,ds$ is bounded above by a constant, independent of y.

Once we have verified this, we calculate the H^p norm of G by

$$\left(\int_{-\infty}^\infty |G(t)|^p\,dt \right)^{1/p} = \left(\int_\Gamma |F(z)|^p\,ds \right)^{1/p}.$$

The mapping taking F to G is isometric and thus extends to an isometry between $H^p(\Omega)$ and $H^p(P)$.

To show that this closed subspace is the whole of $H^p(P)$, we use duality. We let $g(x) \in L^q(\mathbb{R}, dx)$ denote a function such that

(3.13) $$\int_{-\infty}^\infty (F \circ \Phi)\Phi'^{1/p} g(x)\,dx = 0, \qquad F \in H^p(\Omega).$$

We now change this integral into an integral along Γ and (3.13) becomes $\int_\Gamma F(z)h(z)\,dz = 0$, where $(h \circ \Phi)\Phi' = \Phi'^{1/p}g(x)$, which gives $(h \circ \Phi)\Phi'^{1/q} = g(x)$. Theorem 1 now tells us that $h(z)$ is in $H^q(\Omega)$. By the direct part of our argument, $g(x)$ belongs to $H^q(P)$, and hence the functions $(F \circ \Phi)\Phi'^{1/p}$ are dense in $H^p(P)$.

We have established the following result.

Theorem 4. *Let $1 < p < \infty$. If $\Phi : P \to \Omega$ is a conformal representation of the upper half-plane onto the open set Ω above a Lipschitz graph Γ, normalized as above, then $F \mapsto (F \circ \Phi)\Phi'^{1/p}$ is an isometric isomorphism of $H^p(\Omega)$ (considered as a closed subspace of $L^p(\Gamma)$) with $H^p(P)$.*

We shall get a more complete statement by also considering the extreme cases $p = 1$ and $p = \infty$.

When $p = 1$, the arguments we used to prove Theorem 1 no longer work.

Recall that $F \in H^1(\Omega)$ if F is holomorphic in Ω and there exists a constant C such that $\sup_{y>0} \int_\Gamma |F(z + iy)| \, ds \le C$.

If $F \in H^1(\Omega)$, we form $F_m(z) = F(z + i/m)$, $m \ge 1$, and these functions F_m are in $H^1(\Omega)$. Furthermore, they have an obvious trace on Γ and satisfy

$$F_m(z) = \frac{1}{2\pi i} \int_{\Gamma - i/2m} \frac{F_m(\zeta)}{\zeta - z} \, d\zeta \,,$$

if $z \in \Omega \cup \Gamma$. This can be established by following the proof we gave for the corresponding assertion of Theorem 1. From this relationship, we can deduce that $F_m \in H^\infty(\Omega)$, but the L^∞ norms of the functions F_m are clearly not bounded. We then set $f_m = (F_m \circ \Phi)\Phi'$, which is holomorphic on P, and, for each $\varepsilon > 0$, we put

$$f_{m,\varepsilon}(z) = f_m(z)(1 + \varepsilon\Phi'(z))^{-1}(1 - i\varepsilon z)^{-2} \,.$$

Since $\Phi'(z) \in \Sigma$, we have $|\Phi'(z)(1 + \varepsilon\Phi'(z))^{-1}| \le C/\varepsilon$ and it follows that $|f_{m,\varepsilon}(z)| \le C(m, \varepsilon)|1 - i\varepsilon z|^{-2}$. Thus $f_{m,\varepsilon}(z) \in H^1(P)$, which is a closed subspace of $L^1(\mathbb{R})$.

But $(F_m \circ \Phi)\Phi'$ belongs to $L^1(\mathbb{R})$ and $|(1 + \varepsilon\Phi')^{-1}(1 - i\varepsilon z)^{-2}| \le 1$. Lebesgue's dominated convergence theorem shows that $(F_m \circ \Phi)\Phi' \in H^1(P)$.

Having got this far, we use a classical result ([239], Theorem (7.22) of Chapter 7), namely, the factorization of a function in $H^1(P)$ into the product of two functions in $H^2(P)$. This gives $(F_m \circ \Phi)\Phi' = g_m h_m$, where $g_m, h_m \in H^2(P)$, and where $\|g_m\|_2^2 = \|h_m\|_2^2 = \|(F_m \circ \Phi)\Phi'\|_1 = \int_\Gamma |F_m(z)| \, ds$.

We then define G_m and H_m by $(G_m \circ \Phi)\Phi'^{1/2} = g_m$ and $(H_m \circ \Phi)\Phi'^{1/2} = h_m$. We thus have $F_m = G_m H_m$ and both G_m and H_m belong to $H^2(\Omega)$. Further,

$$\int_\Gamma |G_m(z)|^2 \, ds = \int_\Gamma |H_m(z)|^2 \, ds = \int_\Gamma |F_m(z)| \, ds \,.$$

Inequality (2.10) shows us that, if $G_m^*(z) = \sup_{\tau>0} |G_m(z+i\tau)|$, $z \in \Gamma$, then $\|G_m^*\|_{L^2(\Gamma)} \le C\|G_m\|_{L^2(\Gamma)}$. This gives $\|F_m^*\|_{L^1(\Gamma)} \le C\|F_m\|_{L^1(\Gamma)}$.

On the one hand, $\|F_m\|_{L^1(\Gamma)} \le \|F\|_{H^1(\Omega)}$. On the other, $F_m^*(z) = \sup_{\tau \ge 1/m} |F(z+i\tau)|$ increases to $F^*(z) = \sup_{\tau>0} |F(z+i\tau)|$. Hence, by simply taking the limit, $\|F^*\|_{L^1(\Gamma)} \le C\|F\|_{L^1(\Gamma)}$. From this point, all the conclusions of Theorem 1 extend without difficulty to $p = 1$.

The case $p = \infty$ has no interesting features, because the mapping taking $F \in H^\infty(\Omega)$ to $F \circ \Phi \in H^\infty(P)$ is clearly an isometric isomorphism.

4 The operators associated with complex analysis

We again let Γ denote a Lipschitz graph and let Ω_1 be the open set above Γ. In what follows, we can always replace Ω_1 by the open set Ω_2 underneath Γ, as long as we make the obvious alterations.

Let Γ be the graph of the Lipschitz function a. We let $M = \|a'\|_\infty$ and let S be the sector $y \geq M'|x|$, where $M' > M$. Let $K(z, w)$ be a function of the two complex variables z and w, which is holomorphic in the open set $W \subset \mathbb{C}^2$ defined by $w - z \notin S$, and which satisfies

(4.1) $|K(z, w)| \leq |z - w|^{-1}$ for $w - z \notin S$.

Theorem 5. *With the above hypotheses, for each $f \in L^2(\Gamma, ds)$, the function $F : \Omega_1 \to \mathbb{C}$ defined by $F(z) = \int_\Gamma K(z, w) f(w) \, dw$ belongs to $H^2(\Omega_1)$.*

Calderón's operator belongs to the more general class of operators described by Theorem 5, so Theorem 5 implies Calderón's theorem.

We shall use the $T(b)$-theorem to prove Theorem 5.

The function $F(z)$ is defined and holomorphic in Ω_1, because of the geometric hypotheses and the Cauchy-Schwarz inequality. Now, if $F(z)$ is holomorphic in the open set defined by $y < M|x|$ and satisfies $|F(z)| \leq C'|z|^{-1}$ on $y < M'|x|$, whenever $M' < M$, then the derivative $F'(z)$ similarly satisfies $|F'(z)| \leq C''|z|^{-2}$, for $y < M'|x|$ and $M' < M$. We can see this by applying Cauchy's formula to a circle, centre z, and radius half the distance from z to the cone $y = M|x|$.

Returning to $K(z, w)$, we thus get

(4.2) $\left| \dfrac{\partial K}{\partial z} \right| + \left| \dfrac{\partial K}{\partial w} \right| \leq C'' |z - w|^{-2}$,

if $w - z \notin S''$, where S'' is the sector $y \geq M''|x|$ and $M < M'' < M'$.

Naturally, we shall identify M' with M'' and can thus suppose that (4.1) and (4.2) are satisfied for $w - z \notin S$.

To prove the theorem, we need to establish that the functions F_τ, defined on Γ, for $\tau > 0$, by

(4.3) $F_\tau(z) = \int_\Gamma K(z + i\tau, w) f(w) \, dw$,

satisfy

(4.4) $\left(\int_\Gamma |F_\tau(z)|^2 \, ds \right)^{1/2} \leq C \left(\int_\Gamma |f(z)|^2 \, ds \right)^{1/2}$,

uniformly in $\tau > 0$. The upper bound of the left-hand side of (4.4) is, in fact, the norm of F in $H^2(\Omega_1)$.

To get (4.4), we just apply the $T(b)$ theorem to the kernel $K_\tau(z, w) =$

$K(z + i\tau, w)$. For this, we need to show that there is a constant C such that, for each interval $I \subset \Gamma$,

$$(4.5) \qquad \int_I \left| \int_I K_\tau(z, w)\, dw \right| ds \le C|I|$$

and

$$(4.6) \qquad \int_I \left| \int_I K_\tau(z, w)\, dz \right| ds \le C|I|.$$

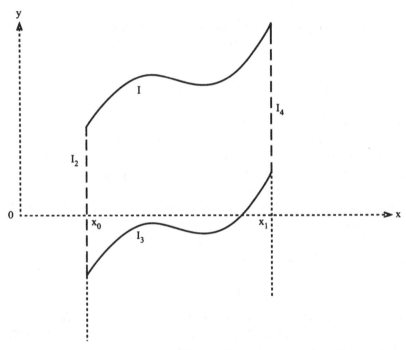

Figure 1

In terms of the kernel $K(z, w)$, inequality (4.5) means that

$$\int_I \left| \int_I K(z + i\tau, w)\, dw \right| ds \le C|I|.$$

We get this inequality by changing the contour, as in Figure 1, that is, by replacing I (defined by $z(x) = x + ia(x)$, $x_0 \le x \le x_1$) by the union $J = I_2 \cup I_3 \cup I_4$, where I_2 is defined by $z = x_0 + iy$, $a(x_0) - l \le y \le a(x_0)$, $l = x_1 - x_0$; I_3 is defined by $z(x) = x + i(a(x) - l)$, $x_0 \le x \le x_1$; and, lastly, I_4 is given by $z(x) = x_1 + iy$, $a(x_1) - l \le y \le a(x_1)$.

Clearly,

$$\int_I K_\tau(z,w)\,dw = \int_{I_2} K_\tau(z,w)\,dw + \int_{I_3} K_\tau(z,w)\,dw + \int_{I_4} K_\tau(z,w)\,dw \,.$$

It then suffices to use the estimate $|K_\tau(z,w)| \le C|z-w|^{-1}$ and the obvious remark that $\int_I \int_{I_j} |z-w|^{-1}\,dw\,ds \le C|I|$, for $j = 2$, 3 and 4.

To get (4.6), we proceed symmetrically, using a contour lying entirely above Γ. Theorem 5 is thus established. Calderón's theorem now follows as a particular case.

We are going to use Theorem 5 to deduce the existence of a commutative algebra of operators $T : L^2(\Gamma) \to L^2(\Gamma)$ for which we have a symbolic calculus. The symbols, denoted by $m(\xi)$, are *holomorphic Marcinkiewicz symbols*. That is, they are defined on $\mathbb{R} \setminus \{0\}$ and satisfy

$$(4.7) \qquad \left| \left(\frac{\partial}{\partial \xi} \right)^k m(\xi) \right| \le C\alpha^k k! |\xi|^{-k}, \qquad k \in \mathbb{N},$$

where $C \ge 0$ and $\alpha > 1$ are constants. This implies that each such $m(\xi)$ is the restriction to $\mathbb{R} \setminus \{0\}$ of a function $m(\zeta)$ which is bounded and holomorphic on the open sector $|\eta| < \beta|\xi|$, where $\zeta = \xi + i\eta$ and $0 < \beta < (\alpha^2 - 1)^{-1/2}$. Conversely, if $\beta > (\alpha^2 - 1)^{-1/2}$ and if $m(\zeta)$ is holomorphic and bounded on $|\eta| < \beta|\xi|$, then (4.7) is satisfied.

In our application to the theory of operators, we must suppose that $(\alpha^2 - 1)^{-1/2} > M = \|a'\|_\infty$ (recall that the Lipschitz curve Γ is the graph of the Lipschitz function $a(x)$). This will enable us to choose β so that $(\alpha^2 - 1)^{-1/2} > \beta > M$. The algebra of symbols we consider is the union, over all α satisfying $1 < \alpha < (1 + M^2)^{1/2}/M$, of the classes defined by (4.7). This union can also be regarded as the algebra of the functions which are holomorphic and bounded on a sector G, defined by $|\eta| < \beta|\xi|$, where $|\beta| > M$. We denote this algebra of symbols by S_M.

With each symbol $m \in S_M$ we associate a distribution S which is the inverse Fourier-Laplace transform of m. Computing S is simplified by the approximation technique which we now describe.

For each $\varepsilon > 0$ and each $m \in S_M$, we put $m_\varepsilon(\xi) = e^{-\varepsilon|\xi|} m(\xi)$. For the holomorphic extension $m(\zeta)$ of $m(\xi)$, if we write $\zeta = \xi + i\eta$, then $m_\varepsilon(\zeta) = e^{-\varepsilon\zeta} m(\zeta)$ when $\xi > 0$, whereas $m_\varepsilon(\zeta) = e^{\varepsilon\zeta} m(\zeta)$ when $\xi < 0$. It follows immediately that the $m_\varepsilon(\xi)$, $\varepsilon > 0$, form a bounded subset of S_M.

If $m \in S_M$, we now define $S_\varepsilon(x)$ by $S_\varepsilon(x) = (2\pi)^{-1} \int_{-\infty}^{\infty} e^{ix\xi} m_\varepsilon(\xi)\,d\xi$. The symbols $m_\varepsilon(\xi)$ converge to $m(\xi)$ in the sense of tempered distributions and the S_ε tend to S, the inverse Fourier transform of m.

We shall characterize the distributions S, obtained in this way, as the boundary values of certain holomorphic functions.

For $\gamma > 0$, we define H_γ^+ and H_γ^- as follows: $F \in H_\gamma^+$ if $F(z)$ is holomorphic in $y > \gamma|x|$ and satisfies $|F(z)| \leq C|z|^{-1}$, for a certain constant C; the space H_γ^- is similarly composed of the functions which are holomorphic in $y < \gamma|x|$ and satisfy the same growth condition.

The characterization of the distributions S is based on the following proposition.

Proposition 1. *Let $m \in S_M$ be a symbol supported by $[0, \infty)$. Then its inverse Fourier transform S belongs to H_γ^+ for some $\gamma > M$. Conversely, if $S \in H_\gamma^+$, for some $\gamma > M$, then the functions $S(x+i\varepsilon)$, $\varepsilon > 0$, converge, in the sense of tempered distributions, to a distribution $S(x)$ which is the inverse Fourier transform of a symbol $m \in S_M$.*

To verify the first statement, we put $S(x) = (2\pi)^{-1} \int_0^\infty e^{ix\xi} m(\xi)\, d\xi$. For this we have to require that $m \in L^1(0, \infty)$, but we can reduce to this case by the approximation procedure described above. Now, $S(x)$ is the restriction to the real axis of $S(z) = \int_0^\infty e^{iz\xi} m(\xi)\, d\xi$, defined for $\Im m\, z > 0$. Fixing z for the moment, we change the contour of integration $[0, \infty)$ into the half-line $\zeta = (1 + ip)t$, $t \geq 0$. We can do this as long as $y + px > 0$ $(z = x + iy)$ and $|p| < \beta$: if $x \geq 0$, we can make an arbitrary choice of $p \in [0, \beta)$ and when $x \leq 0$, p is an arbitrary number in $(-\beta, 0]$. Once this choice of contour has been made, $S(z)$ can automatically be continued analytically to the union of open half-planes $y + px > 0$, $|p| < \beta$. This union is precisely the open set given by $y > -\beta|x|$.

The inequality $|S(z)| \leq C|z|^{-1}$ is a consequence of the following lemma.

Lemma 8. *Let $m(\xi) \in L^\infty(\mathbb{R})$ be such that $|m(\xi)| \leq C_0$. Further, suppose that $m(\xi)$ is of class C^2 on $\mathbb{R} \setminus \{0\}$, with second derivative satisfying $|m''(\xi)| \leq C_1 \xi^{-2}$. Then the inverse Fourier transform S of m is continuous on $\mathbb{R} \setminus \{0\}$ and satisfies $|S(x)| \leq C|x|^{-1}$.*

The proof, which is elementary, consists of integrating the oscillatory integral $\int m(\xi) e^{ix\xi}\, d\xi$ twice, by parts. This poses some problems at infinity, but the details are left to the reader.

The lemma admits a converse in which we start with a distribution $S \in \mathcal{D}'(\mathbb{R})$ with the following properties:

(4.8)   the product $xS(x)$ is locally integrable;

(4.9)   there is a constant C_0 such that $|\int_{-T}^T S(x)\, dx| \leq C_0$, for each $T > 0$;

(4.10) $S(x)$ is of class C^1 on $\mathbb{R} \setminus \{0\}$, vanishes at infinity, and satisfies $|S'(x)| \leq C_1 x^{-2}$.

Then the Fourier transform of S is in $L^\infty(\mathbb{R})$.

This converse leads to the second part of Proposition 1. We leave the details to the reader.

Now we have all the tools needed to construct a commutative algebra A_M of operators $T : L^2(\Gamma) \to L^2(\Gamma)$. The algebra will be isomorphic to the algebra of symbols S_M.

Let $m \in S_M$ be a symbol with support in $[0, \infty)$. The distribution S, which is the inverse Fourier transform of m, can be extended as a holomorphic function on $y > -\beta|x|$, for some $\beta > M$. We then form the kernel $K(z, w) = S(z - w)$. It satisfies the conditions of Theorem 5 and defines a bounded operator on $L^2(\Gamma)$. We let T_m denote this operator.

Let m_1 and m_2 be two symbols in S_M with supports in $[0, \infty)$. We denote the corresponding operators by T_1 and T_2 and their kernels by $S_1(z - w)$ and $S_2(z - w)$, respectively. To compute the operator $T_3 = T_2 T_1$, we first of all replace $m_1(\xi)$ and $m_2(\xi)$ by $m_1(\xi)e^{-\varepsilon\xi}$ and $m_2(\xi)e^{-\varepsilon\xi}$. This amounts to replacing $S_1(z - w)$ by $S_1(z + i\varepsilon - w)$, and similarly for S_2. Then $T_2^\varepsilon T_1^\varepsilon$ is defined by the kernel $K_3(z, w) = \int_\Gamma S_1(z + i\varepsilon - \zeta)S_2(\zeta + i\varepsilon - w)\, d\zeta$, $z, w \in \Gamma$. We may write

$$K_3(z, w) = \int_{\Gamma + i\varepsilon} S_1(z + 2i\varepsilon - \zeta)S_2(\zeta - w)\, d\zeta = S_3(z + 2i\varepsilon - w)$$

and a simple deformation of the contour of integration allows us to verify that the Fourier transform of $S_3(x)$ is the product $m_1(\xi)m_2(\xi)$.

This shows that T_3 is the operator corresponding to the product of symbols $m_3(\xi) = m_1(\xi)m_2(\xi)$.

The other cases are similar and we get the claimed isomorphism between the algebra of operators A_M and the algebra of symbols S_M.

The oblique projection operator of $L^2(\Gamma, ds)$, with kernel $H^2(\Omega_2)$, onto the Hardy space $H^2(\Omega_1)$ has as symbol the characteristic function of the half-line $[0, \infty)$. Similarly, the oblique projection operator of $L^2(\Gamma, ds)$ onto $H^2(\Omega_2)$, with kernel $H^2(\Omega_1)$, has as symbol the characteristic function of $(-\infty, 0]$. If we let χ_+ and χ_- denote these characteristic functions, then the symbol $e^{-\varepsilon\xi}\chi_+(\xi)$, $\varepsilon > 0$, corresponds to the operator which takes $f = F_+ + F_-$ (where $F_+ \in H^2(\Omega_1)$ and $F_- \in H^2(\Omega_2)$) to $F_+(z + i\varepsilon)$ which we consider as a function in $L^2(\Gamma, ds)$.

A further application of the $T(b)$ theorem gives the following theorem of C. Kenig ([157]).

Theorem 6. *Let Ω_1 be the open set above the graph Γ of a Lipschitz function $a : \mathbb{R} \to \mathbb{R}$. Then the norms $(\int_\Gamma |F(z)|^2\, ds)^{1/2}$ and $(\iint_{\Omega_1} |F'(z)|^2(y - a(x))\, dx\, dy)^{1/2}$ are equivalent on $H^2(\Omega_1)$.*

We first check that there exists a constant $C = C(M)$, $M = \|a'\|_\infty$, such that

$$(4.11) \quad \left(\iint_{\Omega_1} |F'(z)|^2 (y - a(x)) \, dx \, dy\right)^{1/2} \le C(M)\left(\int_\Gamma |F(z)|^2 \, ds\right)^{1/2},$$

for $F \in H^2(\Omega_1)$.

We shall then show that a strong version of this inequality automatically implies the converse inequality.

To establish (4.11), we consider the kernel taking values in $H = L^2(0, \infty)$, which is defined by $\mathbb{K}(z, w) = K(z, w, t) = t^{1/2}(z + it - w)^{-2}$, $z, w \in \Gamma$. Then, if f belongs to $H^2(\Omega_1)$,

$$\int_\Gamma K(z, w, t) f(w) \, dw = -2\pi i t^{1/2} f'(z + it).$$

The inequality (4.11) thus follows from a stronger property, namely, the continuity of the operator $T : L^2(\Gamma) \to L^2_H(\Gamma)$, whose kernel is $\mathbb{K}(z, w)$. To avoid using singular integrals, it is convenient to replace T by a sequence of truncated operators defined by the kernels $K_N(z, w, t) = K(z, w, t)$, when $N^{-1} \le t \le N$, and $K_N(z, w, t) = 0$, otherwise.

The continuity of T is then an immediate consequence of the Hilbert space version of the $T(b)$ theorem. The verification of conditions (14.2) and (14.3) of Chapter 11 is done by the same deformation of contour as was used earlier in this section and illustrated by figure 1. The details are left to the reader.

Once the continuity has been established, we prove the converse to inequality (4.11) by using the following remarkable identity.

$$(4.12) \qquad F(z) = \frac{2i}{\pi} \int_0^\infty \int_\Gamma t \frac{F'(w + it)}{(z + it - w)^2} \, dw \, dt,$$

for $F \in H^2(\Omega_1)$.

Let us first prove (4.12). We write $z + it - w = z + 2it - (w + it)$, which gives

$$\int_\Gamma \frac{F'(w + it)}{(z + it - w)^2} \, dw = \int_{\Gamma + it} \frac{F'(\zeta)}{(\zeta - (z + it))^2} \, d\zeta = 2\pi i F''(z + 2it).$$

Then we get

$$\int_0^\infty t F''(z + 2it) \, dt = -\frac{1}{4} F(z)$$

by integrating by parts.

As for the proof of the converse inequality, the continuity of the operator $T : L^2(\Gamma) \to L^2_H(\Gamma)$, gives us, via the adjoint operator, the following result: for every function $f(w, t) \in L^2(\Omega_1)$, $w \in \Gamma$, $t > 0$,

$$g(z) = \int_\Gamma \int_0^\infty t^{1/2}(w + it - z)^{-2} f(w, t) \, dw \, dt$$

is in $L^2(\Gamma)$ and

(4.13) $$\|g\|_{L^2(\Gamma)} \leq C\|f\|_{L^2(\Omega_1)},$$

where C depends only on $M = \|a'\|_\infty$.

Naturally, the same result holds for Ω_2, the open set underneath Γ. This amounts to changing i into $-i$ in the definition of g and putting

$$h(z) = \mathcal{L}(f)(z) = \int_\Gamma \int_0^\infty t^{1/2}(w - it - z)^{-2} f(w,t)\, dw\, dt.$$

We now return to (4.12) and write $f(w,t) = t^{1/2} F'(w+it)$. This gives $F = 2i\pi^{-1}\mathcal{L}(f)$ and

$$\|F\|_{L^2(\Gamma)} \leq \left(C \int\int_{\Omega_1} t|F'(z+it)|^2\, dx\, dy \right)^{1/2},$$

as required.

Before leaving Kenig's theorem, we reformulate it in the language of wavelets. For each $\zeta \notin \Gamma$, we consider the function $\psi_\zeta \in L^2(\Gamma, ds)$, defined by $\psi_\zeta(z) = (\mathrm{dist}(\zeta, \Gamma))^{1/2}(\zeta - z)^{-2}$. This function has the regularity and localization properties that we have come to expect of wavelets. But the cancellation of $\psi_\zeta(z)$ is expressed by $\int_\Gamma \psi_\zeta(z)\, dz = 0$. Moreover, the functions $\psi_\zeta(z)$ are not normalized in the usual way: it would be necessary to replace $(\mathrm{dist}(\zeta, \Gamma))^{1/2}$ by $(\mathrm{dist}(\zeta, \Gamma))^{3/2}$ to achieve a normalization such that $c_1 \leq \|\psi_\zeta\|_{L^2(\Gamma)} \leq c_2$.

Our wavelets are indexed by $\zeta \in \mathbb{C}$ and we consider (continuous) linear combinations of wavelets of the form

(4.14) $$g(z) = \int\int \psi_\zeta(z)\alpha(\zeta)\, d\xi\, d\eta,$$

where the coefficients $\alpha(\zeta)$ satisfy

(4.15) $$\int\int |\alpha(\zeta)|^2\, d\xi\, d\eta < \infty.$$

We then have our reformulation of Kenig's theorem.

Theorem 7. *With the above notation, there is a constant $C = C(M)$, $M = \|a'\|_\infty$, such that $\|g\|_{L^2(\Gamma)} \leq C(\int\int |\alpha(\zeta)|^2\, d\xi\, d\eta)^{1/2}$. Further, the continuous linear operator $\mathcal{L} : L^2(\mathbb{R}^2) \to L^2(\Gamma)$, which takes α to g, is surjective.*

Before proving this theorem, we remark that it provides a particularly simple way of decomposing an arbitrary function $g \in L^2(\Gamma)$ into a sum $g_1 + g_2$, where $g_1 \in H^2(\Omega_1)$ and $g_2 \in H^2(\Omega_2)$.

Indeed, $g = \mathcal{L}(\alpha)$ and it is enough to split α into $\alpha_1 + \alpha_2$, where α_1 has support in Ω_1 and α_2 has support in Ω_2. We then put $g_1 = \mathcal{L}(\alpha_1)$ and $g_2 = \mathcal{L}(\alpha_2)$.

Let us proceed to the proof of Theorem 7.

The continuity of $\mathcal{L}$ is an immediate consequence of (4.13). It thus follows from Theorem 6, together with the identity $L^2(\Gamma) = H^2(\Omega_1) + H^2(\Omega_2)$. The fact that $\mathcal{L}$ is surjective uses the same ingredients. We first decompose g into $g_1 + g_2$, where $g_1 \in H^2(\Omega_1)$ and $g_2 \in H^2(\Omega_2)$. The identity (4.12) then lets us expand g_1 and g_2 as wavelets.

Kenig's theorem and Calderón's theorem thus imply Theorem 7.

The approach of this section has been to use the $T(b)$ theorem to prove everything. But the $T(b)$-theorem was proved only in 1985, some time after both Calderón's and Kenig's theorems were known. The following diagram summarizes the relationships between the theorems.

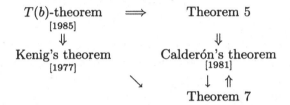

$$T(b)\text{-theorem} \quad \Longrightarrow \quad \text{Theorem 5}$$
$$[1985]$$
$$\Downarrow \qquad\qquad\qquad \Downarrow$$
$$\text{Kenig's theorem} \qquad \text{Calderón's theorem}$$
$$[1977] \qquad\qquad\qquad [1981]$$
$$\searrow \qquad \downarrow\ \Uparrow$$
$$\text{Theorem 7}$$

5 The "shortest" proof

We now want to give another proof of Calderón's theorem. In the Spring of 1987, P. Jones and S. Semmes ([64]) discovered that it was possible to prove Calderón's theorem very simply, using only Theorem 6. Kenig discovered Theorem 6 in 1977 (it was published a year later). So, in a sense, Calderón's theorem lay unnoticed in Kenig's Ph.D. dissertation. Let us state when the theorems were discovered:-

I Calderón's theorem (1977),

II Kenig's theorem (1977),

III Coifman, McIntosh, and Meyer (1981): Calderón's theorem in its strongest form,

IV the $T(b)$-theorem (1985).

Diagramatically,

$$\text{Calderón (1977)}$$
$$\Downarrow \text{[G.David]}$$
$$\text{Kenig (1977)} \quad \Longrightarrow \quad \text{full Calderón theorem.}$$

After either Calderón's 1977 theorem, or Kenig's theorem of the same year, the obstacles in the way of a proof of the full form of Calderón's theorem were mainly of a psychological nature. We were amazed by this fact, which is why we are giving an account of all the approaches to Calderón's theorem.

To appreciate the "shortest" proof of Jones and Semmes, we must make the proof of Kenig's theorem independent of the $T(b)$ theorem, since Calderón's theorem is also an immediate corollary of the $T(b)$ theorem.

But there have "always" been several direct proofs of Kenig's theorem. The first is due to B. Dahlberg ([81] and [82]). Another can be found in Kenig's thesis. Using either of these methods, a simple proof of the L^2 continuity of the Cauchy kernel on Lipschitz curves can be obtained. We shall describe that proof now.

To start with, we establish the continuity of the operator $\mathcal{L}$ of Theorem 7, without using Calderón's theorem, namely, the decomposition $L^2(\Gamma) = H^2(\Omega_1) + H^2(\Omega_2)$. That is, we begin with a function $f(z,t) \in L^2(\Omega_1)$, where $z \in \Gamma$, $t > 0$, and where $(z,t) \in \Gamma \times (0,\infty)$ is identified with $z + it \in \Omega_1$. We next form

$$(5.1) \qquad g_\pm(w) = \int_\Gamma \int_0^\infty \frac{t^{1/2}}{(z - w \pm it)^2} f(z,t)\, dz\, dt$$

and claim that

$$(5.2) \qquad \|g_\pm\|_{L^2(\Gamma)} \leq C \|f\|_{L^2(\Omega_1)}\,.$$

We shall deal only with the case of the kernel $t^{1/2}(z-w+it)^{-2}$, writing $g(w)$ for $g_+(w)$: since the other case is similar, we shall leave it to the reader. Now $g(w)$ is holomorphic in Ω_2 and, to apply Theorem 6, we need to know, a priori, that $g(w)$ belongs to $H^2(\Omega_2)$. But it is sufficient, for that purpose, to suppose that $f(z,t)$ is a function of compact support contained in Ω_1. The general case is given by simply passing to the limit.

We therefore calculate $\int_\Gamma |g(w)|^2 \, ds$ using Kenig's theorem, that is, by estimating $\int\int_{\Omega_2} \operatorname{dist}(w,\Gamma)|g'(w)|^2 \, du\, dv$, where we have put $w = u + iv$.

We write $|g'(w)|^2 = g'(w)\overline{g'(w)}$, which leads to finding an upper bound for the six-fold integral

$$I = \int\int_{\Omega_1} \int\int_{\Omega_1} \int\int_{\Omega_2} \frac{(\operatorname{dist}(z,\Gamma))^{1/2}(\operatorname{dist}(\zeta,\Gamma))^{1/2}}{|z - w|^3 |\zeta - w|^3}$$
$$\times |f(z)||f(\zeta)| \operatorname{dist}(w,\Gamma) \, dx\, dy\, d\xi\, d\eta\, du\, dv\,,$$

where, with a slight change of notation, $z = x + iy \in \Omega_1$, $\zeta = \xi + i\eta \in \Omega_1$ and $w = u + iv \in \Omega_2$.

Clearly, $z \in \Omega_1$ and $w \in \Omega_2$ lead to

$$|z - w| \geq |z - w|/2 + \operatorname{dist}(z,\Gamma)/2\,.$$

Putting $a \wedge b = \inf(a, b)$, we observe that, for $z, \zeta \in \Omega_1$ and $w \in \Omega_2$,

$$\text{dist}(w, \Gamma) \leq |w - z| \wedge |w - \zeta|.$$

Lastly, for every $\alpha > 0$, there is a constant $C(\alpha)$ such that, for $a, b \in \mathbb{C}$, $r > 0$, and $R > 0$,

(5.3)
$$\iint_{\Omega_2} \frac{|w - a| \wedge |w - b|}{(|w - a|^3 + R^3)(|w - b|^3 + r^3)} \, du \, dv$$
$$\leq C(\alpha) \frac{(r \wedge R)^{-\alpha}}{(|b - a| + R + r)^{3-\alpha}}.$$

We are left with estimating

$$\iint_{\Omega_1} \iint_{\Omega_1} R(z, \zeta) |f(z)| |f(\zeta)| \, dx \, dy \, d\xi \, d\eta$$

where

$$R(z, \zeta) = \frac{(\text{dist}(z, \Gamma))^{1/2} (\text{dist}(\zeta, \Gamma))^{1/2} (\text{dist}(z, \Gamma) \wedge \text{dist}(\zeta, \Gamma))^{-\alpha}}{(|z - \zeta| + \text{dist}(z, \Gamma) + \text{dist}(\zeta, \Gamma))^{3-\alpha}}.$$

To do this, it is enough to apply Schur's lemma, that is, to verify that there exists a constant C, such that $\iint R(z, \zeta) \, dx \, dy \leq C$. The symmetry of the kernel $R(z, \zeta)$ will then let us deduce the L^2 continuity which we require. The verification is simple, because it is enough to use x and $t = y - a(x)$ as variables to reduce the expression to

$$\int_{-\infty}^{\infty} \int_0^{\infty} \frac{t^{1/2} \tau^{1/2} (t \wedge \tau)^{-\alpha}}{(|x - \xi| + t + \tau)^{3-\alpha}} \, dx \, dt$$

and it is immediate that this is bounded above.

We have just verified that, if

$$g_\pm(w) = \int_\Gamma \int_0^{\infty} \frac{t^{1/2}}{(z - w \pm it)^2} f(z, t) \, dz \, dt,$$

then

(5.4)
$$\|g_\pm\|_{L^2(\Gamma)} \leq C \left(\int_\Gamma \int_0^{\infty} |f(z, t)|^2 \, ds \, dt \right)^{1/2}.$$

In fact, the L^2 continuity of the operator defined by the Cauchy kernel is a dual form of (5.4), as the following argument shows.

We start with an arbitrary function $h \in L^2(\Gamma, ds)$ and evaluate $J = \int_\Gamma g_\pm(w) h(w) \, dw$. By (5.4), for each $f \in L^2(\Omega_1)$,

$$|J| \leq C \|h\|_{L^2(\Gamma)} \|f\|_{L^2(\Omega_1)}.$$

Taking the least upper bound as f runs through the unit ball of $L^2(\Omega_1)$ and writing $F_\pm(z, t) = t^{1/2} \int_\Gamma (z - w \pm it)^{-2} h(w) \, dw$, we get

(5.5)
$$\|F_\pm\|_{L^2(\Omega_1)} \leq C \|h\|_{L^2(\Gamma)}.$$

We now interpret $F_\pm$, using Cauchy's formula. We put

(5.6)
$$H_+(\zeta) = \frac{1}{2\pi i} \int_\Gamma \frac{h(w)}{w - \zeta} \, dw, \qquad \zeta \in \Omega_1,$$

so that

$$F_+(z,t) = 2\pi i t^{1/2} H'_+(z+it), \qquad z \in \Gamma, \quad t > 0.$$

Inequality (5.5), together with Theorem 6, provides

$$(5.7) \quad \|H_+\|_{L^2(\Gamma)} \le C \left(\int_\Gamma \int_0^\infty t |H'_+(z+it)|^2 \, ds \, dt \right)^{1/2} \le C' \|h\|_{L^2(\Gamma)}.$$

The continuity of the Cauchy operator has thus been established.

In conclusion, it would be interesting to find two discrete subsets $S_1 \subset \Omega_1$ and $S_2 \subset \Omega_2$ such that the two collections ψ_ζ, $\zeta \in S_1$ and $\zeta \in S_2$, once normalized in L^2 norm, form a Riesz basis of $L^2(\Gamma, ds)$. This would make it possible to construct new operators on $L^2(\Gamma, ds)$, by imitating what we have done for orthonormal wavelets. These operators would leave the Hardy spaces $H^2(\Omega_1)$ and $H^2(\Omega_2)$ invariant.

6 Statement of David's theorem

We devote the rest of this chapter to the statement and proof of David's theorem, which, in one sense, brings the problem of the continuity of the Cauchy kernel to a close.

Let us start by a geometric approach, based on generalized Hardy spaces. Let D be a domain (that is, a connected, bounded, open set) in the complex plane, whose boundary is a rectifiable Jordan curve Γ. We shall assume, for convenience, that the total length of Γ is 1, and we shall write $z(s)$, $0 \le s \le 1$, for the parametric representation of Γ, with respect to arc-length s.

For $1 < p < \infty$, we put $L^p(\Gamma) = L^p(\Gamma, ds)$, which is isometrically isomorphic to $L^p[0,1]$.

Following Keldysh, Lavrentiev, and Smirnov, we let $H^p(D) \subset L^p(\Gamma)$ denote the closure in $L^p(\Gamma)$ of the polynomials in z. We shall use $\mathcal{H}^p(\Gamma)$ to denote the, possibly larger, space formed of functions in $L^p(\Gamma)$ such that $\int_\Gamma z^k f(z) \, dz = 0$, for all $k \in \mathbb{N}$.

If f is in $\mathcal{H}^p(\Gamma)$, then the function $F : D \to \mathbb{C}$, defined by $F(z) = (2\pi i)^{-1} \int_\Gamma f(\zeta)(\zeta - z)^{-1} \, d\zeta$, is holomorphic in D and f is the trace of F on Γ, in the sense that, for almost all $z_0 \in \Gamma$, $f(z_0)$ is the non-tangential limit of $F(z)$ as $z \to z_0$.

To verify this property, we let z^* denote the reflection of z in z_0, and note that

$$(6.1) \qquad \int_\Gamma \frac{f(\zeta)}{\zeta - z^*} \, d\zeta = 0,$$

because $g(z) = (\zeta - z^*)^{-1}$ is in $H^q(D)$ and $\int_\Gamma f(\zeta)g(\zeta)\,d\zeta = 0$ if $f \in \mathcal{H}^p(D)$ and $g \in H^q(D)$.

We now replace the Cauchy kernel $(2\pi i)^{-1}(\zeta - z)^{-1}$ by

$$K(z, \zeta) = \frac{1}{2\pi i} \left(\frac{1}{\zeta - z} - \frac{1}{\zeta - z^*} \right).$$

We suppose that $z_0 = z(s_0)$ is a point such that $z'(s_0)$ exists. Then, if $|s - s_0| < \varepsilon$, $z(s)$ is in a sector defined by $\zeta = z_0 + tz'(s_0)e^{i\phi}$, where $|\phi| \leq \phi(\varepsilon)$, $t \in \mathbb{R}$, and $\phi(\varepsilon) \to 0$ as $\varepsilon \to 0$. It is then easy to verify that, if z belongs to a sector $S(z_0)$ contained in D, whose boundary does not contain the tangent to Γ at z_0, then $K(z, \zeta)$ satisfies (for ζ sufficiently close to z_0) the same estimates that we used in the proof of Theorem 1.

Suppose now that $0 \in D$ and let us write Ω for the complement of $\bar{D}$ in the complex plane. We define $H^p(\Omega)$ as the closure in $L^p(\Gamma)$ of the polynomials in z^{-1} which vanish at infinity and, similarly, $\mathcal{H}^p(\Gamma)$ as the set of functions $f \in L^p(\Gamma)$ such that $\int_\Gamma f(z)z^{-k}\,dz = 0$, for $k \geq 1$, $k \in \mathbb{N}$.

We can, moreover, do this directly on the Riemann sphere. We let Γ be a rectifiable Jordan curve traced on the sphere and separating the two connected components D_1 and D_2 of its complement. We define $H^p(D_1)$ as the closure in $L^p(\Gamma)$ of the rational functions $P(z)/Q(z)$ whose poles lie in D_2 and vice versa for $H^p(D_2)$. To avoid the constant functions belonging to both $H^p(D_1)$ and $H^p(D_2)$, we normalize the functions in $H^p(D_2)$ by requiring them to vanish at some fixed point $z_0 \in D_2$.

Let us return to the case of the complex plane. Then every rational function $P(z)/Q(z)$, whose poles do not lie on Γ, can be written, uniquely, as

$$(6.2) \qquad \frac{P(z)}{Q(z)} = \frac{P_1(z)}{Q_1(z)} + \frac{P_2(z)}{Q_2(z)},$$

where P_1/Q_1 is holomorphic on D (and continuous on $\bar{D}$) and P_2/Q_2 is holomorphic on Ω, continuous on $\bar{\Omega}$, and zero at infinity. We let $\mathcal{C}_1$ and $\mathcal{C}_2$ denote the operators defined by $\mathcal{C}_1(P/Q) = P_1/Q_1$ and $\mathcal{C}_2(P/Q) = P_2/Q_2$.

The Hartogs-Rosenthal theorem tells us that, if $K \subset \mathbb{C}$ is a compact set of measure zero (for Lebesgue measure $dx\,dy$ on $\mathbb{R}^2$) in the complex plane, then the rational functions $P(z)/Q(z)$, whose poles do not belong to K, are dense in the continuous functions on K.

To prove this, we apply the Hahn-Banach theorem, arguing by contradiction that there is a bounded, regular, Borel measure μ, supported by K and orthogonal to the rational functions in question. Replacing μ by $\bar{\mu}$, this condition can be written as $\mu \star z^{-1} = 0$ on the complement of K. But μ is a compactly supported measure and z^{-1} is locally integrable in $\mathbb{R}^2$. As a consequence, $\mu \star z^{-1}$ is in $L^1_{\text{loc}}(\mathbb{R}^2)$. It is also zero almost everywhere in $\mathbb{R}^2$ and thus zero as an element of $L^1_{\text{loc}}(\mathbb{R}^2)$. To conclude, we see that $\mu \star z^{-1} = 0$ in the sense of distributions and, on applying the operator $\partial/\partial\bar{z}$, we get $\mu = 0$.

Our rectifiable curve Γ is a compact set of measure zero in the complex plane, so it follows from the Hartogs-Rosenthal theorem that the rational functions $f(z) = P(z)/Q(z)$, which are continuous on Γ, are dense in $L^p(\Gamma)$, for $1 < p < \infty$. We shall use $R(\Gamma) \subset L^p(\Gamma)$ to denote this dense subspace.

For almost all $z_0 \in \Gamma$ and $f = P/Q \in R(\Gamma)$, we have

$$(6.3) \qquad \frac{P_1(z_0)}{Q_1(z_0)} - \frac{P_2(z_0)}{Q_2(z_0)} = \frac{1}{\pi i} \, \text{PV} \int_\Gamma \frac{f(z)}{z - z_0} \, dz$$

and the principal value exists at every point $z_0 = z(s_0)$ where $z'(s_0)$ exists. This can be established by an elementary calculation, left to the reader.

Thus

$$\frac{P_1(z_0)}{Q_1(z_0)} = \frac{1}{2} f(z_0) + \frac{1}{2\pi i} \, \text{PV} \int_\Gamma \frac{f(z)}{z - z_0} \, dz \, .$$

Our problem is to find out whether the operators $\mathcal{C}_1$ and $\mathcal{C}_2$ extend as continuous linear operators on $L^2(\Gamma)$. The Hartogs-Rosenthal theorem has shown us that the question makes sense.

We can now state David's theorem ([93]).

Theorem 8. *Let Γ be a rectifiable Jordan curve of total length l. With the above notation, the following five properties of Γ are equivalent:*

(6.4) *there exists a constant $C \geq 2$ such that, for every $r > 0$ and every $z_0 \in \Gamma$, we have $|\{s \in [0, l] : |z(s) - z_0| \leq r\}| \leq Cr$;*

(6.5) $L^2(\Gamma) = H^2(D) + H^2(\Omega)$, *a direct sum which is only orthogonal when Γ is a circle;*

(6.6) $L^p(\Gamma) = H^p(D) + H^p(\Omega)$, *for some $1 < p < \infty$;*

(6.7) $L^p(\Gamma) = H^p(D) + H^p(\Omega)$, *for all $1 < p < \infty$;*

(6.8) $L^p(\Gamma) = H^p(D) + H^p(\Omega)$, *for all $1 < p < \infty$, furthermore*
$$H^p(D) = \mathcal{H}^p(D) \text{ and } H^p(\Omega) = \mathcal{H}^p(\Omega) \, .$$

Condition (6.4) was introduced by Ahlfors ([1]) and, for that reason, such curves are called *Ahlfors regular curves*. Ahlfors' condition means that, for each disk in the complex plane, the part of the curve Γ which lies inside the disk has total length not greater than a fixed constant C times the diameter of the disk.

Clearly, there is a version of the theorem in which Γ passes through the point at infinity. We then suppose that $\int_\Gamma |z(s) - z_0|^{-p}\, ds < \infty$, for $z_0 \notin \Gamma$ and for all $p > 1$. This condition guarantees that the rational functions which vanish at infinity, and are continuous on Γ, still belong to $L^p(\Gamma)$. We let Ω_1 and Ω_2 denote the two connected open sets whose common boundary is Γ. Then the generalized Hardy spaces $H^p(\Omega_1)$ and $H^p(\Omega_2)$ are defined as the closures in $L^p(\Gamma, ds)$ of the rational functions f_1 and f_2, respectively, which are holomorphic in Ω_1 or Ω_2, respectively, continuous on Γ, and vanishing at infinity. The problem we want to resolve is whether $L^p(\Gamma)$ is the direct sum of $H^p(\Omega_1)$ and $H^p(\Omega_2)$, for a particular p, or for all p $(1 < p < \infty)$.

In this case, once again, the problem does not depend on p and is equivalent to condition (6.4) where, this time, $0 < r < \infty$ and large values of r play a part.

An example of an Ahlfors regular curve is a Lavrentiev curve. This is a rectifiable Jordan curve $z(s)$, $-\infty < s < \infty$, for which there is a constant $C \geq 1$ such that, for every $s \in \mathbb{R}$ and all $t > s$, we have $t - s \leq C|z(t) - z(s)|$. In the case of Lavrentiev curves, the Cauchy kernel $\mathrm{PV}(z(s) - z(t))^{-1}$ conveniently satisfies the Calderón-Zygmund estimates. One of the difficulties we shall encounter in the proof of David's theorem is that the Cauchy kernel no longer satisfies these estimates: the singularity of $(z(s) - z(t))^{-1}$ may be significantly stronger than that of the Calderón-Zygmund kernels.

In this section we shall just show that (6.4) is necessary for the continuity of the operators $\mathcal{C}_1$ and $\mathcal{C}_2$ on any $L^p(\Gamma)$ space.

We consider a disk Δ, centre $z_0 \in \Gamma$, radius $r > 0$, and we let E be the set $\{s \in \mathbb{R} : z(s) \in \Delta\}$. Then we let $z_1 = z(s_1)$ be a point of Γ such that $|z_0 - z_1| = 10r$. Such an s_1 can be found, since $|z(s) - z_0|$ takes all the values between 0 and ∞ when s runs through the real line. Having done this, we consider an interval I, centre s_1, such that, for $s \in I$, we have $|z(s) - z(s_1)| \leq r/2$. Let α be the argument of $z_0 - z_1$. Then, if $z \in \Gamma \cap \Delta$ and $\zeta = z(s)$, $s \in I$, the modulus of $z - \zeta$ lies between $8r$ and $12r$, while the argument of $z - \zeta$ is in $[\alpha - \pi/4, \alpha + \pi/4]$. This implies that, for every function $f(s) \geq 0$ with support in E and for every $t \in I$,

$$(6.9) \qquad \left| \int_E \frac{f(s)}{z(s) - z(t)}\, ds \right| \geq \frac{C}{r} \int_E f(s)\, ds,$$

where $C > 0$ is a constant which is easy to calculate.

We apply this remark to the characteristic function f of E and apply the hypothesis that

$$g(t) = \int_E \frac{1}{z(s) - z(t)}\, ds$$

restricted to I has an L^p norm dominated by that of f. These inequalities give $r^{-1+1/p}|E| \leq C'|E|^{1/p}$, or $|E| \leq C''r$, as required.

The rest of this chapter is devoted to the converse, that is, the proof that (6.4) $\Rightarrow$ (6.8).

Here is the outline of the proof.

The geometric definition of Ahlfors regular curves implies that, at every scale, a regular curve Γ has *good Lipschitz copies*. More precisely, when (6.4) is satisfied, there exists a constant M such that, for every interval $I \subset \mathbb{R}$, we can find an orthonormal co-ordinate system R_I and a Lipschitz function a_I, satisfying $\|a_I'\|_\infty \leq M$ and such that the parametric representation of the graph Γ_I of a_I in the co-ordinate system R_I, denoted by $z_I(s)$, satisfies $z_I(s) = z(s)$, for $s \in E \subset I$, with

$$(6.10) \qquad |E| \geq \gamma|I|,$$

$\gamma > 0$, like M, being a constant which depends only on the constant $C \geq 2$ which appears in (6.4).

The second notable point of the proof is the possibility of *transferring* the estimates which apply in the case of Lipschitz curves *onto* the general curves described by Theorem 8. This "transference" method depends on techniques of real analysis, invented by Calderón and Zygmund and then developed by Burkholder and Gundy. These methods are similar to those described in Chapter 7. The next section is about this line of attack.

7 Transference

Let $\mathbb{C}^*$ denote $\mathbb{C} \setminus \{0\}$ and let K be a function defined on $\mathbb{C}^*$ which is homogeneous of degree -1 and odd. We also suppose that K is infinitely differentiable on $\mathbb{C}^*$.

Let μ be a positive, regular, Borel measure on the complex plane. For every $\varepsilon > 0$, we form the truncated operator T_ε^μ defined by

$$(7.1) \qquad T_\varepsilon^\mu f(z) = \int_{|z-w| \geq \varepsilon} K(z-w)f(w)\,d\mu(w),$$

when f is a continuous function of compact support.

We define the maximal operator T_*^μ by

$$(7.2) \qquad T_*^\mu f(z) = \sup_{\varepsilon > 0} |T_\varepsilon^\mu f(z)|.$$

The problem of characterizing the measures μ such that, for every kernel K satisfying the hypotheses above, the operator T_*^μ is bounded on $L^2(d\mu)$, is still open.

We intend to show that the operator is bounded when μ is the arc-length measure on an Ahlfors regular curve Γ.

We know that this is the case when Γ is a Lipschitz graph. Furthermore, the hypotheses on K and the inequality we intend to establish are invariant under rotation and translation.

We can therefore allow arbitrary displacements on Lipschitz graphs. The corresponding arc-length measures $d\mu$ will uniformly be "good measures" for which $\|T_*^\mu f\|_2 \leq C\|f\|_2$. The constant C depends only on the Lipschitz norm we use.

We shall restrict our attention to a collection Σ of positive, regular, Borel measures μ. To say that $\mu \in \Sigma$ will mean there are constants C_1 and C_2 such that, for each disk D of radius r in the complex plane, we have $\mu(D) \leq C_2 r$ with, further, $\mu(D) \geq C_1 r$, when the centre z_0 of D lies in the support of μ.

Proposition 2. *Let μ and σ be measures in Σ. Suppose that, for every $1 < p < \infty$, the operator T_*^μ is bounded on $L^p(d\mu)$. Then $T_*^\mu : L^p(d\mu) \to L^p(d\sigma)$ and $T_*^\sigma : L^p(d\sigma) \to L^p(d\mu)$ are both continuous.*

The proof of the proposition depends on a series of lemmas very similar to those used in Chapter 7.

If $\mu \in \Sigma$ and $f \in L^1(d\mu)$, we define the maximal function $M_\mu(f)$ by

$$(7.3) \qquad M_\mu(f)(z) = \sup_{r>0} \left\{ \frac{1}{r} \int_{|z-w| \leq r} |f(w)|\, d\mu \right\}.$$

This definition has the following property: on the support of μ, the maximal function $M_\mu f$ dominates $|f|$ pointwise (up to a multiplicative constant). But, outside the the support of μ, there is no relationship at all between $M_\mu f$ and $|f|$. The next lemma can be understood in the light of this remark. Indeed, if μ_1 and μ_2 are measures in Σ with disjoint supports, it is impossible to get $\|f|_{L^p(d\mu_2)} \leq C\|f\|_{L^p(d\mu_1)}$ for every $f \in L^p(d\mu_1)$. But, since it is not true that $M_{\mu_1} f \geq |f|$ everywhere, we can hope to get the inequality

$$\|M_{\mu_1} f\|_{L^p(d\mu_2)} \leq C\|f\|_{L^p(d\mu_1)}, \qquad 1 < p \leq \infty.$$

Lemma 9. *Let μ_1 and μ_2 be measures in Σ. Then, for every $p > 1$, there exists a constant $C = C(\mu_1, \mu_2, p)$ such that, for every function $f \in L^p(d\mu_1)$,*

$$(7.4) \qquad \|M_{\mu_1} f\|_{L^p(d\mu_2)} \leq C\|f\|_{L^p(d\mu_1)}.$$

The proof of this lemma is a variant of the proof of the Hardy-Littlewood theorem. For the convenience of the reader, we shall give it here.

We first observe that the sub-linear operator $M_{\mu_1} : L^\infty(d\mu_1) \to L^\infty(d\mu_2)$ is continuous, because $\mu_1 \in \Sigma$. If we can show that M_{μ_1} is continuous from $L^1(d\mu_1)$ to weak-$L^1(d\mu_2)$, then the Marcinkiewicz interpolation theorem will give the result we want.

To obtain the weak-type estimate, we suppose that $\int |f| d\mu_1 = 1$ and we let $\lambda > 0$ be a threshold value for testing the weak continuity.

Let $m \geq 1$ be an integer (which will be an index of truncation, tending to infinity) and $\Omega_m = \{z : |z| < m \quad \text{and} \quad M_{\mu_1} f(z) > \lambda\}$. Ω_m is open, because M_{μ_1} is lower semi-continuous.

Putting $\nu_1 = |f| d\mu_1$, the definition of the maximal function shows that, for every $z \in \Omega_m$, there exists a compact disk D_z, centre z, radius $r(z) > 0$, such that $\nu_1(D_z) > \lambda r(z)$.

From Besicovitch's theorem ([126], p. 2), we get a (possibly finite) sequence $D_k = D_{z_k}$ of disks D_z such that Ω_m is contained in the union of the disks D_k and such that each point belongs to at most C_0 of these disks.

Since $\mu_2 \in \Sigma$, we get

$$\mu_2(\Omega_m) \leq \sum_{k \geq 0} \mu_2(D_k) \leq C \sum r_k < C\lambda^{-1} \sum \nu_1(D_k)$$

$$\leq CC_0 \lambda^{-1} \nu_1(1) = CC_0 \lambda^{-1}.$$

To conclude, it is enough to let m tend to infinity.

Lemma 10. *If $\mu \in \Sigma$, then there exists a constant $C = C(\mu)$ such that, for every $z_0 \in \mathbb{C}$ and each $r > 0$,*

$$(7.5) \qquad r \int_{|z-z_0| \geq r} \frac{|f(z)|}{|z - z_0|^2} \, d\mu(z) \leq CM_\mu f(z_0).$$

The proof is immediate. It is enough to decompose $|z - z_0| \geq r$ into dyadic annuli $2^j r \leq |z - z_0| < 2^{j+1} r$ and to apply the definition of $M_\mu f$ to each of these.

The significance of the next lemma is that it is enough to know the maximal operator T_*^μ on the support of μ to be able to evaluate it everywhere. This confirms our intuition, because it is on the support of μ that the situation is at its worst.

Lemma 11. *Let $\mu \in \Sigma$. There exist constants C_1 and C_2 (depending on μ and K) such that, for every $z \in \mathbb{C}$ and every continuous function f of compact support,*

$$(7.6) \qquad T_*^\mu f(z) \leq C_1 M_\mu(T_*^\mu f)(z) + C_2 M_\mu f(z).$$

To establish (7.6), we first replace z by z_0 and the left-hand side of (7.6) by $T_*^\mu(z_0)$. We now want to find an estimate, uniform in $\varepsilon > 0$, of $|T_\varepsilon^\mu f(z_0)|$.

Let $D(\varepsilon)$ be the open disk $|z - z_0| < \varepsilon$ and let $D(\varepsilon/2)$ be the disk of half the radius. We write $f = f_1 + f_2$, where $f_1(z) = f(z)$, on $D(\varepsilon)$, and $f_1(z) = 0$, otherwise. Then

$$T_\varepsilon^\mu f(z_0) = T^\mu f_2(z_0) = \int K(z_0 - w) f_2(w) \, d\mu(w).$$

We then observe that, for every $z \in D(\varepsilon/2)$,

(7.7) $$|T^\mu f_2(z) - T^\mu f_2(z_0)| \le C M_\mu f(z_0)$$

(we could even replace the right-hand side by $C M_\mu f(z_1)$, where $z_1 \in D(\varepsilon/2)$, a remark that we shall use later).

The above inequality follows immediately from Lemma 10 and the fact that

$$|K(z_0 - w) - K(z - w)| \le C|z - z_0||z - w|^{-2},$$

if $z \in D(\varepsilon/2)$ and $w \notin D(\varepsilon)$.

The inequality (7.7) leads to

$$T^\mu f_2(z_0) = T^\mu f_2(z) + O((M_\mu f)(z_0)) = T_\varepsilon^\mu f(z) + O((M_\mu f)(z_0))$$

and then to

(7.8) $$|T_\varepsilon^\mu f(z_0)| \le T_\star^\mu f(z) + C M_\mu f(z_0),$$

for $z \in D(\varepsilon/2)$.

We shall compare ε with the distance d from z_0 to the support E of μ.

If $\varepsilon \ge 4d$, we take the mean of the inequalities (7.8) with respect to the restriction of μ to $D(\varepsilon/2)$. By the definition of Σ, $\mu(D(\varepsilon/2))$ and ε are of the same order of magnitude and we get

$$|T_\varepsilon^\mu f(z_0)| \le C M_\mu T_\star^\mu f(z_0) + C' M_\mu f(z_0).$$

If $d/2 \le \varepsilon < 4d$, we observe that $|T_\varepsilon^\mu f(z_0) - T_{4d}^\mu f(z_0)| \le C M_\mu f(z_0)$, and we are back to the preceding case.

Finally, if $0 < \varepsilon < d/2$, we necessarily have $T_\varepsilon^\mu = T_{4d}^\mu$, which brings us back, once again, to the first case.

We have proved Lemma 11.

We now return to the proof of Proposition 2. We suppose that, for $1 < p < \infty$, $T_\star^\mu : L^p(d\mu) \to L^p(d\mu)$ is continuous. Then (7.6) and Lemma 9 give the continuity of $T_\star^\mu : L^p(d\mu) \to L^p(d\sigma)$.

In some sense, the continuity of $T_\star^\sigma : L^p(d\sigma) \to L^p(d\mu)$ is the dual of the preceding result. The truncated operators $T_\varepsilon^\mu : L^p(d\mu) \to L^p(d\sigma)$ are uniformly bounded, because the maximal operator $T_\star^\mu$ is continuous as an operator from $L^p(d\mu)$ to $L^p(d\sigma)$. By duality, it follows that the truncated operators $T_\varepsilon^\sigma : L^q(d\sigma) \to L^q(d\mu)$ are also uniformly bounded $(1/p + 1/q = 1)$. We still have to pass to the maximal operator $T_\star^\sigma$, for which we again need some "real-variable" techniques.

Let ε_j be a sequence such that $T^\sigma_{\varepsilon_j}$ converges weakly to an operator that we shall denote by T^σ. So, for $f \in L^q(d\sigma)$ and $g \in L^p(d\mu)$,

$$\langle T^\sigma f, g \rangle = \lim_{j \to \infty} \langle T^\sigma_{\varepsilon_j} f, g \rangle \, .$$

We intend to verify that the pointwise inequality

$$(7.9) \qquad T^\sigma_* f(z_0) \le C M_\mu(T^\sigma f)(z_0) + C_r (M_\sigma |f|^r (z_0))^{1/r}$$

is satisfied, for $1 < r < \infty$ and for all z_0 in the support of μ. This inequality, for $1 < r < p < \infty$, together with Lemma 9, will give the continuity of $T^\sigma_* : L^p(d\sigma) \to L^p(d\mu)$.

To establish (7.9), we again use the decomposition $f = f_1 + f_2$ of Lemma 11, together with the associated notation. For $z \in D(\varepsilon/2)$,

$$T^\sigma_\varepsilon f(z_0) = T^\sigma f_2(z) + R_1(z) = T^\sigma f(z) - T^\sigma f_1(z) + R_1(z) \, ,$$

where

$$|R_1(z)| \le C M_\sigma f(z_0) \, .$$

So

$$(7.10) \qquad |T^\sigma_\varepsilon f(z_0)| \le |T^\sigma f(z)| + |T^\sigma f_1(z)| + C M_\sigma f(z_0)$$

and we compute the mean of the right-hand side of (7.10) with respect to the restriction of μ to $D(\varepsilon/2)$.

Since z_0 is in the support of μ, $\mu(D(\varepsilon/2))$ and ε are of the same order of magnitude. We thus get

$$(7.11) \qquad |T^\sigma_\varepsilon f(z_0)| \le C M_\mu(T^\sigma f)(z_0) + C M_\sigma f(z_0) + R_\varepsilon(z_0) \, ,$$

where

$$\begin{aligned}
R_\varepsilon(z_0) &= \frac{1}{\mu(D(\varepsilon/2))} \int_{D(\varepsilon/2)} |T^\sigma f_1(z)| \, d\mu(z) \\
&\le \left(\frac{1}{\mu(D(\varepsilon/2))} \int_{D(\varepsilon/2)} |T^\sigma f_1(z)|^r \, d\mu(z) \right)^{1/r} \\
&\le C(M_\sigma |f|^r(z_0))^{1/r} \, .
\end{aligned}$$

The last inequality comes from the continuity of $T^\sigma : L^r(d\sigma) \to L^r(d\mu)$.

8 Calderón-Zygmund decomposition of Ahlfors regular curves

We recall that Γ is an oriented, rectifiable Jordan curve in the complex plane, passing through the point at infinity. The arc-length measure on Γ, denoted by $d\sigma$, is defined by $\int f \, d\sigma = \int_{-\infty}^{\infty} f(z(s)) \, ds$, where $z(s)$ is, of course, the arc-length parametric representation of Γ.

We suppose, once and for all, that there exists a constant $C \ge 2$ such that, for every disk D in the complex plane, of radius r,

$$(8.1) \qquad \sigma(D) \le C r \, .$$

The measure σ thus belongs to the class Σ defined in the previous section.

The object of the present section is to show that every Ahlfors regular curve has good Lipschitz copies, of uniform quality, at all scales.

Proposition 3. *There exist constants M and $\nu > 0$, depending only on the constant C of (8.1), such that the following property is satisfied:*

for every interval $I \subset \mathbb{R}$, of length $|I|$, there exist a compact subset $E \subset I$, a Lipschitz function $a_I(x)$ and an orthonormal co-ordinate system R_I such that, if we let Γ_I denote the graph of $a_I(x)$ in R_I and let $z_I(s)$ denote the arc-length parametric representation of Γ_I,

$$(8.2) \qquad \qquad \|a_I'\|_\infty \leq M \,,$$

$$(8.3) \qquad \qquad |E| \geq \nu|I| \,,$$

and, for all $s \in E$,

$$(8.4) \qquad \qquad z_I(s) = z(s) \,.$$

The proof of this proposition depends on a version of the rising sun lemma (Chapter 9, Lemma 7).

Lemma 12. *Let $I = [a, b] \subset \mathbb{R}$ be an interval of length $|I|$ and let $f : I \to \mathbb{R}$ be a continuous function such that $f(b) - f(a) = l > 0$. Then there is a compact subset $E \subset I$ such that, for $x, x' \in E$,*

$$(8.5) \qquad x' \geq x \Rightarrow f(x') - f(x) \geq \frac{l}{2|I|}(x' - x)$$

and

$$(8.6) \qquad \qquad |f(E)| \geq \frac{l}{2} \,.$$

Lemma 12 means that, if the fluctuations of f have a total outcome l which is positive, then f must be increasing, with a slope of at least $l/(2|I|)$, on an appreciable subset of I, whose size is measured by (8.6).

In our application of the lemma, the function $f(x)$ will also be Lipschitz and satisfy $\|f'\|_\infty \leq 1$. Then (8.6) will give $|E| \geq |f(E)| \geq l/2$. In what follows, we shall extend the restriction of f to E as a function g, defined on the whole real line and constrained to be linear on the contiguous intervals of E (and to be of the form $(l/2|I|)x + c$ on each of the remaining half-lines). In this case, we clearly get

$$(8.7) \qquad x' \geq x \Rightarrow g(x') - g(x) \geq \frac{l}{2|I|}(x' - x) \,,$$

for all $x', x \in \mathbb{R}$, and

$$(8.8) \qquad \qquad g(x) = f(x) \qquad \text{for } x \in E.$$

We return to Proposition 3.

Let $I = [a, b]$ and let $K \subset \Gamma$ be the arc of Γ between $z(a)$ and $z(b)$. We denote the diameter of the compact set K by d_K. Since Γ is an Ahlfors regular curve, $|I| \leq C d_K$. Let a_1 and b_1 be two points of I such that $|z(b_1) - z(a_1)| = d_K$. Since $|z(b_1) - z(a_1)| \leq b_1 - a_1$, it follows that $b_1 - a_1 \geq c(b - a)$ for $c = C^{-1} > 0$. We now forget about the interval I, replacing it with $[a_1, b_1]$, and look for the set E in $[a_1, b_1]$. We can further simplify the notation and omit the subscripts, putting $a = a_1$ and $b = b_1$. We may suppose, from now on, that there is a constant $c' > 0$ such that $|z(b) - z(a)| \geq c'(b - a)$.

A translation followed by a rotation now lets us assume that $z(a) = 0$ and $z(b) = l \geq c'(b - a) > 0$.

We finally write $z(s) = x(s) + iy(s)$, in order to apply Lemma 12 to the function $x(s)$.

So we have an increasing function $\xi(s)$ and a constant $c'' > 0$ such that $\xi(s') - \xi(s) \geq c''(s' - s)$ whenever $s', s \in \mathbb{R}$ satisfy $s' > s$. Further, $\xi(s) = x(s)$ for $s \in E$, where $|E| \geq \nu |I|$, with $\nu > 0$ being a constant. We put $\zeta_I(s) = \xi(s) + iy(s)$, where $z(s) = x(s) + iy(s)$. It is no longer true that s is the arc-length of the curve represented parametrically by $\zeta_I(s)$. But the curve Γ_I, given by $\zeta_I(s)$, is the graph of a Lipschitz function and $\zeta_I(s) = z(s)$ when $s \in E$. By the construction of $\xi(s)$, we have $c''(s' - s) \leq \xi(s') - \xi(s) \leq s' - s$, so that the arc-length t of the curve Γ_I and the parameter s are related by $c_1(s' - s) \leq t' - t \leq c_2(s' - s)$, where $c_2 \geq c_1 > 0$ are constants.

All that is left to do is to modify Γ_I at the points $s \notin E$ so that, for each complementary interval (s_j, s'_j), the arc-lengths of Γ_I and of Γ between $z_j = z(s_j)$ and $z'_j = z(s'_j)$ are the same. These arc-lengths are both of the same order of magnitude as $s'_j - s$ or $x'_j - x_j$ (where, again, $x_j = x(s_j)$ and $x'_j = x(s'_j)$).

The modification is very simple. It consists of replacing Γ_I, for $x_j \leq x \leq x'_j$, by a polygonal line $z_j w_j z'_j$, where $w_j = u_j + iv_j$, $u_j = (x_j + x'_j)/2$, and where v_j is chosen appropriately. Doing this does not alter the Lipschitz character of Γ_I, and we have completely proved the proposition.

9 The proof of David's theorem

Let Γ be an Ahlfors regular curve. We let $K(z)$ denote a homogeneous kernel on $\mathbb{C}^*$, of degree -1, odd, and infinitely differentiable. We know that, if $a : \mathbb{R} \to \mathbb{R}$ is a Lipschitz function satisfying $\|a'\|_\infty \leq M$, then the kernel

$$K(a, s, t) = K(s + ia(s) - t - ia(t)), \qquad s, t \in \mathbb{R},$$

defines a continuous linear operator on $L^p(\mathbb{R})$, for $1 < p < \infty$, and that the norm of this operator depends on the function a only to the extent that M does. If we put $z(s) = s + ia(s)$, for $s \in \mathbb{R}$, we see that the same conclusion holds for $K(\lambda(z(s) - z(t)))$, when $|\lambda| = 1$, because $K(z)$ and $K(\lambda z)$ satisfy the same hypotheses.

By the general theory of Calderón-Zygmund operators of Chapter 7, the maximal operator corresponding to $K(a, s, t)$ is also bounded on $L^p(\mathbb{R})$. This maximal operator is not the operator we considered in the earlier sections of this chapter, but Lemma 11 enables us to pass from one to the other.

Lastly, if μ is the arc-length measure along a Lipschitz curve, the operator $T^\mu_\star : L^p(d\mu) \to L^p(d\mu)$ is bounded and the operator norm depends only on M (and K).

Applying Proposition 2, we can transfer this estimate. Let Γ be an Ahlfors regular curve and σ the arc-length measure on Γ. Then

$$(9.1) \qquad \|T^\mu_\star(f)\|_{L^p(d\sigma)} \le C\|f\|_{L^p(d\mu)}$$

and

$$(9.2) \qquad \|T^\sigma_\star(f)\|_{L^p(d\mu)} \le C\|f\|_{L^p(d\sigma)} \,,$$

where $C = C(K, p, M, \Gamma)$.

We want to deduce that

$$(9.3) \qquad \|T^\sigma_\star(f)\|_{L^p(d\sigma)} \le C\|f\|_{L^p(d\sigma)} \,.$$

To do this, we first recall Burkholder and Gundy's lemma, albeit in a slightly different form from that used in Chapter 7.

Lemma 13. *Let p be an exponent, with $1 < p < \infty$, let R be a positive number and u, v be two non-negative functions on $\mathbb{R}$, satisfying the following conditions:*

(9.4) *$u(x)$, restricted to $|x| \ge R$, belongs to L^p;*

(9.5) *there exist $\varepsilon > 0$, $\gamma(\varepsilon) > 0$, and β, with $0 \le \beta < 1$, such that $\beta < (1 + \varepsilon)^{-p}$, and, for every $\lambda > 0$,*

$$|\{x \in \mathbb{R} : u(x) > \lambda + \varepsilon\lambda \text{ and } v(x) \le \gamma(\varepsilon)\lambda\}| \le \beta|\{x \in \mathbb{R} : u(x) > \lambda\}| \,.$$

Then $u \in L^p(\mathbb{R})$ and

$$(9.6) \qquad \|u\|_p \le ((1 + \varepsilon)^{-p} - \beta)^{-1/p}(\gamma(\varepsilon))^{-1}\|v\|_p \,.$$

The proof is identical to that given in Chapter 7, even though the hypotheses are stated in a slightly different way.

In our case, f is a function of compact support (in the complex plane), $u(s) = (T^\sigma_\star f)(z(s))$ and $v(s) = (M_\sigma|f|^r(z(s)))^{1/r}$, where $1 < r < p$. The measure σ is, as we have said, the arc-length measure along the Ahlfors regular curve with arc-length parametric representation $z(s)$.

If we manage to verify (9.5), then David's theorem will follow from (9.6).

By construction, $u(s)$ is lower semi-continuous and vanishes at infinity. The set Ω, defined by $\Omega = \{s : u(s) > \lambda\}$, is thus a bounded open set and hence the union of disjoint open intervals (a_j, b_j).

As usual, we let E_j be the set of $s \in (a_j, b_j)$ such that $u(s) > \lambda + \varepsilon\lambda$, where $\varepsilon > 0$ will be chosen below. We may restrict to the case where there exists $\xi \in (a_j, b_j)$ such that $v(\xi) \leq \gamma\lambda$ and we then intend to show that $|E_j| \leq \beta(b_j - a_j)$. Summing the inequalities then gives (9.5).

As ever, we write $f = f_1 + f_2$, where $f_1(z) = f(z)$, if $|z - z(a_j)| \leq 2l_j$, where $l_j = b_j - a_j$, and $f_1(z) = 0$, otherwise. Then, on repeating the proof of (7.6), we get

$$(9.7) \qquad T_*^\sigma f_2(z) \leq T_*^\sigma f(z(a_j)) + CM_\sigma f(z'),$$

if $|z - z(a_j)| \leq l_j$ and $|z' - z(a_j)| \leq l_j$.

Now $T_*^\sigma f(z(a_j)) \leq \lambda$, by the definition of Ω, and if we choose $z' = z(\xi)$, we get, for $z = z(s)$ with $a_j \leq s \leq b_j$,

$$(9.8) \qquad T_*^\sigma f_2(z) \leq \lambda + C_0\gamma\lambda.$$

We are left with estimating $T_*^\sigma f_1(z)$, which we shall do by replacing the regular curve Γ by a "Lipschitz approximation" Λ. More precisely, there exist a compact subset $K_j \subset [a_j, b_j]$, such that $|K_j| \geq \nu l_j$, and a Lipschitz curve Λ_j, with arc-length parametrization $\zeta_j(s)$, such that $\zeta_j(s) = z(s)$, for $s \in K_j$.

Let ds be arc-length measure on Λ_j. By (9.2),

$$(9.9) \qquad \int_{-\infty}^{\infty} ((T_*^\sigma f_1)(\zeta_j(s)))^r \, ds \leq C \int |f_1|^r \, d\sigma.$$

We get an upper bound for the right-hand side of (9.9) by observing that, if D_j is the disk $|z - z(a_j)| \leq 2l_j$ and if $|z' - z(a_j)| \leq l_j$, then the definition of the operator M_σ gives

$$\int |f_1|^r \, d\sigma = \int_{D_j} |f|^r \, d\sigma \leq C' l_j M_\sigma |f|^r(z').$$

We again take $z' = z(\xi)$ and (9.9) gives

$$(9.10) \qquad \int_{K_j} ((T_*^\sigma f_1)(z(s)))^r \, ds \leq C'' l_j \gamma^r \lambda^r.$$

We choose $\gamma = \gamma(\varepsilon)$ sufficiently small so that $C_0\gamma < \varepsilon/2$ and $C''\gamma^r \leq \nu/2(\varepsilon/2)^r$. By these choices and (9.10), the measure of the set $R_j \subset K_j$ of the s such that $T_*^\sigma f_1(z(s)) \geq \lambda\varepsilon/2$ is not greater than $\nu l_j/2$.

The set $\Delta_j = K_j \setminus R_j$ satisfies $|\Delta_j| \geq \nu l_j/2$ and, if $s \in \Delta_j$, we have

$T^\sigma_* f_1(z(s)) \le \lambda\varepsilon/2$ and $T^\sigma_* f_2(z(s)) \le \lambda + \lambda\varepsilon/2$, giving $T^\sigma_* f(z(s)) \le \lambda + \varepsilon\lambda$.

The number $\beta \in (0,1)$ is thus $1 - \nu/2$ and, taking $0 < \varepsilon < \beta^{-p} - 1$, (9.5) is proved as, indeed, is David's theorem.

10 Further results

David and Semmes are in the process of extending David's theorem on rectifiable curves to hypersurfaces in $\mathbb{R}^{n+1}$.

Here is a result due to Semmes ([212]).

We consider an orientable surface $S \subset \mathbb{R}^{n+1}$ which separates $\mathbb{R}^{n+1}$ into two open connected components Ω_1 and Ω_2 and such that there is a constant $K > 1$ so that, for all $x \in S$ and for all $R > 0$, we have

$$(10.1) \qquad K^{-1}R^n < \sigma\{y : |y - x| \le R\} \le KR^n,$$

where σ is the surface measure on S.

The second hypothesis is a property of accessibility from outside. We suppose that there is a constant $\delta > 0$ such that, for all $x \in S$ and every $R > 0$, we can find $y \in \Omega_1$ and $z \in \Omega_2$ satisfying $|x - y| < R$ and $|x - z| < R$, but such that the balls of centre y and z and of radius δR lie entirely within Ω_1 and Ω_2, respectively.

Then, if $P(x)$ is an odd, homogeneous polynomial of degree l and if $k(x) = P(x)|x|^{-n-l}$, then the operator defined by

$$(10.2) \qquad Tf(x) = \mathrm{PV} \int_S k(x - y)f(y)\, d\sigma(y)$$

is bounded on $L^2(d\sigma)$.

Unlike the one-dimensional case, this result is not optimal and it is unlikely that we can obtain necessary and sufficient conditions for the continuity of all operators of the form (10.2) in terms of the geometry of the surface S.

13

Multilinear operators

1 Introduction

This chapter is devoted to the analysis and construction of multilinear operators. These are given by algorithms which, starting with $k+1$ functions $a_1(x), \ldots, a_k(x), f(x)$, all defined on $\mathbb{R}^n$, give rise to a new function $g(x)$, also defined on $\mathbb{R}^n$. The operators which claim our attention resemble pointwise multiplication $g(x) = a_1(x) \cdots a_k(x) f(x)$ and we shall require them to satisfy Hölder's inequality

$$\|g\|_r \leq C \|a_1\|_{p_1} \cdots \|a_k\|_{p_k} \|f\|_q \,,$$

where $1 < r < \infty$, $1 < p_1, \ldots, p_k \leq \infty$, and $1 < q < \infty$ are exponents satisfying $1/r = 1/p_1 + \cdots + 1/p_k + 1/q$. Further, we shall require that the algorithm which derives g from the sequence $(a_1, \ldots, a_k, f)$ should, just like the usual product, be compatible with translations and positive dilations performed simultaneously on $a_1, \ldots, a_k$ and f. If these conditions (that is, Hölder's inequality and the commutativity rules) are satisfied, then the algorithm is completely defined by a multilinear symbol $\tau(\eta, \xi) \in L^\infty(\mathbb{R}^n \times \mathbb{R}^{nk})$, where $\eta = (\eta_1, \ldots, \eta_k) \in \mathbb{R}^{nk}$, $\xi \in \mathbb{R}^n$, and

$$(1.1) \qquad g(x) = \frac{1}{(2\pi)^{n(k+1)}} \int\int e^{ix \cdot (\xi + \tilde{\eta})} \tau(\eta, \xi) \hat{a}(\eta) \hat{f}(\xi) \, d\eta \, d\xi \,,$$

where $\hat{a}(\eta) = \hat{a}_1(\eta_1) \cdots \hat{a}_k(\eta_k)$, $d\eta = d\eta_1 \cdots d\eta_k$ and $\tilde{\eta} = \eta_1 + \cdots + \eta_k$.

If $\tau(\eta, \xi) = 1$, we get the usual product.

The necessary condition $\tau \in L^\infty(\mathbb{R}^{nk} \times \mathbb{R}^n)$ is not sufficient for (1.1) to define a multilinear operator with the same continuity properties as the usual product (that is, satisfying the Hölder inequality with constant, as given above.)

Multilinear operators arise when we study holomorphic functionals, which themselves appear, for example, in the analysis of *inverse problems*. These are problems where the unknown function $a(x)$ (for example the resistance inside the earth) arises as a coefficient of a partial differential equation. We suppose that, by some means or other, we are equipped with the operator $T(a)[f] = g$ which enables us to solve the differential equation in some appropriate space. The *inverse problem* consists of finding out what $a(x)$ is from knowledge of the operator $T(a) \in \mathcal{L}(H, H)$, where H is a Hilbert space. In many cases, the dependence of $T(a)$ on a is holomorphic and it is tempting to apply the implicit function theorem. A *holomorphic functional* $T(a)$ is one that may be written (for example, in a neighbourhood of 0) as

$$(1.2) \qquad T(a) = T_0 + T_1(a) + \cdots + T_k(a, \ldots, a) + \cdots,$$

where $T_0 \in \mathcal{L}(H, H)$ is independent of a, $T_1(a)$ is linear in a with values in $\mathcal{L}(H, H)$, and, in general, the T_k are "homogeneous polynomials of degree k", that is, restrictions to the diagonal of *multilinear operators* $\tilde{T}_k(a_1, \ldots, a_k)$, with values in $\mathcal{L}(H, H)$.

In order to use the implicit function theorem, it is necessary to have a Banach space B at our disposal, in which the norm of a is equivalent to the operator norm of $T_1(a) : H \to H$. In the examples which follow, the mapping which makes $T_1(a) \in \mathcal{L}(H, H)$ correspond to a cannot be surjective as a mapping from B to $\mathcal{L}(H, H)$. Indeed, the collection of operators $T(a)$ is a closed manifold V in the Banach space $\mathcal{L}(H, H)$ and the *inverse problem* is the study of the function a, regarded as a holomorphic functional on the manifold V.

As the reader will already perceive, it is a question of finding which Banach space B enables the inverse problem to be solved (in the sense we have described), when, as in the majority of cases, the functional $T(a)$ is only defined on a subspace $A \subset B$. The fact that it is a proper subspace corresponds to having to make superfluous hypotheses about regularity to start with, in order to give a meaning to the formalism of (1.2).

We thus have estimates $\|\tilde{T}_k(a_1, \ldots, a_k)\|_{\mathcal{L}(H,H)} \leq C^k \|a_1\|_A \cdots \|a_k\|_A$. These estimates express the fact that $T(a)$ is holomorphic in a neighbourhood of 0 in the Banach space A. But, if $A \neq B$, it is clear that these estimates are not optimal.

The *optimal estimates* are

(1.3) $\|\tilde{T}_k(a_1,\ldots,a_k)\|_{\mathcal{L}(H,H)} \leq C^k\|a_1\|_B \cdots \|a_k\|_B$.

We cannot do better, because $C\|a\|_B \leq \|T_1(a)\|_{\mathcal{L}(H,H)} \leq C'\|a\|_B$. We then say that B is the holomorphy space associated with the holomorphic functional $T(a)$. As we have shown, we cannot attack the inverse problem without knowing the holomorphy space which will parametrize the manifold V of operators $T(a)$.

If $H = L^2(\mathbb{R}^n)$, if $B = L^\infty(\mathbb{R}^n)$ and if the problem has a geometrical significance, then all the algorithms commute with translations and changes of scale. In this case, the multilinear operators

$$\tilde{T}_k(a_1,\ldots,a_k)[f] = g, \qquad a_1,\ldots,a_k \in L^\infty(\mathbb{R}^n), \quad f,q \in L^2(\mathbb{R}^n)$$

which we are trying to construct, will be given by (1.1). Or, more exactly, they will be given by (1.1), subject to supplementary hypotheses on the a_j, $1 \leq j \leq k$. These hypotheses are that the a_j belong to the Wiener algebra $A(\mathbb{R}^n)$ consisting of the Fourier transforms of the functions in $L^1(\mathbb{R}^n)$. Passing from $A(\mathbb{R}^n)$ to $L^\infty(\mathbb{R}^n)$ remains a major difficulty, because we still have only partial results: sufficient conditions on the multilinear symbol which enable us to establish the Hölder inequalities. To show that these conditions are, indeed, sufficient, we apply David and Journé's $T(1)$ theorem.

The same approach will be used in Chapter 14, to deal with Kato's problem about the domain of the square root of an accretive, differential operator, written in a divergent form.

2 The general theory of multilinear operators

Let us begin by recalling the definition and properties of the Wiener algebra $A(\mathbb{R}^n)$. It consists of those functions $\hat{f}$ which are the Fourier transforms of functions $f \in L^1(\mathbb{R}^n)$ and, by definition, the norm of $\hat{f}$ in $A(\mathbb{R}^n)$ is the norm of f in $L^1(\mathbb{R}^n)$. Thus $A(\mathbb{R}^n)$ is a (dense) subalgebra of $C_0(\mathbb{R}^n)$, the algebra of all continuous functions on $\mathbb{R}^n$ which vanish at infinity. Furthermore, the norm of $g \in A(\mathbb{R}^n)$ is invariant under translation and dilation.

The Wiener algebra $A(\mathbb{R}^n)$ is the natural space to start from, for the theory of multilinear operators which commute with translations and dilations.

Indeed, let $R_x : A(\mathbb{R}^n) \to A(\mathbb{R}^n)$ be the operation of translation by x, that is, $R_x f(y) = f(y-x)$. Similarly, for $\delta > 0$, we define $D_\delta : A(\mathbb{R}^n) \to A(\mathbb{R}^n)$ by $D_\delta f(y) = f(\delta^{-1}y)$. In the next proposition, we write A for $A(\mathbb{R}^n)$ and L^2 for $L^2(\mathbb{R}^n)$. Also, in what follows, for

$a \in A$ it will be convenient to write $\hat{a} \in L^1(\mathbb{R}^n)$ for the function whose (inverse) Fourier transform is a.

Proposition 1. *Let $\pi : A^k \times L^2 \to L^2$ be a $(k+1)$-linear operator which is continuous and satisfies*

$$(2.1) \qquad \pi(R_x a_1, \ldots, R_x a_k, R_x f) = R_x \pi(a_1, \ldots, a_k, f),$$

for all $x \in \mathbb{R}^n$, $a_1, \ldots, a_k \in A$, and $f \in L^2$. Then there exists a function $\tau \in L^\infty(\mathbb{R}^{n(k+1)})$, called the symbol of π, such that

$$(2.2) \quad \pi(a_1, \ldots, a_k, f) = \frac{1}{(2\pi)^{n(k+1)}} \int\!\!\int e^{ix\cdot(\xi+\tilde{\eta})} \tau(\eta, \xi) \hat{a}(\eta) \hat{f}(\xi) \, d\eta \, d\xi,$$

where $\eta = (\eta_1, \ldots, \eta_k)$, $\tilde{\eta} = \eta_1 + \cdots + \eta_k$ and $\hat{a}(\eta) = \hat{a}_1(\eta_1) \cdots \hat{a}_k(\eta_k)$.

If, moreover, π commutes with dilations, in the sense that

$$(2.3) \qquad \pi(D_\delta a_1, \ldots, D_\delta a_k, D_\delta f) = D_\delta \pi(a_1, \ldots, a_k, f),$$

for all $\delta > 0$, then τ is homogeneous of degree 0:

$$(2.4) \qquad \tau(\delta\eta, \delta\xi) = \tau(\eta, \xi) \qquad \text{for all } \delta > 0.$$

Conversely, if $\tau \in L^\infty(\mathbb{R}^{n(k+1)})$, then (2.2) defines a multilinear operator which commutes with translations (in the sense of (2.1)) and

$$(2.5) \qquad \|\pi(a_1, \ldots, a_k, f)\|_2 \le \|\tau\|_\infty \|a_1\|_A \cdots \|a_k\|_A \|f\|_2.$$

This means, in particular, that the multilinear theory is trivial if we replace $C_0(\mathbb{R}^n)$ by $A(\mathbb{R}^n)$. The serious problem is to extend (2.5), replacing the right-hand side by $\|\tau\|_E \|a_1\|_\infty \cdots \|a_k\|_\infty \|f\|_2$, where $\|\tau\|_E$ is an appropriate norm (necessarily more precise than the L^∞ norm.) It is, moreover, easy to verify that $\|\tau\|_E$ cannot be the L^∞ norm of τ.

To establish (2.2), we use the following heuristic approach. We replace $a_1(x)$ by $e^{i\alpha_1 \cdot x}, \ldots, a_k(x)$ by $e^{i\alpha_k \cdot x}$, where $\alpha_1, \ldots, \alpha_k \in \mathbb{R}^n$, and we consider the operator $T_{(\alpha_1, \ldots, \alpha_k)} : L^2(\mathbb{R}^n) \to L^2(\mathbb{R}^n)$ defined by

$$(2.6) \qquad T_{(\alpha_1, \ldots, \alpha_k)}(f) = e^{-i(\alpha_1 + \cdots + \alpha_k)\cdot x} \pi(e^{i\alpha_1 \cdot x}, \ldots, e^{i\alpha_k \cdot x}, f).$$

This is not really defined, because the functions $e^{i\alpha \cdot x}$, $\alpha \in \mathbb{R}^n$, do not belong to $A(\mathbb{R}^n)$. But, for all $\delta > 0$, the functions $e^{i\alpha \cdot x} e^{-\delta|x|^2}$ do belong to $A(\mathbb{R}^n)$ and their norm is 1. So what is forbidden appears as a limiting case of what is permitted.

Let us continue to dream. By (2.1), the operators $T_{(\alpha_1, \ldots, \alpha_k)} : L^2 \to L^2$ commute with translations and are thus defined, via the Fourier transform, by a multiplier $\tau(\alpha_1, \ldots, \alpha_k, \xi) \in L^\infty(\mathbb{R}^{n(k+1)})$. If we pretend that $f(x) = e^{i\xi \cdot x}$ is in $L^2(\mathbb{R}^n)$, and if we write $\chi_j(x) = e^{i\alpha_j \cdot x}$, we get

$$(2.7) \qquad \pi(\chi_1, \ldots, \chi_k, f) = \tau \chi_1 \cdots \chi_k f$$

(where the right-hand side of (2.7) is the usual product of the $k+2$ functions $\tau, \chi_1, \ldots, \chi_k$, and f). We then obtain (2.2) from (2.7) "by linearity".

We now give a proper proof of (2.2). Returning to reality, we have $a_1, \ldots, a_k \in A(\mathbb{R}^n)$, $f \in L^2(\mathbb{R}^n)$ and we analyse π by introducing the multilinear form $J : (L^1(\mathbb{R}^n))^k \times (L^2(\mathbb{R}^n))^2 \to \mathbb{C}$, defined by

$$(2.8) \qquad J(b_1, \ldots, b_k, f, g) = \int \pi(\hat{b}_1, \ldots, \hat{b}_k, \hat{f}) \hat{g} \, dx \,,$$

where $\hat{b}_j (= a_j)$ is the Fourier transform of b_j, $1 \leq j \leq k$.

We thus have

$$(2.9) \qquad |J(b_1, \ldots, b_k, f, g)| \leq C \|b_1\|_1 \cdots \|b_k\|_1 \|f\|_2 \|g\|_2$$

and translation-invariance gives

$$(2.10) \qquad J(\chi b_1, \ldots, \chi b_k, \chi f, \bar{\chi} g) = J(b_1, \ldots, b_k, f, g) \,,$$

where $\chi(x) = e^{i\omega \cdot x}$, $\omega \in \mathbb{R}^n$.

To get any further, we restrict J to $(\mathcal{S}(\mathbb{R}^n))^{k+2}$ and use the analysis of multilinear forms given by Schwartz's kernel theorem. This tells us that there exists a distribution S belonging to $\mathcal{S}'(\mathbb{R}^{n(k+2)})$ such that, if $b_1, \ldots, b_k, f$, and g are in $\mathcal{S}(\mathbb{R}^n)$, then

$$(2.11) \qquad J(b_1, \ldots, b_k, f, g) = \langle S, b_1 \otimes \cdots \otimes b_k \otimes f \otimes g \rangle \,.$$

We use (2.10) to get $S = \tilde{\chi} S$, where $\tilde{\chi}(x_1, \ldots, x_k, x_{k+1}, x_{k+2}) = \chi(x_1 + \cdots + x_k + x_{k+1} - x_{k+2})$, for $x_1, \ldots, x_{k+2} \in \mathbb{R}^n$. The distribution S is thus necessarily supported by the set where $\tilde{\chi} = 1$, whatever the choice of ω involved in χ. As a consequence, S is a single-layered distribution, supported by $x_{k+2} = x_1 + \cdots + x_{k+1}$.

Finally, there exists a distribution σ in the variables $x_1, \ldots, x_{k+1}$ of which S is the image, under the mapping which takes $(x_1, \ldots, x_{k+1})$ to $(x_1, \ldots, x_{k+1}, x_1 + \cdots + x_{k+1})$.

We have shown that the identity

$$J(b_1, \ldots, b_k, f, g) = \int \cdots \int b_1(x_1) b_2(x_2) \cdots b_k(x_k)$$
$$\times f(x_{k+1}) g(x_1 + \cdots + x_{k+1}) \, d\sigma(x_1, \ldots, x_{k+1})$$

holds where, by abuse of notation, we have extended the standard form for writing an integral to the case of a distribution.

We shall get at the continuity of J by considering a parameter $\delta > 0$ (whose rôle will be to tend to 0) and the particular choices given by the Gaussian functions

$$b_1(x_1) = \frac{e^{-|x_1 - x_1'|^2/\delta^2}}{\delta^n}, \quad \ldots, \quad b_k(x_k) = \frac{e^{-|x_k - x_k'|^2/\delta^2}}{\delta^n},$$

$$f(x_{k+1}) = \frac{e^{-|x_{k+1} - x_{k+1}'|^2/\delta^2}}{\delta^{n/2}},$$

where $x_1, x_1', \ldots, x_{k+1}, x_{k+1}' \in \mathbb{R}^n$, and, lastly, putting $x_{k+2} = x_1 + \cdots +$

x_{k+1} and $x'_{k+2} = x'_1 + \cdots + x'_{k+1}$,

$$g(x_{k+2}) = \frac{e^{-|x_{k+2}-x'_{k+2}|^2/\delta^2}}{\delta^{n/2}}.$$

Then

$$b_1(x_1)\ldots b_k(x_k)f(x_{k+1})g(x_1 + \cdots + x_{k+1}) = \delta^{-n(k+1)}e^{-Q(x-x')/\delta^2},$$

where $Q(x)$ is the quadratic form $|x_1|^2 + \cdots + |x_{k+1}|^2 + |x_1 + \cdots + x_{k+1}|^2$, $x = (x_1, \ldots, x_{k+1})$, and $x' = (x'_1, \ldots, x'_{k+1})$.

So, for this choice of functions, we see that $J(b_1, \ldots, b_k, f, g)$ is the convolution product of the distribution $\sigma \in \mathcal{S}'(\mathbb{R}^{n(k+1)})$ and an approximate identity G_δ (up to a constant depending only on n and k). We can thus write the inequality expressing the continuity of J as $\|\sigma \star G_\delta\|_\infty \le C$ and, passing to the limit, we get $\sigma \in L^\infty(\mathbb{R}^{n(k+1)})$. We have established (2.2): the symbol τ is none other than σ.

It is now easy to verify that (2.4) follows from (2.3).

Conversely, starting with (2.2), to establish the continuity of the operator defined by that equation, we again look at $\int \pi(a_1, \ldots, a_k, f)g\,dx$, for $g \in L^2(\mathbb{R}^n)$, which is

$$\iint \hat{g}(-\xi - \eta_1 - \cdots - \eta_k)\tau(\eta, \xi)\hat{a}_1(\eta_1)\cdots\hat{a}_k(\eta_k)\hat{f}(\xi)\,d\eta\,d\xi.$$

On integrating with respect to ξ, using the Cauchy-Schwarz inequality, and then integrating with respect to $\eta_1, \ldots, \eta_k$, we establish the continuity. It is a trivial matter to verify (2.1).

Before attempting the difficult case, where we try to replace the Wiener algebra $A(\mathbb{R}^n)$ by the algebra $C_0(\mathbb{R}^n)$ of continuous functions on $\mathbb{R}^n$ which vanish at infinity, we show how to replace $A(\mathbb{R}^n)$ by the algebra $B(\mathbb{R}^n)$ of Fourier-Stieltjes transforms of complex, bounded, regular, Borel measures $\mu \in M(\mathbb{R}^n)$. We follow a line similar to that used in passing from $C_0(\mathbb{R}^n)$ to $L^\infty(\mathbb{R}^n)$.

Let μ be a bounded, regular, Borel measure. We say that a sequence μ_j, $j \in \mathbb{N}$, of bounded, regular, Borel measures *tends strictly* to μ if, for every continuous, bounded function $b(x)$, $\lim_{j\to\infty} \int b(x)\,d\mu_j(x) = \int b(x)\,d\mu(x)$. It comes to the same thing to require this just for every continuous function $b(x)$ which vanishes at infinity, as long as we add the condition

$$\int_{|x|\ge T} d|\mu_j| \le \varepsilon(T),$$

where $\varepsilon(T) \to 0$ as $T \to \infty$, uniformly in j.

We extend the multilinear operator $\pi : A^k \times L^2 \to L^2$ to $(B(\mathbb{R}^n))^k \times L^2(\mathbb{R}^n)$ by continuity. Now, the algebra A is a closed linear subspace (in fact, an ideal) of $B = B(\mathbb{R}^n)$. But A becomes dense in B, when we use the notion of strict convergence. We say that a sequence a_m, $m \in \mathbb{N}$, of functions in A converges strictly to $b \in B$ if the sequence of measures $\hat{a}_m \, d\xi$ converges strictly to μ, where a_m is the Fourier transform of $\hat{a}_m$ and b is the Fourier-Stieltjes transform of the measure μ.

The extra condition that we require of the symbol is that, if $\eta_{j,m} \to \eta_j \in \mathbb{R}^n$, as $m \to \infty$, then $\tau(\eta_{1,m}, \ldots, \eta_{k,m}, \xi - \eta_{1,m} - \cdots - \eta_{k,m}) \to \tau(\eta_1, \ldots, \eta_k, \xi - \eta_1 - \cdots - \eta_k)$, for almost all ξ. This is satisfied, for example, when $\tau(\eta, \xi)$ is continuous on $\mathbb{R}^{nk} \times \mathbb{R}^n \setminus \{(0,0)\}$.

Let us show that, if this continuity property is satisfied, then

$$(2.12) \qquad \lim_{m \to \infty} \|\pi(b, f) - \pi(a_m, f)\|_2 = 0 \,,$$

when $f \in L^2$, $a_m \in A^k$, $b \in B^k$, and whenever a_m tends strictly to b. When we have done this, (2.12) will serve as the definition of $\pi(b, f)$.

Returning to (2.2), we let X_η denote the operation of pointwise multiplication by $e^{i\tilde{\eta} \cdot x}$, where $\eta = (\eta_1, \ldots, \eta_k)$ and $\tilde{\eta} = \eta_1 + \cdots + \eta_k$. We let M_η be the convolution operator which, after applying the Fourier transform, becomes multiplication by $\tau(\eta, \xi - \tilde{\eta})$. Then

$$\pi(a, f) = \frac{1}{(2\pi)^{nk}} \int M_\eta X_\eta(f) \hat{a}(\eta) \, d\eta \,.$$

Now, by the continuity hypotheses on the symbol τ, the $L^2(\mathbb{R}^n)$-valued function of $\eta \in \mathbb{R}^n$ given by $M_\eta X_\eta(f)$ is continuous and bounded. Following on from this remark, (2.12) is just a consequence of the definition of strict convergence, a definition which applies just as well to bounded, continuous functions with values in a separable Hilbert space.

Before concluding this section, we describe some easy results about the collections Γ_k, $k \in \mathbb{N}$, of operators $\pi(a_1, \ldots, a_k)$, where $a_1, \ldots, a_k$ are elements of A or B and where the operators are applied to $f \in L^2$.

Then, if $T \in \Gamma_k$, the same is true for its transpose ${}^t T$. Indeed, if $f, g \in L^2(\mathbb{R}^n)$,

$$\int T(f)g \, dx = \frac{1}{(2\pi)^{n(k+1)}} \iint \hat{g}(-\xi - \tilde{\eta}) \hat{f}(\xi) \hat{a}(\eta) \tau(\eta, \xi) \, d\eta \, d\xi$$

$$= \frac{1}{(2\pi)^{n(k+1)}} \iint \hat{g}(\xi) \hat{f}(-\xi - \tilde{\eta}) \hat{a}(\eta) \tau(\eta, -\xi - \tilde{\eta}) \, d\eta \, d\xi \,.$$

The symbol $\sigma(\eta, \xi)$ of ${}^t T$ is thus $\tau(\eta, -\xi - \tilde{\eta})$.

We observe, similarly, that, if $T_1 \in \Gamma_k$ and $T_2 \in \Gamma_l$, then $T_3 = T_1 \circ T_2$ belongs to Γ_{k+l}. In fact the symbol $\tau_3(\xi, \alpha_1, \ldots, \alpha_k, \beta_1, \ldots, \beta_l)$ of T_3 is given by the product $\tau_1(\xi + \beta_1 + \cdots + \beta_l, \alpha_1, \ldots, \alpha_k) \tau_2(\xi, \beta_1, \ldots, \beta_l)$. The verification is immediate and is left to the reader.

3 A criterion for the continuity of multilinear operators

We are going to investigate the fundamental problem of extending the (L^2-valued) multilinear operators (by continuity) to $(C_0(\mathbb{R}^n))^k \times L^2(\mathbb{R}^n)$ from where they were defined in the first place, that is on $A^k \times L^2$.

Because $\|a\|_\infty \leq \|a\|_A$, for all $a \in A(\mathbb{R}^n)$, the condition $\tau(\eta, \xi) \in L^\infty(\mathbb{R}^{nk} \times \mathbb{R}^n)$ is necessary for the continuity of π from $(C_0)^k \times L^2$ to L^2. But this condition is not sufficient. Here is an example where $n = k = 1$, that is, where the dimension is 1 and the operator is bilinear. We put $\pi(a, f) = (Ha)f$, where H is the Hilbert transform. In this case, $\tau(\eta, \xi) = -i\pi \operatorname{sgn} \eta$, but $\pi(a, f)$ has not got the required continuity property, because H is not bounded as an operator from C_0 into L^∞.

On modifying this example (H being replaced by other convolution operators), we see that we must suppose that a is in the Wiener algebra, if we want to make all the bilinear operators $\pi(a, f)$ continuous, given only that their symbols belong to L^∞.

A sufficient condition for continuity on $(C_0(\mathbb{R}^n))^k \times L^2(\mathbb{R}^n)$ is given by the following theorem.

Theorem 1. *Suppose that the symbol $\tau(\eta, \xi) \in L^\infty(\mathbb{R}^{n(k+1)})$ also satisfies the condition*

$$(3.1) \qquad |\partial^\gamma \tau(\eta, \xi)| \leq C(|\xi| + |\eta|)^{-|\gamma|},$$

where $\gamma = (\gamma_1, \ldots, \gamma_k, \gamma_{k+1})$, for $\gamma_1, \ldots, \gamma_k, \gamma_{k+1} \in \mathbb{N}^n$, and

$$|\gamma| = |\gamma_1| + \cdots + |\gamma_k| + |\gamma_{k+1}| \leq N = n(k+1) + 1.$$

Then the corresponding multilinear operator given by (2.2) satisfies

$$(3.2) \qquad \|\pi(a_1, \ldots, a_k, f)\|_2 \leq C\|a_1\|_\infty \cdots \|a_k\|_\infty \|f\|_2.$$

It follows, from this result, that $\pi(a_1, \ldots, a_k, f)$ extends by continuity to the case where $a_1, \ldots, a_k \in C_0$ and $f \in L^2$.

But we need to extend the operation $\pi(a_1, \ldots, a_k, f)$ to the case where $a_1, \ldots, a_k \in L^\infty$ and $f \in L^\infty$.

If $k = 1$ and we assume that (3.2) holds, then the extension problem has an immediate solution. We know that C_0 is dense in L^∞ for the topology $\sigma(L^\infty, L^1)$. Now, this topology may be defined by the following equivalent condition: a sequence $a_m \in L^\infty$ converges weakly to $b \in L^\infty$ if and only if the operators $A_m : L^2 \to L^2$ of pointwise multiplication by $a_m(x)$ converge weakly to the operator B of pointwise multiplication by $b(x)$, that is, if and only if

$$\lim_{m \to \infty} \langle A_m f, g \rangle = \langle Bf, g \rangle,$$

for all $f, g \in L^2$.

The theorem is then completed by the following proposition.

Proposition 2. *If $k = 1$ and the functions $a_m \in C_0$ converge weakly to $b \in L^\infty$, in the sense of the topology $\sigma(L^\infty, L^1)$, then the operators $T_m : L^2 \to L^2$, defined by $T_m(f) = \pi(a_m, f)$, converge weakly to an operator T, which we denote by $\pi(b, f)$.*

Here is the proof of the above remark.

We consider the trilinear form

$$J(a, f, g) = \int \pi(a, f) g \, dx = \int a \, d\mu_{(f,g)} \,,$$

where the existence of the bounded regular Borel measure $d\mu_{(f,g)}$ follows from the estimate $|J(a, f, g)| \le C\|a\|_\infty \|f\|_2 \|g\|_2$, for $a \in C_0$ and $f, g \in L^2$.

We intend to verify that $d\mu_{(f,g)}$ lies in $L^1(\mathbb{R}^n)$ by approximating general functions f and g in $L^2(\mathbb{R}^n)$ by appropriate simple functions. We shall use functions in $\mathcal{S}(\mathbb{R}^n)$ whose Fourier transforms have compact support away from 0. In that case, $d\mu_{(f,g)} = h(x) \, dx$, where

$$h(x) = \frac{1}{(2\pi)^{2n}} \int\!\!\int e^{-ix\cdot\eta} \hat{g}(-\eta - \xi) \hat{f}(\xi) \tau(\eta, \xi) \, d\eta \, d\xi \,.$$

Integrating at first with respect to ξ, we obtain a function of η, of compact support and whose regularity is enough for $h(x)$ to be continuous and $O(|x|^{-n-1})$ at infinity. By density and continuity, we conclude that $d\mu_{(f,g)}$ belongs to $L^1(\mathbb{R}^n)$, when f and g belong to $L^2(\mathbb{R}^n)$. This gives the required conclusion.

This method, which works if $k = 1$, is no longer applicable if $k \ge 2$, when typically non-linear features appear. An obvious example is the usual product $\pi(a_1, a_2, f) = a_1 a_2 f$. If $n = 1$ and

$$a_{1,m}(x) = e^{imx}(1 + x^2)^{-1/2} \quad \text{and} \quad a_{2,m} = e^{-imx}(1 + x^2)^{-1/2} \,,$$

then both $a_{1,m}$ and $a_{2,m}$ tend weakly to 0, but $a_{1,m} a_{2,m} f$ does nothing of the kind.

We shall therefore abandon this weak topology (for functions or operators) in favour of an idea of convergence (for L^∞ functions) which leads to convergence in the strong operator topology. Recall that a sequence of bounded linear operators $T_m : L^2 \to L^2$, $m \in \mathbb{N}$, converges strongly to $T : L^2 \to L^2$ if and only if, for all $f \in L^2$, the sequence $T_m(f)$ converges to $T(f)$ in norm. At the level of functions, we say that a sequence $a_m(x)$ of $L^\infty(\mathbb{R}^n)$ functions *converges strictly* to $b(x)$ if and only if the operators $A_m(x)$ of pointwise multiplication by $a_m(x)$ converge strongly to the operator B of pointwise multiplication by $b(x)$. By the Banach-Steinhaus theorem, this implies that the sequence $\|a_m\|_\infty$ is bounded and that, for every $T > 0$, $(\int_{-T}^{T} |b(x) - a_m(x)|^2 \, dx)^{1/2}$ tends to 0.

This condition is clearly sufficient, and the following proposition serves as the completion of Theorem 1 for general k.

Proposition 3. *Suppose that, for $1 \leq j \leq k$, the functions $a_{j,m}(x)$ belong to $C_0(\mathbb{R}^n)$ and converge strictly to functions $a_j \in L^\infty(\mathbb{R}^n)$. Then, for every function $f \in L^2(\mathbb{R}^n)$, the functions $\pi(a_{1,m} \ldots, a_{k,m}, f)$ converge in L^2 norm to a limit, denoted by $\pi(a_1, \ldots, a_k, f)$ and independent of the choice of sequences $a_{1,m}, \ldots, a_{k,m}$ used in the approximation. Further, the multilinear operator, thus extended to $(L^\infty)^k \times L^2$, is itself continuous, in the sense that the strict convergence of the functions $a_{j,m} \in L^\infty(\mathbb{R}^n)$ to $a_j \in L^\infty(\mathbb{R}^n)$ gives L^2 convergence of $\pi(a_{1,m} \ldots, a_{k,m}, f)$ to $\pi(a_1, \ldots, a_k, f)$, for every $f \in L^2(\mathbb{R}^n)$.*

We first go back to Theorem 1. Suppose that $f \in \mathcal{S}(\mathbb{R}^n)$ and that $a_1, \ldots, a_k \in A(\mathbb{R}^n)$. We can then compute $g = \pi(a_1, \ldots, a_k, f)$ as the limit, as $\varepsilon > 0$ tends to 0, of

$$g_\varepsilon(x) = \frac{1}{(2\pi)^{n(k+1)}} \int\int e^{ix \cdot (\bar{\eta} + \xi)} e^{-\varepsilon|\eta|^2 - \varepsilon|\xi|^2} \tau(\eta, \xi) \hat{a}(\eta) \hat{f}(\xi) \, d\eta \, d\xi.$$

Lebesgue's dominated convergence theorem gives $g(x) = \lim_{\varepsilon \downarrow 0} g_\varepsilon(x)$. If we know how to establish (3.2) for g_ε, with a constant C independent of ε, then Fatou's lemma will allow us to pass to the limit. We can even truncate τ in a neighbourhood of 0 and from now on we shall discuss such a truncated symbol and also omit the subscript ε.

We let $K(u_1, \ldots, u_k, x)$ denote the inverse Fourier transform of τ. On the one hand, K is continuous on $\mathbb{R}^{n(k+1)}$ and is $O(|u_1| + \cdots + |u_k| + |x|)^{-N})$ at infinity (but the constants are not uniform with respect to $\varepsilon > 0$).

On the other hand, $|K(u, x)| \leq C_0(|u| + |x|)^{-n(k+1)}$ and

$$\left|\frac{\partial K}{\partial u_{j,l}}\right| + \left|\frac{\partial K}{\partial x_j}\right| \leq C_1(|u| + |x|)^{-n(k+1)-1}.$$

A straightforward calculation gives

$$(3.3) \qquad \pi(b, f)x = \int \tilde{K}(b, x, y) f(y) \, dy,$$

where

$$\tilde{K}(b, x, y) = \int \cdots \int K(x - u_1, \ldots, x - u_k, x - y) b_1(u_1) \cdots b_k(u_k) \, du_1 \cdots du_k.$$

Then it is easy to verify that

$$(3.4) \qquad |\tilde{K}(b, x, y)| \leq C'_0 \|b_1\|_\infty \cdots \|b_k\|_\infty |x - y|^{-n},$$

$$(3.5) \qquad \left|\frac{\partial \tilde{K}(b, x, y)}{\partial x_j}\right| \leq C'_1 \|b_1\|_\infty \cdots \|b_k\|_\infty |x - y|^{-n-1},$$

and

$$(3.6) \qquad \left| \frac{\partial \tilde{K}(b, x, y)}{\partial y_j} \right| \leq C_1' \|b_1\|_\infty \cdots \|b_k\|_\infty |x - y|^{-n-1},$$

for $1 \leq j \leq n$.

This is an invitation to use David and Journé's $T(1)$ theorem to establish the continuity of the linear operator $T^{(k)}$ defined by $T^{(k)}(f) = \pi(b_1, \ldots, b_k, f)$.

We proceed by induction on k. We have $T^{(k)}(1) = L(b_1, \ldots, b_{k-1})[b_k]$, where the symbol of L is $\tau(\eta_1, \ldots, \eta_k, 0)$. This symbol satisfies the condition (3.1) and, thus, defines a Calderón-Zygmund operator $T^{(k-1)}$ which is continuous on $L^2(\mathbb{R}^n)$, by the induction hypothesis.

As a consequence, the operator $T^{(k-1)}$ is continuous from L^∞ to BMO and we have $T^{(k)}(1) \in$ BMO. In fact, we are never dealing with singular integrals, because τ has been replaced by τ_ϵ and the qualitative results should really be written as quantitative estimates. We have not done so, in order to keep this section within reasonable bounds.

As far as ${}^tT^{(k)}$ is concerned, the argument is the same, once we have worked out the symbol of the adjoint, which we did at the end of Section 2.

To prove the weak continuity, we may apply Proposition 6 of Chapter 8. Let χ_ω denote the function $e^{i\omega \cdot x}$, $x \in \mathbb{R}^n$. We shall show that there exists a constant C such that $\|T^{(k)}\|_{\text{BMO}} \leq C$, for all $\omega \in \mathbb{R}^n$. This will verify the weak continuity property.

Now, $T^{(k)}(\chi_\omega) = (2\pi)^{-nk} \int \cdots \int e^{ix \cdot (\eta_1 + \cdots + \eta_k)} \tau(\eta_1, \ldots \eta_{k-1}, \eta_k - \omega, \omega)$ $\times \hat{b}_1(\eta_1) \cdots \hat{b}_{k-1}(\eta_{k-1}) \hat{b}_k(\eta_k - \omega) \, d\eta_1 \cdots d\eta_k = L_\omega(b_1, \ldots, b_{k-1})[\chi_\omega b_k]$, where the symbol of L_ω is $\tau(\eta_1, \cdots, \eta_{k-1}, \eta_k - \omega, \omega)$. These symbols satisfy condition (3.1), uniformly in ω, and, once again, the induction hypothesis allows us to conclude.

We now turn to Proposition 3, which we again prove by induction on k. We let $T_m = T_{k,m}$ denote the operators defined, on $L^2(\mathbb{R}^n)$, by $T_m(f) = \pi(a_{1,m}, \ldots, a_{k,m}, f)$. We know that the norms of the operators $T_m : L^2 \to L^2$ form a bounded sequence. To establish the strong convergence of the sequence T_m, it is thus enough to verify that $T_m(f)$ converges in $L^2(\mathbb{R}^n)$ for a dense linear subspace of $L^2(\mathbb{R}^n)$. The subspace V we use is defined by the following conditions: $f \in \mathcal{S}(\mathbb{R}^n)$ and the Fourier transform $\hat{f}$ of f has compact support which does not include 0.

We begin by expanding $\tau(\eta, \xi)\hat{f}(\xi)$ as a series $\sum_0^\infty p_j(\xi)q_j(\eta)$, which will be absolutely convergent with respect to all the norms we use. We get the series by letting $T > 0$ be a sufficiently large number for the support of $\hat{f}(\xi)$ to be contained in $|\xi| \leq T/3$. Then we extend the function of ξ given by $\tau(\eta,\xi)\hat{f}(\xi)$ by periodicity, of period T, in each co-ordinate $\xi_1,\dots,\xi_n$. We compute the Fourier coefficients

$$T^{-n}\int \tau(\eta,\xi)\hat{f}(\xi)e^{-2\pi i\xi\cdot k/T}\,d\xi = q_k(\eta)$$

of this function and arrive at

$$\tau(\eta,\xi)\hat{f}(\xi) = \sum_{k\in\mathbb{Z}^n} q_k(\eta)e^{2\pi i\xi\cdot k/T}\chi(\xi)\,,$$

where $\chi(\xi)$ is a "cut-off" function, whose rôle is to separate the main term $\tau(\eta,\xi)\hat{f}(\xi)$ from the parasitic terms obtained through periodification. A simple change of notation gives the expansion we want.

We now work term by term, forgetting the index j, so we reduce to the case where $\tau(\eta,\xi)\hat{f}(\xi)$ has become $p(\xi)q(\eta)$ with

$$(3.7) \qquad\qquad \sup_{0\le|\alpha|\le N}\|(1+|\eta|)^{|\alpha|}\partial^\alpha q(\eta)\|_\infty < \infty\,.$$

Then

$$\pi(a_{1,m},\dots,a_{k,m},f) = \omega(x)Q(a_{1,m},\dots,a_{k,m})\,,$$

where

$$Q(a_{1,m},\dots,a_{k,m})$$
$$= \int\cdots\int e^{ix\cdot\bar{\eta}}q(\eta_1,\dots,\eta_k)\hat{a}_{1,m}(\eta_1)\cdots\hat{a}_{k,m}(\eta_k)\,d\eta_1\cdots d\eta_k$$

and $\omega(x)$ is continuous and $O(|x|^{-n})$ at infinity. The required conclusion will follow, if we can show that the functions $g_m(x) = Q(a_{1,m},\dots,a_{k,m})$ converge to $g_\infty(x) = Q(a_1,\dots,a_k)$ in $L^2(\mathbb{R}^n,(1+|x|)^{-2n}\,dx)$ norm.

To establish that convergence, consider the operators $L_m : L^2(\mathbb{R}^n) \to L^2(\mathbb{R}^n)$ defined by $L_m(f) = Q(a_{1,m},\dots,a_{k-1,m},f)$. By Theorem 1, these operators are uniformly continuous. Further, if the distribution $R \in \mathcal{S}'(\mathbb{R}^{nk})$ is the inverse Fourier transform of the symbol $q(\eta)$, the kernel $L_m(x,y)$ of L_m is given by

$$L_m(x,y)$$
$$= \int\cdots\int R(x-u_1,\dots,x-u_{k-1},x-y)a_{1,m}(u_1)\cdots a_{k-1,m}(u_{k-1})\,du\,.$$

Since $q(\eta)$ is regular, R decreases sufficiently quickly at infinity for $|L_m(x,y)| \le C|x-y|^{-n-1}$ to hold, where C is independent of m. Moreover, the strict convergence of $a_{j,m}$ to a_j implies that $L_m(x,y)$ converges simply to $L(x,y)$, which convergence is defined by passing to the limit under the integral sign.

By the induction hypothesis on the conclusion of Proposition 3,

$$(3.8) \qquad\qquad \|L_m(f) - L(f)\|_2 \to 0\,,$$

for $f \in L^2(\mathbb{R}^n)$. We also know that

$$(3.9) \qquad\qquad |L_m(x,y)| \le C|x-y|^{-n-1}$$

as well as

$$L_m(x, y) \to L(x, y) \qquad \text{for } x \neq y.$$

These three properties alone imply that $L_m(b_m)$ converges to $L(b)$ in $L^2(\mathbb{R}^n, (1 + |x|)^{-2n} dx)$ norm, whenever b_m converges strictly to b. Indeed, if we decompose L_m into $A_m + B_m$, where B_m is defined by the kernel $L_m(x, y)\chi_{\{|x-y|\geq 1\}}$ (χ_E denoting the characteristic function of E), then $B_m(b_m)$ converges uniformly to $B_\infty(b)$ and $\|B_m(f) - B_\infty(f)\|_2 \to 0$, for $f \in L^2$. This means that we can replace L_m by A_m. It now is enough to observe that the functions $A_m(b_m)$ converge to $A_\infty(b)$ in L^2_{loc} and that these functions satisfy the condition

$$(3.10) \qquad \sup_{x \in \mathbb{R}^n} \left(\int_{|x-y|\leq 1} |A_m(b_m)|^2 \, dy \right)^{1/2} < \infty.$$

This is enough to establish convergence in $L^2(\mathbb{R}^n, (1 + |x|)^{-2n} dx)$.

We conclude this section by another proof of the weak continuity property of the operators $T_k(f) = \pi(a_{1,m}, \ldots, a_{k,m}, f)$ of Theorem 1.

We use Proposition 5 of Chapter 8. To check the weak continuity of T_k, it is enough to show that $\|T_k(\psi_\lambda)\|_2 \leq C$, when ψ_λ, $\lambda \in \Lambda$, is an orthonormal wavelet basis. We observe that the conditions on the multilinear symbol τ are all dilation-invariant. This means that the collection of all the operators T_k given by choices of τ and $a_1, \ldots, a_k$, with $\|a_1\|_\infty \leq 1, \ldots, \|a_k\|_\infty \leq 1$, is globally invariant under translation and dilation. This remark lets us reduce the calculations to a mere $2^n - 1$ calculations involving the "mother" wavelet. We choose the Littlewood-Paley wavelets, so that $\hat{\psi}_\lambda$ has support in the dyadic shell

$$\frac{2\pi}{3} \leq \sup(|\xi_1|, \ldots, |\xi_n|) \leq \frac{8\pi}{3}.$$

Omitting the index λ, we are left with considering

$$\int \int e^{ix \cdot (\xi + \bar{\eta})} \tau(\eta, \xi) \hat{a}(\eta) \hat{\psi}(\xi) \, d\eta \, d\xi$$

and verifying that this function belongs to $L^2(\mathbb{R}^n)$. It is enough to use the expansion $\tau(\eta, \xi) = \sum_0^\infty p_j(\xi) q_j(\eta)$ above to obtain $\|T_k(\psi_\lambda)\|_2 \leq C$.

This proof has the advantage of working in the case where L^∞ is replaced by BMO, a situation that we shall study in the next section.

4 Multilinear operators defined on $(\text{BMO})^k$

We shall try to replace L^∞ by BMO wherever possible in Theorem 1. This question gives rise to two problems. The first is about the possibility of replacing the estimate

$$\|\pi(a_1, \ldots, a_k, f)\|_2 \leq C\|a_1\|_\infty \cdots \|a_k\|_\infty \|f\|_2$$

by a more precise estimate, namely

(4.1) $\|\pi(a_1, \ldots, a_k, f)\|_2 \leq C\|a_1\|_{\text{BMO}} \cdots \|a_k\|_{\text{BMO}}\|f\|_2 \,,$

when $a_1, \ldots, a_k$ are elements of the Wiener algebra $A(\mathbb{R}^n)$.

The second question consists of extending the multilinear operator $\pi(a_1, \ldots, a_k, f)$, well-defined when $a_j \in A(\mathbb{R}^n)$, to the general case when $a_j \in$ BMO. This extension cannot be done naively, because $A(\mathbb{R}^n)$ is not dense in BMO. Nor can we use the $\sigma(\text{BMO}, H^1)$ topology, which is not fine enough for our purpose. We use a middle way, in which weak convergence is reinforced and adapted to non-linear problems.

Definition 1. *Let b_m, $m \in \mathbb{N}$, be a sequence of functions in* $\text{BMO}(\mathbb{R}^n)$. *We say that b_m converges strictly to b if and only if the two following conditions hold:*

(4.2) $\|b_m\|_{\text{BMO}} \leq C$

and there exists a sequence c_m of constants such that, for all $R > 0$,

(4.3) $\displaystyle \lim_{m \to \infty} \int_{|x| \leq R} |b(x) - b_m(x) - c_m|^2 \, dx = 0 \,.$

Here is the reason for the definition. We consider the linear action of BMO on L^2 given by the paraproduct $\pi(\beta, f) = \sum_{-\infty}^{\infty} \Delta_k(\beta) S_{k-1}(f)$ of Chapter 8, section 11. Then we have

Proposition 4. *A sequence β_m, $m \in \mathbb{N}$, of BMO functions, converges strictly to $\beta \in$ BMO if and only if the operators $T_m : L^2 \to L^2$, defined by $T_m(f) = \pi(\beta_m, f)$, converge strongly to the operator T, defined by $T(f) = \pi(\beta, f)$.*

To verify this equivalence, we begin by observing that (4.2) is given by the Banach-Steinhaus theorem. If a sequence $T_m : L^2 \to L^2$ is strongly convergent, then the norms of the operators are bounded. But these norms are equivalent to the BMO norms of the functions β_m.

The operators T_m thus form a bounded sequence of Calderón-Zygmund operators and, if T_m converges strongly to T, then $T_m(1)$ converges to $T(1)$ in the $\sigma(\text{BMO}, H^1)$ topology, as we showed in Lemma 3 of Chapter 7. Thus β_m converges to β in the $\sigma(\text{BMO}, H^1)$ topology and it follows that $\Delta_k(\beta_m)$ converges uniformly to $\Delta_k(\beta)$, for fixed k.

We test the strong convergence on functions $f \in \mathcal{S}(\mathbb{R}^n)$ whose Fourier transform is supported by $2^{l-1} \leq |\xi| \leq 2^{l+1}$, $l \in \mathbb{Z}$. This is sufficient, because these functions form a total subset of $L^2(\mathbb{R}^n)$. Then $S_{k-1}(f) = 0$, if $k \leq l - 1$, and $S_{k-1}(f) = f$, if $k \geq l + 2$. The two other values of k cause no trouble, because we already know that $\Delta_k(\beta_m)$ converges uniformly to $\Delta_k(\beta)$.

We thus arrive at $\sum_{k\geq l+2} f(\Delta_k \beta_m) = f(\beta_m - S_{l+2}(\beta_m))$, which has to converge to $f(\beta - S_{l+2}(\beta))$, in L^2 norm.

Let $R > 0$. We can find a function f, with the above properties, which does not vanish on $|x| \leq R$. Further, the weak convergence of β_m to β lets us replace $S_{l+2}(\beta_m)$ by $S_0(\beta_m)$ and then by the constant $c_m = S_0(\beta_m)(0)$. The error terms, introduced in this way, can be written as a scalar product of β_m with an H^1 function. We thus arrive at condition (4.3).

Conversely, if (4.2) and (4.3) hold, then we easily replace (4.3) by the convergence, in $L^2(\mathbb{R}^n, (1 + |x|)^{-n-1}\, dx)$, of the functions $\beta_m - S_0(\beta_m)$ to the function $\beta - S_0(\beta)$. We then conclude that the functions $\pi(\beta_m, f)$ converge to $\pi(\beta, f)$ in L^2, when f has the special properties as above.

We are now in a position to state the fundamental theorem.

Theorem 2. *Suppose that, together with the notation and hypotheses of Theorem 1, we have* $\tau(\eta_1, \ldots, \eta_k, \xi) = 0$, *whenever one of the* $\eta_j = 0$ *($1 \leq j \leq k$). Then the multilinear operator* $\pi : A^k \times L^2 \to L^2$, *defined by (2.2), satisfies*

$$(4.4) \qquad \|\pi(b_1, \ldots, b_k, f)\|_2 \leq C\|b_1\|_{\mathrm{BMO}} \cdots \|b_k\|_{\mathrm{BMO}} \|f\|_2$$

and extends, by continuity, as a multilinear operator $\pi : (\mathrm{BMO})^k \times L^2 \to L^2$. *This extension is obtained by approximating to general functions* b_j *in* BMO *by sequences* $b_{j,m}$ *of functions in the Wiener algebra* A *which converge strictly to* b_j.

The proof follows that of Theorem 1, step by step. The new facet is the estimate of the kernel of the operator $T : L^2 \to L^2$, defined by $T(f) = \pi(b_1, \ldots, b_k, f)$. The kernel is

$$(4.5) \qquad \tilde{K}(b, x, y) = \int \cdots \int K(x - u_1, \ldots, x - u_k, x - y)$$
$$\times\, b_1(u_1) \cdots b_k(u_k)\, du_1 \cdots du_k\,,$$

where the Fourier transform of K in the sense of distributions is $\tau(\eta, \xi)$.

The expression $(|x_2| + \cdots + |x_k| + |z|)^{nk} K(u_1, x_2, \ldots, x_k, z)$, where $z \in \mathbb{R}^n$ is non-zero, is a molecule, when considered as a function of the variable u_1, centred about 0, of width $|x_2| + \cdots + |x_k| + |z|$. The cancellation condition is ensured by $\tau(0, \eta_2, \ldots, \eta_k, \xi) = 0$. We can thus multiply this molecule by $b_1 \in$ BMO and integrate. We then argue step by step, checking that at each stage we are integrating a molecule multiplied by a BMO function. This calculation, of which we have given only an outline, leads to

$$|\tilde{K}(b, x, y)| \leq C\|b_1\|_{\mathrm{BMO}} \cdots \|b_k\|_{\mathrm{BMO}} |x - y|^{-n}\,.$$

The estimates of the partial derivatives of $|\tilde{K}(b, x, y)|$ with respect to

x and y are established similarly. The continuity expressed by (4.4) is arrived at by applying David and Journé's $T(1)$ theorem. Of course, we need to carry out a process of induction on k, with $T^{(k)}(1)$ being given by $T^{(k-1)}(b_k)$, where $T^{(k-1)}$ is, by induction, a Calderón-Zygmund operator. This operator takes BMO to itself, if and only if $T^{(k-1)}(1) = 0$, modulo the constant functions, and it is this condition which is ensured by $\tau(\eta_1, \ldots, \eta_{k-1}, 0, \xi) = 0$.

The weak continuity property cannot be established by the first of the proofs we gave, because the product of a BMO function by a character χ_ω does not, in general, belong to BMO. On the other hand, the second proof extends very easily and we leave the reader the task of doing this.

Before finishing this section, let us observe that the conditions

$$\tau(0, \eta_2, \ldots, \eta_k, \xi) = \cdots = \tau(\eta_1, \ldots, \eta_{k-1}, 0, \xi) = 0$$

are necessary for extending multilinear operators to $(\mathrm{BMO})^k$. Indeed, let us fix $a_1, \ldots, a_{k-1}$ and f in the subspace V used above and replace a_k by the sequence $a_{k,m} = \theta(m^{-1}x)$, where the Fourier transform $\hat\theta$ of θ is a function of compact support. We further suppose that $\hat\theta$ is positive and $\theta(0) = 1$. Then

$$\pi(a_1, \ldots, a_{k-1}, a_{k,m}, f)(x)$$
$$= \frac{1}{(2\pi)^{n(k+1)}} \int \cdots \int e^{ix\cdot(\eta_1 + \cdots + \eta_k + \xi)} \tau(\eta, \xi)$$
$$\times \hat a_1(\eta_1) \cdots \hat a_{k-1}(\eta_{k-1}) m^n \hat\theta(m\eta_k) \hat f(\xi)\, d\eta\, d\xi$$
$$\to \frac{1}{(2\pi)^{nk}} \int \cdots \int e^{ix\cdot(\eta_1 + \cdots + \eta_{k-1} + \xi)} \tau(\eta_1, \cdots, \eta_{k-1}, 0, \xi)$$
$$\times \hat a_1(\eta_1) \cdots \hat a_{k-1}(\eta_{k-1}) \hat f(\xi)\, d\eta\, d\xi\,.$$

The convergence takes place in $L^2(\mathbb{R}^n)$. But the sequence $\theta(m^{-1}x)$ tends to 0 strictly and we must therefore have $\tau(\eta_1, \ldots, \eta_{k-1}, 0, \xi) = 0$.

5 The general theory of holomorphic functionals

We now are going to relate the theory of multilinear operators of the last few sections to the theory of a certain type of holomorphic functionals. These functionals are holomorphic on the Banach space $L^\infty(\mathbb{R}^n)$, taking values in the algebra $\mathcal{A} = \mathcal{L}(L^2(\mathbb{R}^n), L^2(\mathbb{R}^n))$ of bounded linear operators. We let $\Omega \subset L^\infty(\mathbb{R}^n)$ denote the open unit ball; the functions of $L^\infty(\mathbb{R}^n)$ may take real or complex values.

Let G be the group of affine transformations $g(x) = \delta x + h$, $\delta > 0$, $h \in \mathbb{R}^n$. We define the unitary action U_g of G on $L^2(\mathbb{R}^n)$ by $U_g f(x) =$

$\delta^{-n/2}f(\delta^{-1}(x-h))$. For the action of G on $L^\infty(\mathbb{R}^n)$, it will be appropriate to use the operators $V_g b(x) = b(\delta^{-1}(x-h))$, $\delta > 0$, $h \in \mathbb{R}^n$.

We are interested in the holomorphic functionals

(5.1) $$F : \Omega \to \mathcal{A}$$

which obey the commutativity rule

(5.2) *for every $b \in \Omega$ and $g \in G$, $F(V_g b) = U_g F(b) U_g^{-1}$*

and the continuity rule

(5.3)   *if $0 < r < 1$ and if the $b_j \in \Omega$ satisfy $\|b_j\|_\infty \leq R$ and converge strictly to b, then the operators $F(b_j)$ converge strongly to $F(b)$.*

We can express the latter condition more symmetrically by requiring that, if the operators $B_j : L^2 \to L^2$, given by pointwise multiplication by the functions $b_j(x)$, converge strongly to the operator B (pointwise multiplication by $b(x)$), then $F(b_j)$ converges strongly to $F(b)$.

This condition means that, in a certain sense, $F(b)$ resembles the operation of pointwise multiplication by $b(x)$. As in the multilinear case, the condition is indispensable if we want to start our analysis of $F(b)$ by restricting to the case where b is in the Wiener algebra $A(\mathbb{R}^n)$.

The holomorphic functionals we have just defined are analysed in the following theorem. Synthesis, on the other hand, poses difficult problems which we shall describe later.

Theorem 3. *A holomorphic functional $F : \Omega \to \mathcal{A}$, which satisfies (5.2) and (5.3), has the form*

(5.4) $$F(b) = T_0 + T_1(b) + \cdots + T_k(b, \ldots, b) + \cdots,$$

where, for $b_1, \ldots, b_k$ in the Wiener algebra $A(\mathbb{R}^n)$,

(5.5) $T_k(b_1, \ldots, b_k)(f)(x)$
$$= \frac{1}{(2\pi)^{n(k+1)}} \int \cdots \int e^{ix \cdot (\xi + \eta_1 + \cdots + \eta_k)} \tau_k(\eta_1, \ldots, \eta_k, \xi)$$
$$\times \hat{a}_1(\eta_1) \ldots \hat{a}_k(\eta_k) \hat{f}(\xi) d\eta_1 \ldots d\eta_k \, d\xi.$$

The corresponding multilinear symbol τ_k satisfies

(5.6) $$\tau_k \in L^\infty(\mathbb{R}^{n(k+1)}),$$

and, for all $\lambda > 0$,

(5.7) $$\tau_k(\lambda\eta_1, \ldots, \lambda\eta_k, \lambda\xi) = \tau_k(\eta_1, \ldots, \eta_k, \xi).$$

Further, there exists a constant $C \geq 0$ such that, for all $k \in \mathbb{N}$, every choice of $b_1, \ldots, b_k$ in $L^\infty(\mathbb{R}^n)$, and each $f \in L^2(\mathbb{R}^n)$,

(5.8) $$\|T_k(b_1, \ldots, b_k)(f)\|_2 \leq C^k \|b_1\|_\infty \cdots \|b_k\|_\infty \|f\|_2.$$

Finally, the operators T_k inherit the continuity property (5.3) from the functional F.

Before proving Theorem 3, we must remind the reader that holomorphic functionals F on an infinite Banach space B may have the following pathological behaviour. It is possible to find functionals which are entire on the space $L^\infty(\mathbb{R})$ but which are not bounded on a given ball centre 0. Here is an example. We consider the linear forms

$$\lambda_k(f) = \int_0^1 f(x)e^{ikx}\,dx\,,$$

for $k \in \mathbb{N}$ and $f \in L^\infty(\mathbb{R})$, and we put

$$F(f) = \sum_0^\infty (\lambda_k(f))^k\,.$$

This functional is entire, because $\lambda_k(f) \to 0$ as $k \to \infty$. But it is not bounded on the ball $\|f\|_\infty \le 2$. If it were, $\int_0^{2\pi} e^{-ik\theta} F(e^{i\theta} f)\,d\theta$ would also be bounded (uniformly in k). Thus

$$\sup\left\{ \left| \int_0^{2\pi} e^{-ik\theta} F(e^{i\theta} f)\,d\theta \right| \ : \ \|f\|_\infty \le 2 \right\}$$

would be a bounded sequence. But this sequence is $2\pi 2^k$.

We return to Theorem 3. A holomorphic functional F on the unit ball $\|b\|_\infty < 1$ is continuous at 0 and thus bounded on a neighbourhood of 0, on some ball $\|b\|_\infty < \delta$. But we have no control over the value of $\delta > 0$.

We put

$$T_k(b) = \frac{1}{2\pi} \int_0^{2\pi} e^{-ik\theta} F(e^{i\theta} b)\,d\theta\,,$$

when $\|b\|_\infty < \delta$, and thus have $\|T_k(b)\|_\mathcal{A} \le C$, a constant which is independent of k. By homogeneity, for any b,

$$\|T_k(b)\|_\mathcal{A} \le C\delta^{-k}\|b\|_\infty^k\,.$$

The $T_k(b)$ are homogeneous polynomials in b of degree k and thus the restrictions to $b_1 = \cdots = b_k = b$ of symmetric multilinear operators. This gives (5.5), and (5.6).

From F, the T_k inherit the commutativity identities (5.2) and the continuity property (5.3). It follows that the T_k can be analysed by applying Proposition 1, thus completing the proof of Theorem 3.

This is as far as we can get, because we do not, at the moment, have any workable continuity criteria to get the growth of operator norms required by (5.8) which depend only on the form of the multilinear symbols τ_k. Another difficulty arises when we try to add together the estimates (5.8). It can happen, as we have already indicated, that $F(b)$ is holomorphic in $\|b\|_\infty < 1$, while the optimal constant C in (5.8) is 10^{10}. Then, we can only add the inequalities for $\|b\|_\infty < 10^{-10}$.

Let us examine an example to illustrate the preceding considerations.

Let $n = 1$ and let $b(t) \in L^\infty(\mathbb{R})$ be a real- or complex-valued function with $\|b\|_\infty < 1$. Let $B(t)$ be a primitive of $b(t)$ and put $z(t) = t + B(t)$, $-\infty < x < \infty$. Then $z(t)$, $-\infty < t < \infty$, is a parametrization of a Lipschitz graph Γ. Indeed, we write $b = b_1 + ib_2$, where b_1 and b_2 are real-valued and $\|b_1\|_\infty \leq r < 1$. We put $x = t + B_1(t)$, $y = B_2(t)$ and observe that the mapping which takes t to x is bi-Lipschitz. Thus $y = A(x)$, where $\|A'\|_\infty \leq \|b\|_\infty/(1 - \|b\|_\infty)$. We consider the operator defined by the Cauchy kernel on Γ, acting on $L^2(\Gamma)$. Or, rather, we adopt the equivalent realization given by $K(x, y) = \mathrm{PV}(z(x) - z(y))^{-1}$ acting on $L^2(\mathbb{R}, dx)$. We let $T(b)$ denote this last operator.

Theorem 4. *With the above notation, $T(b)$ is a holomorphic functional on $\|b\|_\infty < 1$, taking values in $\mathcal{L}(L^2(\mathbb{R}), L^2(\mathbb{R}))$, and satisfying conditions (5.2) and (5.3).*

Before proving the theorem, let us remark that, when $b(x)$ is real-valued, then $T(b) = \pi M_h U_h H U_h^{-1} M_h$, where H is the Hilbert transform, $h(x) = x + B(x)$, U_h is the isomorphism induced by the bi-Lipschitz change of variable h, that is, $U_h(f) = (f \circ h)\sqrt{h'(x)}$. Lastly, M_h is the operation of pointwise mutiplication by $(h'(x))^{-1/2}$. When $b(x)$ is real-valued, $T(b)$ is thus obviously bounded on $L^2(\mathbb{R})$. *The Cauchy kernel on a curve thus appears as the analytic continuation of the operators which we get from the Hilbert transform by conjugation with bi-Lipschitz changes of variable.*

To establish Theorem 4, let $K_{\varepsilon,R}(x, y) = (z(x) - z(y))^{-1}$, when $\varepsilon < |x - y| \leq R$ $(R > \varepsilon > 0)$, and $K_{\varepsilon,R} = 0$ otherwise, be the truncated kernels. Since $T(b)$ is a Calderón-Zygmund operator, the truncated operators $T_{\varepsilon,R}$ defined by $K_{\varepsilon,R}$ are uniformly bounded on $L^2(\mathbb{R})$ and converge strongly to $T(b)$. These truncated operators $T_{\varepsilon,R}(b)$ are clearly holomorphic on $\|b\|_\infty < 1$ and are uniformly bounded on $\|b\|_\infty \leq r < 1$. Since the truncated operators converge strongly to $T(b)$, the latter is also holomorphic on $\|b\|_\infty < 1$.

We expand the operator $T(b)$ as the series $\sum_0^\infty (-1)^k \Gamma_k(b)$, where the $\Gamma_k(b)$ are defined by the kernels $\mathrm{PV}((B(x) - B(y))^k/(x - y)^{k+1})$. The kernels of the corresponding multilinear operators are

$$\mathrm{PV} \frac{(B_1(x) - B_1(y)) \cdots (B_k(x) - B_k(y))}{(x - y)^{k+1}},$$

where $B_1' = b_1 \in L^\infty(\mathbb{R}), \ldots, B_k' = b_k \in L^\infty(\mathbb{R}^n)$. The corresponding multilinear symbol is

$$\tau(\eta_1, \ldots, \eta_k) = \frac{\pi}{k!} \frac{1}{\eta_1 \cdots \eta_k} \Delta_{\eta_1} \cdots \Delta_{\eta_k}(\xi^k \operatorname{sgn} \xi),$$

where

$$\Delta_\alpha f(\xi) = f(\xi + \alpha) - f(\xi) \,.$$

This multilinear approach was the basis of all subsequent work on Calderón's programme, but did not give the estimate (5.8), which Calderón had obtained in 1977 by complex variable methods ([38]).

In fact, we can even improve the estimate (5.8) if we have a bound for the norm $T(b) : L^2 \to L^2$, when $\|b\|_\infty < 1$ and $\|b\|_\infty$ is close to 1. Using the results of Chapter 9, we get

$$\|T(b)\| \le C(1 - \|b\|_\infty)^{-5} \,.$$

Since

$$\Gamma_k(b) = \frac{1}{2\pi} \int_0^{2\pi} e^{-ik\theta} T(e^{i\theta} b) \, d\theta \,,$$

we get $\|\Gamma_k(b)\| \le C(1 - \|b\|_\infty)^{-5}$, if $\|b\|_\infty < 1$. We then bring the homogeneity of Γ_k into play and suppose that $\|b\|_\infty \le 1 - (k+1)^{-1}$. In this case, we get $\|\Gamma_k(b)\|_\infty \le C(1 + k)^5$, and this estimate extends to the case $\|b\|_\infty \le 1$, since $(1 - (k+1)^{-1})^{-k} \le C'$.

This estimate is not optimal ([53]). We do not know what the best estimate is.

On the other hand, the optimal estimate of $\|T(b)\|$ is known and is given by $C(1 - \|b\|_\infty)^{-1/2}$, as has been shown by David ([94]) and Murai ([194]).

We conclude the proof of Theorem 4 by verifying that, if the b_j converge strictly to b, and if $\|b_j\|_\infty \le r < 1$, then the corresponding operators $T(b_j)$ converge strongly to $T(b)$.

We again write $T(b) = \sum_0^\infty (-1)^k \Gamma_k(b)$, where we have shown that

$$\|\Gamma_k(b)\| \le C(1 + k)^5 \|b\|_\infty^k \,.$$

The series of operators which appear in $T(b_j)$ and $T(b)$ are absolutely convergent, uniformly in j. It is thus sufficient to verify that, for each fixed k, the operators $\Gamma_k(b_j)$ converge strongly to $\Gamma_k(b)$. Since we already have uniform estimates for the norms of the operators involved, we only need verify that, for each C^1 function f, of compact support, $\Gamma_k(b_j)[f]$ converges to $\Gamma_k(b)[f]$ in L^2 norm.

Let $[-T, T]$ be an interval containing the support of f. If $|x| \ge T + 1$, we can write

$$\Gamma_k(b_j)[f](x) = \int \frac{(B_j(x) - B_j(y))^k}{(x - y)^{k+1}} f(y) \, dy \,.$$

This gives

$$|\Gamma_k(b_j)[f](x)| \le C(f)|x|^{-1}$$

and the functions $\Gamma_k(b_j)[f]$ converge simply to $\Gamma_k(b)[f]$. All this gives L^2 convergence on $|x| \geq T + 1$.

To deal with the L^2 convergence on $[-T - 1, T + 1]$, we integrate by parts, which gives the two terms

$$\Gamma_{k-1}(b_j)[b_j f] \qquad \text{and} \qquad \int \frac{(B_j(x) - B_j(y))^k}{(x - y)^k} f'(y)\, dy\,.$$

The first is dealt with by the induction hypothesis on k allied to the following remark: if the operators $T_j : L^2 \to L^2$ converge strongly to T and the functions $u_j \in L^2$ converge to u in L^2 norm, then $T_j(u_j)$ converges to $T(u)$ in L^2 norm.

The second term is susceptible to Lebesgue's dominated convergence theorem, because the kernel is no longer singular.

In the example that we have examined, $L^\infty(\mathbb{R})$ is the *domain of holomorphy* of the functional $T(b)$, in the sense that we cannot extend the holomorphic functional $T(b)$ to an open subset of a larger Banach space. To see this, it is enough to observe that $\|\Gamma_1(b)\| \geq \pi\|b\|_\infty$, for every $b \in L^\infty(\mathbb{R})$. This inequality may be proved by means of the following lemma.

Lemma 1. *Let $T : L^2(\mathbb{R}^n) \to L^2(\mathbb{R}^n)$ be a bounded linear operator with distribution-kernel $K(x, y)$. For $x_0 \in \mathbb{R}^n$ and $\delta > 0$, write $g(x) = \delta^{-1}(x - x_0)$. Then $\delta^n K(\delta x + x_0, \delta y + x_0)$, is the distribution-kernel of $T_{(x_0, \delta)} = U_g T U_g^{-1}$, whose norm is the same as that of T.*

In our application, $K(x, y) = \mathrm{PV}((B(x) - B(y))/(x - y)^2)$, where B is a primitive of b. We let $\delta > 0$ tend to 0: the weak limit of the operators $T_{(x_0, \delta)}$ is $\pi b(x_0) H$, where H is the Hilbert transform. This weak limit exists at each x_0 for which $b(x_0)$ is the derivative of $B(x)$. We thus get $\pi\|b\|_\infty \leq \|\Gamma_1(b)\|$, as claimed.

6 Application to Calderón's programme

The sharper form of Calderón's symbolic calculus has the following goal: to achieve precise results about the commutator of a classical pseudo-differential operator (essentially a convolution operator) with an operator of pointwise multiplication by a function with prescribed regularity.

We shall give examples where the optimal results can be obtained by the multilinear calculus developed in section 3.

Let us start with a one-dimensional example, in which the pseudo-differential operator is the Hilbert transform H.

Theorem 5. Let $b(x)$ be a locally integrable function of the real variable x and let B be the operator of pointwise multiplication by $b(x)$. Then the commutator $[H, B]$ is bounded on $L^2(\mathbb{R})$ if and only if $b(x)$ is in BMO. In that case, if the functions $b_j \in$ BMO converge to b in the $\sigma(\text{BMO}, H^1)$ topology, then the corresponding commutators $[H, B_j]$ converge to $[H, B]$ in the weak operator topology.

Let us start by verifying the necessity of the condition $b \in$ BMO. Let $I = [a, b]$ be an interval of length l and let J be the interval $[b+l, b+2l]$. We let $K(x, y)$ be the kernel of $T = [B, H]$, which we suppose to be bounded on $L^2(\mathbb{R})$. One way of testing the L^2 continuity is to consider the auxiliary function $g(x) = \int_J K(x, y)(x - y)\, dy$, $x \in I$, and to verify that $\|g\|_{L^2(I)} \leq Cl^{3/2}$. Indeed, if we write $x - y = x - x_0 + x_0 - y$ and $g(x) = g_1(x) + g_2(x)$, where $g_1(x) = (x - x_0)T(\chi_J)$ and $g_2(x) = T((x_0 - \cdot)\chi_J)$, then the continuity of T on L^2 gives the stated inequality.

In our situation,

$$g(x) = \int_J (b(x) - b(y))\, dy = l(b(x) - c_I)$$

and

$$\|g\|_{L^2(I)} \leq Cl^{3/2}$$

is equivalent to the condition $b \in$ BMO.

We now show that the condition $b \in$ BMO is sufficient. We start with the special case where b is in the Wiener algebra $A(\mathbb{R})$. The general case then follows via the weak continuity property described in Theorem 5.

If $f \in L^2$ and if $b \in A(\mathbb{R})$, then

$$[H, B]f = -\pi i \iint e^{ix(\xi+\eta)}(\text{sgn}(\xi + \eta) - \text{sgn}\,\xi)\hat{b}(\eta)\hat{f}(\xi)\, d\eta\, d\xi$$

$$= -\pi i \iint e^{ix(\xi+\eta)}(\text{sgn}(\xi + \eta) - \text{sgn}\,\xi)\hat{b}(\eta)\phi_0\left(\frac{\xi}{\eta}\right)\hat{f}(\xi)\, d\eta\, d\xi,$$

where $\phi_0 \in \mathcal{D}(\mathbb{R})$ equals 1 on $[-1, 1]$. The reason why this identity is valid is that $\text{sgn}(\xi + \eta) = \text{sgn}\,\xi$ if $|\eta| < |\xi|$, and it is only in this case that $\phi_0(\xi/\eta) \neq 1$.

We let $\pi(b, f) = T_b(f)$ be the bilinear operator defined by the symbol $\phi_0(\xi/\eta)$, which is clearly homogeneous of degree 0 and infinitely differentiable except at $(0, 0)$. By Theorem 2, T_b is bounded on $L^2(\mathbb{R})$, when b is in BMO($\mathbb{R}$). Further, $[H, B] = [H, T_b]$, by the bilinear identity above. Thus

(6.1) $\|[H, B]\| \leq C\|b\|_{\text{BMO}}$,

for $b \in A(\mathbb{R})$.

To pass to the general case, we consider the trilinear form

$$\int [H, B](f)\, g\, dx = \int H(bf)\, g\, dx - \int bH(f)\, g\, dx$$
$$= -\int bf\, H(g)\, dx - \int b\, H(f)\, g\, dx$$
$$= -\int bh\, dx$$

where

$$h = gH(f) + fH(g)\,.$$

If $f, g \in L^2(\mathbb{R})$, then h lies in $L^1(\mathbb{R})$ and $|\int bh\, dx| \leq C\|b\|_{\mathrm{BMO}}$, for $b \in A(\mathbb{R})$. As a consequence, h belongs to the Hardy space $H^1(\mathbb{R})$ and we have obtained the estimate

(6.2) $$\|gH(f) + fH(g)\|_{H^1} \leq C\|f\|_2 \|g\|_2\,.$$

This estimate can be connected with the usual properties of the holomorphic Hardy space. Indeed, let us take f and g to be real-valued (which is enough to establish the general case of (6.2)). Then $f + iH(f) = F$ is holomorphic in the upper half-plane and belongs to the (holomorphic) Hardy space $\mathbb{H}^2$. The same is true for $g + iH(g) = G$ and thus the imaginary part of the product FG is in the real Hardy space H^1, of Stein and Weiss.

We see here that the multilinear calculus, as developed in this chapter, extends those algebraic manipulations which the subject's pioneers effected in the context of classical Hardy spaces.

Now for a second example, wherein the Hilbert transform (an operator of order 0 in dimension 1) is replaced by a convolution operator T whose symbol $\tau(\xi)$, $\xi \in \mathbb{R}^n$, is homogeneous, of degree $m \in \mathbb{N}$ and infinitely differentiable in $\mathbb{R}^n \setminus \{0\}$. Thus T is a classical pseudo-differential operator of order m (and the differential operators, with constant coefficients, which are homogeneous of degree m, belong to this category.)

We let $b(x)$ denote a function defined on $\mathbb{R}^n$, whose regularity will be prescribed below, and we let B denote the operator of pointwise multiplication by $b(x)$.

We then have the classical identity of the pseudo-differential calculus

(6.3) $$TB = \sum_{|\alpha| \leq m} B_\alpha T_\alpha + R_m\,,$$

where B_α is the operator of pointwise multiplication by $(\alpha!)^{-1}\partial^\alpha b(x)$, where T_α has symbol $i^{-|\alpha|}\partial^\alpha \tau(\xi)$, and where R_m is bounded on $L^2(\mathbb{R})$, when $b(x)$ is sufficiently regular.

As we have already emphasized, the pseudo-differential approach does not give the optimal results which we now present. In fact, that approach does not even get us as far as proving the continuity of R_m under the (non-optimal) hypothesis that $b(x)$ is a C^m function.

The solution is given by the following statement.

Theorem 6. *Let Λ denote $(-\Delta)^{1/2}$. With the preceding hypotheses on T we have*

$$(6.4) \qquad \|R_m(f)\|_2 \leq C(m,n,T)\|\Lambda^m b\|_{\mathrm{BMO}}\|f\|_2 .$$

If m is odd and if $T = \Lambda^m$, the norm of the operator $R_m : L^2 \to L^2$ and the BMO norm of $\Lambda^m b$ are equivalent.

Before proving this result, we should throw a little light on it. We may replace the condition $\Lambda^m b \in$ BMO by more manageable conditions when m is odd (when m is even, Λ^m is a power of the Laplacian and the condition is obvious). If m is odd, Λ^m is a pseudo-differential operator, acting on a general function b. Indeed, the BMO norm of $\Lambda^m b$ is equivalent to the sum of the BMO norms of the functions $\partial^\alpha b$, $|\alpha| = m$. This is due to the fact that the operators $\partial^\alpha \Lambda^{-m}$ are Calderón-Zygmund operators which are bounded on BMO (Chapter 7, section 4.)

The error term R_m is identically zero if $b(x)$ is a polynomial of degree not exceeding m. In this case, the derivatives of order m of $b(x)$ are constants whose BMO norm is zero.

We come to the proof of (6.4).

As in the case of Theorem 5, we shall begin by dealing with the case where b is sufficiently regular for all the calculations to make sense. In fact, we shall take $b \in \mathcal{S}(\mathbb{R}^n)$. The general case will then be obtained by the density argument used in the proof of Theorem 5.

We consider $R_m(f)$ as a bilinear operator in b and f. The corresponding bilinear symbol is

$$a_m(\eta,\xi) = \tau(\xi + \eta) - \sum_{|\alpha| \leq m} \frac{\eta^\alpha}{\alpha!} \partial^\alpha \tau(\xi)$$

and we need to write this in the form $a_m(\eta,\xi) = |\eta|^m \tau_m(\eta,\xi)$ and to show that the operator defined by the symbol $\tau_m(\eta,\xi)$ satisfies the BMO $\times L^2 \to L^2$ estimate. This prompts us to apply Theorem 2, but we run up against the difficulty that $\tau_m(\eta,\xi)$ does not satisfy condition (3.1). On the other hand, we do have $\tau_m(0,\xi) = 0$, because of the regularity of τ. As for condition (3.1), everything works well for $|\eta| \leq |\xi|/2$, but the estimates do not hold if, for example, $|\xi + \eta|$ is very much smaller than $|\xi|$ (which tends to infinity).

To get round this difficulty, we let θ_0 and θ_1 denote even functions in $\mathcal{D}(\mathbb{R})$ such that $\theta_0(t) + \theta_1(1/t) = 1$, for all $t \in \mathbb{R}$, and $\theta_0(t) = 1$ if

$0 \le t \le 1/3$, $\theta_0(t) = 0$ if $t \ge 1/2$ (so that $\theta_1(t) = 0$ if $t \ge 3$ and $\theta_1(t) = 1$ if $0 \le t \le 2$).

We form the symbols

$$b_m(\eta, \xi) = \tau_m(\eta, \xi)\theta_0(|\eta|/|\xi|) \quad \text{and} \quad c_m(\eta, \xi) = \tau_m(\eta, \xi)\theta_1(|\xi|/|\eta|).$$

Then $b_m(\eta, \xi)$ satisfies the hypotheses of Theorem 2, as a straightforward calculation shows. The corresponding bilinear operator is bounded as an operator from BMO $\times L^2$ to L^2.

As for $c_m(\eta, \xi)$, we separately examine each term

$$\tau(\xi + \eta)\theta_1(|\xi|/|\eta|), \ldots, \frac{\eta^\alpha}{\alpha!}\partial^\alpha \tau(\xi)\theta_1(|\xi|/|\eta|), \ldots,$$

where $0 \le |\alpha| \le m$. We write

$$\tau(\xi + \eta)\theta_1\left(\frac{|\xi|}{|\eta|}\right) = \frac{\tau(\xi + \eta)}{|\xi + \eta|^m}\frac{|\xi + \eta|^m}{|\eta|^m}\theta_1\left(\frac{|\xi|}{|\eta|}\right)|\eta|^m$$

and observe that $\tau(\eta + \xi)/|\xi + \eta|^m$ is a symbol of order 0, corresponding to a bounded operator on $L^2(\mathbb{R})$ which acts on the result of the application of the bilinear operator with symbol $|\xi + \eta|^m|\eta|^{-m}\theta_1(|\xi|/|\eta|)$. This symbol satisfies the hypotheses of Theorem 2 precisely.

We write the other terms in the form

$$\eta^\alpha \partial^\alpha \tau(\xi)\theta_1\left(\frac{|\xi|}{|\eta|}\right) = \frac{\partial^\alpha \tau(\xi)}{|\xi|^{m-|\alpha|}}\left(\frac{|\xi|}{|\eta|}\right)^{m-|\alpha|}\frac{\eta^\alpha}{|\eta|^{|\alpha|}}\theta_1\left(\frac{|\xi|}{|\eta|}\right)|\eta|^m$$

and we apply Theorem 2 to the bilinear symbol

$$(\eta^\alpha/|\eta|^{|\alpha|})(|\xi|/|\eta|)^{m-|\alpha|}\theta_1(|\xi|/|\eta|).$$

To pass from the case where $b(x)$ is in $\mathcal{S}(\mathbb{R}^n)$ to the general case where $b(x)$ is in BMO$(\mathbb{R}^n)$, we use the same argument as in the proof of Theorem 5. The details present no difficulty, so we omit them.

We now come to the last part of Theorem 6. The proof is by determining the restriction to $x \ne y$ of the kernel $R_m(x, y)$ of R_m. This restriction is given by

$$R_m(x, y) = c(m, n)|x - y|^{-n-m}\left(b(y) - \sum_{|\alpha| \le m}\frac{(y - x)^\alpha}{\alpha!}\partial^\alpha b(x)\right).$$

We let B and B' be two balls, of radius $r > 0$, in $\mathbb{R}^n$, with centres x_0 and x'_0, respectively. We suppose that $|x_0 - x'_0| = 3r$. We shall make the operator R_m act on certain functions supported by B' and calculate the L^2 norm of the restriction to B of the function obtained thereby.

But, to simplify what follows, we shall replace R_m by the kernel S_m defined by

$$S_m(x, y) = |x - y|^{n+m}\theta\left(\frac{|x - y|}{r}\right)R_m(x, y),$$

where θ is an even, regular function, of compact support, equal to 0 if

$t \leq 1/2$ or $t \geq 6$ and equal to 1 if $1 \leq t \leq 5$. So, if $x \in B'$ and $y \in B$, we have $S_m(x - y) = |x - y|^{n+m} R_m(x, y)$. Now,

$$|x - y|^{n+m} \theta(|x - y|/r) = r^{n+m} \gamma((x - y)/r),$$

where $\gamma \in \mathcal{D}(\mathbb{R}^n)$. We then write

$$\gamma(r^{-1}(x - y)) = \frac{1}{(2\pi)^n} \int e^{ir^{-1}\xi \cdot (x-y)} \hat{\gamma}(\xi) \, d\xi$$

and this identity lets us separate the variables x and y, since

$$e^{ir^{-1}\xi \cdot (x-y)} = e^{ir^{-1}\xi \cdot x} e^{-ir^{-1}\xi \cdot y}.$$

Multiplication by unimodular functions does not affect the operator norms and, by convex combinations, the norm of S_m is not greater than $Cr^{m+n} \| R_m \|$.

To finish, we consider a function g supported by the unit ball of $\mathbb{R}^n$, which is positive, of mean 1 and of sufficient regularity. Then $\int x^\beta \partial^\alpha g(x) \, dx = 0$, if $|\beta| \leq |\alpha|$ and $\alpha \neq \beta$, while the integral is $(-1)^{|\alpha|} \alpha!$, if $\beta = \alpha$. Encouraged by this remark, we let S_m act on the function $(\partial^\beta g)(r^{-1}(y - x_0'))$ and test the result on the ball B. With $|\beta| = m$, all the terms, apart from $\partial^\beta b(x)$ and the integral $I_B^{(\beta)} = \int \partial^\beta g((y - x_0')/r) b(y) \, dy$, vanish. We do not try to evaluate the integral. Finally, the continuity of $S_m : L^2(B') \to L^2(B)$ enables us to write

$$\int_B |\partial^\beta b(x) - I_B^{(\beta)}|^2 \, dx \leq C|B|$$

for $|\beta| = m$. This is what we had to prove.

If m is an even integer, Λ^m is a differential operator and $R_m = 0$. But the interesting conclusion would consist of replacing T systematically by $\Lambda^m R_1, \ldots, \Lambda R_n$, where $R_j = (\partial/\partial x_j)\Lambda^{-1}$, $1 \leq j \leq n$.

7 McIntosh's theory of multilinear operators

Let $\phi(x) \in L^1(\mathbb{R}^n)$ be a function of mean 1, let $\psi(x) \in L^1(\mathbb{R}^n)$ be a function of mean 0, and, for all $t > 0$, let P_t and Q_t be the operators of convolution with ϕ_t and ψ_t, respectively, where $\phi_t(x) = t^{-n}\phi(x/t)$ and similarly for ψ_t. We let $b_1, \ldots, b_k$ be functions in $L^\infty(\mathbb{R}^n)$ and we let $B_1, \ldots, B_k$ denote the corresponding operators of pointwise multiplication by those functions. We intend to construct bounded operators on $L^2(\mathbb{R}^n)$ by McIntosh's formal calculus ([66]). That is, we begin with a function $m(t) \in L^\infty(0, \infty)$ and we aim to define the operators

$$(7.1) \qquad L_k = \int_0^\infty Q_t B_1 P_t B_2 P_t \cdots B_k P_t m(t) \frac{dt}{t}$$

as the limit, in the strong operator topology, of

$$L_k^{\varepsilon,R} = \int_\varepsilon^R Q_t B_1 P_t \cdots B_k P_t m(t) \, \frac{dt}{t},$$

as $\varepsilon \to 0$ and $R \to \infty$.

It is possible to show that this programme has no chance of success unless we make more restrictive hypotheses on ϕ and ψ. However, the following are sufficient conditions:

(7.2) $|\phi(x)| \le \dfrac{C}{(1+|x|)^{n+1}}$ and $|\nabla\phi(x)| \le \dfrac{C}{(1+|x|)^{n+1}}$;

(7.3) $|\psi(x)| \le \dfrac{C}{(1+|x|)^{n+1}}$, $\displaystyle\int \psi(x)\,dx = 0$, $|\nabla\psi(x)| \le \dfrac{C}{(1+|x|)^{n+1}}$.

Using the theory of Calderón-Zygmund operators and David and Journé's $T(1)$ theorem, we shall establish the following theorem.

Theorem 7. *Assuming (7.2) and (7.3), the operators $L_k^{(\varepsilon,R)}$ converge to L_k in the strong operator topology and L_k is a Calderón-Zygmund operator.*

We first of all remark that the "truncated operators" $L_k^{(\varepsilon,R)}$ are special cases of the general operators L_k, because the truncation can be obtained by replacing $m(t)$ by 0 outside $[\varepsilon, R]$. To simplify the notation, we shall therefore work directly with L_k, but with the proviso that $m(t)$ has support in the interval $[\varepsilon, R]$

The kernel $L_k(x,y)$ of L_k is given by

$$L_k(x,y) = \int_0^\infty \psi_t(x - u_1)b_1(u_1)\phi_t(u_1 - u_2)b_2(u_2)\cdots$$

$$\cdots b_k(u_k)\phi_t(u_k - y)m(t)\,\frac{dt}{t}$$

and we thus have

$$|L_k(x,y)| \le \|b_1\|_\infty \cdots \|b_k\|_\infty \|m\|_\infty \int_0^\infty (|\psi_t| \star |\phi_t| \star \cdots \star |\phi_t|)(x - y)\,\frac{dt}{t}.$$

The estimate of $|L_k(x,y)|$ finishes by remarking that the convolution product of k copies of $(1+|x|)^{-n-1}$ is bounded above by $C(k,n)(1+|x|)^{-n-1}$ and then applying the inequality

$$\int_0^\infty \frac{1}{t^n}\frac{1}{(1+t^{-1}|x-y|)^{n+1}}\,\frac{dt}{t} \le \frac{C(n)}{|x-y|^n}.$$

To estimate $|L_k(x,y) - L_k(x',y)|$, we use the inequality

$$|\psi_t(v) - \psi_t(u)| \le C\frac{|v-u|^\alpha}{t^\alpha}\left(\frac{1}{t^n(1+t^{-1}|u|)^{n+1}} + \frac{1}{t^n(1+t^{-1}|v|)^{n+1}}\right),$$

for $0 < \alpha < 1$, and finish as above. This gives

$$|L_k(x,y) - L_k(x',y)| \leq C_{k,\alpha}|x' - x|^\alpha(|x - y|^{-n-\alpha} + |x' - y|^{-n-\alpha}).$$

The same argument works for $|L_k(x,y') - L_k(x,y)|$.

We have not yet used the hypothesis $\int \psi(x)\,dx = 0$. This comes in when we use David and Journé's $T(1)$ theorem and an induction argument to prove the continuity of the operators L_k. We have $P_t(1) = 1$ and thus $L_k(1) = L_{k-1}(b_k) \in$ BMO, since, by induction, L_{k-1} is a Calderón-Zygmund operator. On the other hand, ${}^tQ_t(1) = 0$ and thus ${}^tL_k(1) = 0$.

The weak continuity is established by showing that

$$\|L_k(\chi_\omega)\|_{\mathrm{BMO}} \leq C_k,$$

where $\chi_\omega(x) = e^{i\omega \cdot x}$. Here, again, $L_k(\chi_\omega) = L_{k-1}^{(\omega)}(b_k\chi_\omega)$, where $m(t)$ is replaced in $L_{k-1}^{(\omega)}$ by $m(t)\hat{\phi}(t\omega)$.

The starting point for these induction arguments is the fact that $L_0 = \int_0^\infty Q_t m(t)t^{-1}\,dt$ is bounded on $L^2(\mathbb{R})$. We show this by observing that L_0 is a convolution operator whose symbol $\int_0^\infty \hat{\psi}(t\xi)m(t)t^{-1}\,dt$ belongs to $L^\infty(\mathbb{R}^n)$. Indeed, $|\hat{\psi}(\xi)| \leq C(\alpha)|\xi|^\alpha$, for $0 < \alpha < 1$, because $|\psi(x)| \leq C(1 + |x|)^{-n-1}$ and $\int \psi(x)\,dx = 0$. For large values of $|\xi|$, we integrate $\int e^{-ix\cdot\xi}\psi(x)\,dx$ by parts and (if $|\xi_1| \geq |\xi_2| \geq \cdots \geq |\xi_n|$) get

$$-\frac{i}{\xi_1}\int e^{-ix\cdot\xi}\frac{\partial\psi}{\partial x_1}\,dx,$$

whose modulus is bounded above by $C|\xi|^{-1}$.

Let us now show that $L_k^{(\varepsilon,R)}$ converges strongly to L_k, when $\varepsilon \to 0$ and $R \to \infty$.

We suppose that f lies in the linear subspace V of $L^2(\mathbb{R}^n)$ which consists of those functions $f \in \mathcal{S}(\mathbb{R}^n)$ whose Fourier transforms vanish in a neighbourhood of 0. By Plancherel's identity, we immediately see that $\|P_tf\|_2 = O(t^{-1})$ as $t \to \infty$ and that $\|P_tf - f\|_2 = O(t^\alpha)$, when $0 < \alpha < 1$ and $t \to 0$.

We split $\int_0^\infty Q_t\ldots B_kP_t m(t)t^{-1}\,dt$ into $\int_0^1 + \int_1^\infty$ and do the same for $L_k^{(\varepsilon,R)}$. In the first integral, we replace P_tf by f, which reduces the problem to studying $L_{k-1}^{(\varepsilon,R)}(b_kf)$. We then use the induction hypothesis. The second integral and the error term introduced into the first are absolutely convergent, so passing to the limit poses no problems.

We have established Theorem 7. Now for some variants. For $0 \leq j \leq k$, we consider the operators

$$(7.4) \qquad L_{j,k} = \int_0^\infty P_tB_1\cdots P_tB_jQ_tB_{j+1}P_t\cdots B_kP_t m(t)\,\frac{dt}{t},$$

where $L_{0,k}$ is defined by (7.1) and $L_{k,k} = \int_0^\infty P_tB_1\cdots P_tB_kQ_t m(t)t^{-1}\,dt$.

With the hypotheses (7.2) and (7.3), all these operators are Calderón-Zygmund operators, bounded on $L^2(\mathbb{R}^n)$.

We are going to compare the operators constructed in this way with those defined by Theorem 1. To simplify the discussion, we shall strengthen the hypotheses in both cases. That is, we shall suppose that (3.1) holds, whatever the integer N (but the constant C will depend on N). We further suppose that, in McIntosh's formal calculus, ϕ and ψ belong to the Schwartz class $\mathcal{S}(\mathbb{R}^n)$ and that all the moments of ψ are zero. Then we have

Proposition 5. *The operators $L_{j,k}$, defined by (7.4), are special cases of the operators defined by Theorem 1, in which $\pi(\eta,\xi) = 0$, if $\eta_{j+1} + \cdots + \xi = 0$ (and $\pi(\eta,\xi) = 0$ for $\xi = 0$, when $j = k$).*

Conversely, every operator arising out of Theorem 1 whose multilinear symbol satisfies this condition may be written as a convergent series of operators $L_{j,k}$ constructed by means of different choices of the functions ψ, ϕ and m.

Lastly, every operator defined by Theorem 1 may be written as a convergent series composed of operators of the form $L_{j,k}$ ($0 \le j \le k$).

The last assertion is a consequence of the first two. Indeed, let $\chi_0(\eta,\xi), \ldots, \chi_k(\eta,\xi)$ be $k+1$ functions which are homogeneous of degree zero, infinitely differentiable except at $(0,0)$, and such that $\chi_j(\eta,\xi) = 0$ on a conical neighbourhood of $\eta_{j+1} + \cdots + \eta_k + \xi = 0$, $0 \le j < k$, and $\chi_k(\eta,\xi) = 0$ on a conical neighbourhood of $\xi = 0$. The significance of these conditions and the fact of the existence of $\chi_0, \ldots, \chi_k$ become obvious if we restrict attention to the sphere $|\eta|^2 + |\xi|^2 = 1$. The intersection of the compact sets $\xi = 0$, $\eta_k + \xi = 0, \ldots, \eta_1 + \eta_k + \xi = 0$ on the sphere is empty and the existence of the partition of unity is classical. We then take a multilinear symbol $\pi(\eta,\xi)$ which satisfies (3.1) and decompose it as $\chi_0(\eta,\xi)\pi(\eta,\xi) + \cdots + \chi_k(\eta,\xi)\pi(\eta,\xi)$. We can apply the second paragraph of Proposition 5 to each term.

The first paragraph of Proposition 5 is obvious. For example, the multilinear symbol of L_k is

$$(7.5) \quad \pi(\eta,\xi) = \int_0^\infty \hat{\psi}(t(\eta_1 + \cdots + \eta_k + \xi))\hat{\phi}(t(\eta_2 + \cdots + \eta_k + \xi)) \cdots$$

$$\cdots \hat{\phi}(t(\eta_k + \xi))\hat{\phi}(t\xi)m(t)\,\frac{dt}{t}$$

and (3.1) follows immediately. It is worth remarking that $\pi(\eta,\xi)$ is homogenous of degree 0 if $m(t)$ is a constant function.

We obtain the second statement of Proposition 5 by a method of separation of variables which we have used before, in Section 3. We shall use it now in a slightly sharper form, described in the following lemma, for which we need to fix some preliminary notation.

We let $\mathcal{D}(\mathbb{R}^p \times \mathbb{R}^q)$ denote the Schwarz space of compactly supported, infinitely differentiable functions. We give it the usual topology. A bounded subset $\mathcal{B}$ of $\mathcal{D}(\mathbb{R}^p \times \mathbb{R}^q)$ is a set of functions $f \in \mathcal{D}$ whose compact supports are contained in a fixed compact set and whose successive derivatives satisfy uniform bounds.

Lemma 2. *For each bounded set $\mathcal{B} \subset \mathcal{D}(\mathbb{R}^p \times \mathbb{R}^q)$, there exist two sequences $g_j(x)$ and $h_j(y)$, $j \in \mathbb{N}$, belonging to bounded subsets $\mathcal{B}_1$ and $\mathcal{B}_2$, respectively, of $\mathcal{D}(\mathbb{R}^p)$ and $\mathcal{D}(\mathbb{R}^q)$ such that every function $f \in \mathcal{B}$ may be written in the form*

$$(7.6) \qquad f(x,y) = \sum_0^\infty \omega_j(f) g_j(x) h_j(y)\,,$$

where $\omega_j(f)$ is a rapidly decreasing sequence.

As we have already had occasion to remark, we get (7.6) by the following three operations. Firstly, we periodify f in each variable, using a sufficiently large period T. Secondly, we expand the periodified function as a Fourier series. Finally, we use a cut-off function to separate $f(x,y)$ from the parasitic terms accompanying the periodification.

The three steps are all linear, so, if $f(0,y) = 0$, for all y, then $g_j(0) = 0$, for all $j \in \mathbb{N}$.

Let us now analyse the multilinear operators of Theorem 1, using McIntosh's formal calculus. We let $\omega(\eta, \xi)$ be a function in $\mathcal{D}(\mathbb{R}^{n(k+1)})$ taking positive or zero values, which we shall suppose to be radial. We further require $\omega(\eta, \xi)$ to vanish in a neighbourhood of 0, but not, of course, to be identically zero outside that neighbourhood. Multiplying ω by a positive constant, if necessary, we can assume that $\int_0^\infty \omega(t\eta, t\xi) t^{-1}\, dt = 1$, for $(\eta, \xi) \neq (0,0)$. We now write

$$\pi(\eta, \xi) = \int_0^\infty \pi_t(t\eta, t\xi) \frac{dt}{t}, \quad \text{where} \quad \pi_t(\eta, \xi) = \pi(t^{-1}\eta, t^{-1}\xi)\omega(\eta, \xi)\,.$$

Since (3.1) holds for all $N \geq 1$, the functions $\pi_t(\xi, \eta)$ form a bounded subset of $\mathcal{D}(\mathbb{R}^{n(k+1)})$. We shall analyse these functions, using Lemma 2 and the new variables $\eta_1 + \cdots + \eta_k + \xi, \eta_2 + \cdots + \eta_k + \xi, \ldots, \eta_k + \xi, \xi$. More accurately, we shall use the obvious generalization of Lemma 2 in which the two variables x and y are replaced by the $k+1$ variables which we have just defined.

Looking at the case $\pi(\eta, \xi) = 0$ when $\eta_1 + \cdots + \eta_k + \xi = 0$, we get the expansion

$$(7.7) \quad \pi_t(\eta, \xi) = \sum_0^\infty m_j(t) a_{1,j}(\eta_1 + \cdots + \eta_k + \xi) \cdots a_{k,j}(\eta_k + \xi) b_j(\xi),$$

where

(7.8) $\|m_j(t)\|_{L^\infty(0,\infty)}$ decreases rapidly,

(7.9) $a_{1,j}(0) = 0$, and all the derivatives of $a_{1,j}$ vanish at 0, and

(7.10) the functions $a_{1,j}, \ldots, a_{k,j}$ and b_j run through a bounded subset of $\mathcal{D}(\mathbb{R}^n)$.

If we now put

$$\hat{\psi}_j(\xi) = a_{1,j}(\xi), \ \hat{\phi}_{1,j}(\xi) = a_{2,j}(\xi), \ldots, \hat{\phi}_{k,j} = b_j(\xi),$$

we obtain $\pi(\eta, \xi)$ as a series of operators arising out of the McIntosh formal calculus, except for needing k different functions ϕ.

The cases where $\pi(\eta, \xi) = 0$ for $\eta_2 + \cdots + \xi = 0, \ldots, \xi = 0$ follow by the same method applied to the terms $\pi(\eta, \xi)\chi_1(\eta, \xi), \ldots, \pi(\eta, \xi)\chi_k(\eta, \xi)$. This concludes the proof of Proposition 5.

It thus seems that Theorem 1 and McIntosh's approach are equivalent routes to the same results. This is not quite the case. Going back and forth between functions and Fourier transforms always requires too many hypotheses: for example, one cannot get the correct estimates for the Calderón commutators in the formalism of Theorem 1, because the corresponding multilinear symbols are not sufficiently regular.

On the other hand, the success of McIntosh's programme first became apparent in the analysis of these commutators. During his visit to Paris in 1980-91, McIntosh suggested studying them by using the identity

$$(7.11) \quad \text{PV} \int_{-\infty}^\infty \frac{(\beta(x) - \beta(y))^k}{(x-y)^{k+1}} f(y) \, dy$$

$$= \text{PV} \int_{-\infty}^\infty [(I + itD)^{-1} B]^k (I + itD)^{-1} f \, \frac{dt}{t},$$

where $D = -i(d/dx)$, $b(x) \in L^\infty(\mathbb{R})$, B is multiplication by $b(x)$, and $\beta(x)$ is a primitive of $b(x)$.

This identity was the starting point for the first proof of the L^2 continuity of the Cauchy kernel on an arbitrary Lipschitz curve ([65]). In fact $(I + itD)^{-1} = P_t - iQ_t$, where P_t and Q_t are defined, as above, using $\phi(x) = e^{-|x|}/2$ and $\psi(x) = -\operatorname{sgn} x e^{-|x|}/2$. The function ψ does not, however, have the regularity that we imposed above. Now, note that P_t is an even function of t, Q_t is an odd function, and PV $\int_{-\infty}^\infty [(P_t - iQ_t)B]^k (P_t - iQ_t) t^{-1} \, dt$ can thus be split into a sum of terms each of which contains the operator Q_t at least once.

We recognize our old friends, the operators of Theorem 7. When the original proof was written, we did not have the $T(1)$ theorem and the analysis of the operator on the right-hand side of (7.11) was carried out by relating the operator, using certain tricks, to quadratic functionals of the form $(\int_0^\infty |Q_t(BP_t)^k f|^2 t^{-1}\, dt)^{1/2}$. These latter were estimated using appropriate "Carleson measures". This means that all the recipes to get the L^2 continuity of the Cauchy kernel on Lipschitz curves use essentially the same basic ingredients: the Carleson measures. If we go back to the proof of the $T(1)$ theorem, they appear in the proof of the continuity of the pseudo-product of a BMO function with an L^2 function.

8 Conclusion

Despite the results described in this chapter, we have only a rudimentary understanding of the multilinear operators $\pi : (L^\infty)^k \times L^2 \to L^2$ which commute with translations and dilations.

To illustrate this, let us consider, for $k = n = 1$, the bilinear symbol $\pi(\eta, \xi) = \text{sgn}(\eta - \xi)$. We do not know whether the corresponding bilinear operator is bounded, as an operator from $L^\infty \times L^2$ to L^2. If we were to restrict ourselves to functions $\pi(\eta, \xi)$, defined on $\mathbb{R}^2 \setminus \{0\}$, and homogeneous of degree zero, we might dream of finding a Banach space E of 2π-periodic functions such that the norm of $\pi(\cos\theta, \sin\theta)$ in the Banach space E was equivalent to the norm of the operator $\pi : L^\infty \times L^2 \to L^2$ corresponding to the symbol.

Even using Theorem 1, we know only that $E \subset L^\infty(\mathbb{R}/2\pi\mathbb{Z})$ and that, for sufficiently large regularity $r > 0$, C^r is a subspace of E.

Going back to the special case $\pi(\eta, \xi) = \text{sgn}(\eta - \xi)$ and using a duality argument (as we did in Theorem 5) we may reformulate this problem in the following way. If f and g are in $S(\mathbb{R})$, we put

$$h(x) = \text{PV} \int_{-\infty}^\infty f(x - t)g(x + t)\, \frac{dt}{t} .$$

and try to see whether there is a constant C such that $\|h\|_1 \le C\|f\|_2\|g\|_2$.

We end by remarking that the multilinear operators whose symbols satisfy the hypotheses of Theorem 1 have other noteworthy continuity properties, namely

$$\|\pi(a_1, \ldots, a_k, f)\|_r \le C\|a_1\|_{p_1} \cdots \|a_k\|_{p_k}\|f\|_q$$

where $1 < p_1, p_2, \ldots, p_k, q, r < \infty$ and $1/r = 1/p_1 + \cdots + 1/p_k + 1/q$.

The proof of these inequalities is much easier than in the extreme case $p_1 = \cdots = p_k = \infty$, given above, and the reader is referred to [68].

14

Multilinear analysis of the square roots of accretive differential operators

1 Introduction

In this chapter we shall give an application of the general methods presented in the previous chapter. We consider a second order differential operator $L = -\sum_1^n \sum_1^n (\partial/\partial x_j)(a_{j,k}(x)\partial/\partial x_k)$, where the functions $a_{j,k}(x)$ are in $L^\infty(\mathbb{R}^n)$ and the matrix $A(x) = (a_{j,k}(x))_{1 \leq j,k \leq n}$ satisfies the following condition: there exists a $\delta > 0$ such that, for all $\xi \in \mathbb{C}^n$ and almost all $x \in \mathbb{R}^n$,

$$(1.1) \qquad \Re e \sum_1^n \sum_1^n a_{j,k}\xi_j\bar{\xi}_k \geq \delta(|\xi_1|^2 + \cdots + |\xi_n|^2).$$

We can see how well-founded this condition is when we consider the bilinear form $J(f,g) = \int (Lf)g\,dx$. This form is defined, for f and g belonging to the Sobolev space H^1, by $J(f,g) = \int (A(x)\nabla f(x)) \cdot \nabla g(x)\,dx$, where $A(x) = (a_{j,k}(x))_{1 \leq j,k \leq n}$ and $\nabla f = (\partial f/\partial x_1, \ldots, \partial f/\partial x_n)$. We can then write (1.1) in the equivalent form

$$(1.2) \qquad \Re e\, J(f,\bar{f}) \geq \delta\|\nabla f\|_2^2.$$

T. Kato suggested studying the accretive square root S of the operator L. We shall define S in the following section. The problem (still open) is to show that the domain of S is H^1. This result is obvious if L is a self-adjoint operator, that is, if the matrix $A(x)$ is self-adjoint. Indeed, S is then self-adjoint and positive and we can write

$$\|S(f)\|_2^2 = \langle S(f), S(f) \rangle = \langle L(f), f \rangle \geq \delta\|\nabla f\|_2^2.$$

We shall establish Kato's conjecture in the special case in which

$\|A(x) - I\|_\infty < \varepsilon(n)$, where $\varepsilon(n) > 0$ depends only on the dimension. To do this, we shall write the operator $\sqrt{L}$ in a form using the resolvent of L, together with a series of multilinear operators which, after some work, can be analysed using David and Journé's $T(1)$ theorem ([96]). Towards the end, we shall return to dimension 1, where a version of Kato's conjecture gives precisely the operator defined by the Cauchy kernel on a Lipschitz curve.

2 Square roots of operators

We start with the self-adjoint case. Let H be a Hilbert space, with inner product $\langle \cdot, \cdot \rangle$. Let $V \subset H$ be a dense linear subspace and $T : V \to H$ a linear operator. We say that T is symmetric if $\langle Tf, g \rangle = \langle f, Tg \rangle$, for all $f, g \in V$. The operator T is self-adjoint, with domain V, if one of the two following equivalent conditions is also satisfied:

(2.1) V is a complete normed space for the norm $(\|T(f)\|^2 + \|f\|^2)^{1/2}$;

(2.2) $T + iI : V \to H$ is an isomorphism.

Suppose, further, that $\langle Tf, f \rangle \geq 0$ for all $f \in V$. Then there exists a unique, positive, self-adjoint operator S such that $S^2 = T$. The domain W of S is the completion of V for the norm defined by $(\langle Tf, f \rangle + \|f\|^2)^{1/2}$. Lastly, in a sense that we shall clarify,

$$(2.3) \qquad S = \frac{1}{\pi} T \int_0^\infty (T + \lambda I)^{-1} \lambda^{-1/2} \, d\lambda.$$

To make the meaning of (2.3) clear, we consider a much more general situation, in which T is no longer self-adjoint and positive, but has the following properties:

(2.4) T is defined on a dense linear subspace $V \subset H$ and takes values in H;

(2.5) for all $\lambda > 0$, $T + \lambda I : V \to H$ is an isomorphism;

(2.6) there is a constant $C \geq 1$ such that, for all $\lambda > 0$, we have $\|(T + \lambda I)^{-1}\| \leq C\lambda^{-1}$.

Under these conditions, there exists an open sector $\Omega \subset \mathbb{C}$ of angle $2\alpha < \pi$ and axis of symmetry $(-\infty, 0)$, such that, for all $\zeta \in \Omega$, $T - \zeta I : V \to H$ is an isomorphism and

$$\|(T - \zeta I)^{-1}\| \leq C|\xi|^{-1} \qquad \text{where} \quad \zeta = \xi + i\eta \in \Omega.$$

What is more, the function $F(\zeta) = (T + \zeta I)^{-1}$ is holomorphic in Ω, and takes values in the algebra $\mathcal{L}(H, H)$ of bounded linear operators on H.

The spectrum $\sigma(T)$ of T is then contained in the complement Σ of Ω, given by $|\theta| \leq \pi - \alpha$.

Our intention is to define the operator $(T + \varepsilon I)^{-1/2}$, for all $\varepsilon > 0$, in the sense of the holomorphic symbolic calculus.

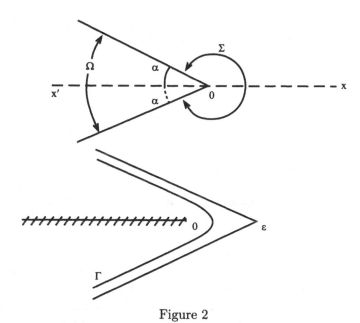

Figure 2

We observe that the spectrum of $T + \varepsilon I$ is contained in $\varepsilon + \Sigma$. We define the holomorphic function $z^{1/2}$ in $\mathbb{C} \setminus (-\infty, 0]$ (the cut is indicated in Figure 2) and we let Γ be a contour contained in $(\Omega + \varepsilon) \setminus (-\infty, 0]$, as suggested by Figure 2. Then, putting $T_\varepsilon = T + \varepsilon I$, we get

$$(2.7) \qquad (T + \varepsilon I)^{-1/2} = \frac{1}{2\pi i} \int_\Gamma (\zeta I - T_\varepsilon)^{-1} \zeta^{-1/2} \, d\zeta .$$

The integral is absolutely convergent, with values in $\mathcal{L}(H, H)$. It does not depend on the choice of contour Γ, because $(\zeta I - T_\varepsilon)^{-1}$ is holomorphic in $\Omega + \varepsilon$.

Let us verify that $(T + \varepsilon I)^{-1/2}(T + \varepsilon I)^{-1/2} = (T + \varepsilon I)^{-1}$.

To do this, we use two distinct contours Γ_1 and Γ_2, which are of the same type as Γ and satisfy the condition $|\zeta_1 - \zeta_2| > \delta > 0$, as ζ_1 describes Γ_1 and ζ_2 describes Γ_2.

We write down (2.7) for each of Γ_1 and Γ_2 and multiply the right-hand sides together. This gives

$$-\frac{1}{4\pi^2} \int_{\Gamma_2} \int_{\Gamma_1} (\zeta_1 I - T_\varepsilon)^{-1}(\zeta_2 I - T_\varepsilon)^{-1}\zeta_1^{-1/2}\zeta_2^{-1/2}\, d\zeta_1\, d\zeta_2\,.$$

To compute this integral, we use the identity

$$(\zeta_1 I - T_\varepsilon)^{-1}(\zeta_2 I - T_\varepsilon)^{-1} = (\zeta_2 - \zeta_1)^{-1}\big[(\zeta_1 I - T_\varepsilon)^{-1} - (\zeta_2 I - T_\varepsilon)^{-1}\big]\,.$$

This leads to two integrals. If, as we may suppose, Γ_1 is contained in an open convex region bounded by Γ_2, we have

$$\frac{1}{2\pi i} \int_{\Gamma_2} (\zeta_2 - \zeta_1)^{-1}\zeta_2^{-1/2}\, d\zeta_2 = 0\,,$$

while

$$\frac{1}{2\pi i} \int_{\Gamma_1} (\zeta_2 - \zeta_1)^{-1}\zeta_1^{-1/2}\, d\zeta_1 = -\zeta_2^{-1/2}\,.$$

So we are left with

$$\frac{1}{2\pi i} \int_{\Gamma_2} \zeta_2^{-1}(\zeta_2 I - T_\varepsilon)^{-1}\, d\zeta_2 = T_\varepsilon^{-1}\,.$$

Since $T_\varepsilon^{-1} : H \to V$ is an isomorphism, $T_\varepsilon^{-1/2} : H \to H$ is an injective operator. Indeed, if $T_\varepsilon^{-1/2}(x) = 0$, then $T_\varepsilon^{-1}(x) = T_\varepsilon^{-1/2}\big(T_\varepsilon^{-1/2}(x)\big) = 0$, so $x = 0$.

To simplify the notation, we set $A = T_\varepsilon^{-1/2}$ and $B = T_\varepsilon T_\varepsilon^{-1/2}$. We know that $BA = I$ and from this we shall deduce that the domain of the (unbounded) operator B coincides with the image of A. We already know that $\operatorname{Im} A \subset \operatorname{Dom} B$. Let $y \in \operatorname{Dom} B$. We put $By = z = (BA)z = Bx$, where $x = Az$. Thus $B(y - x) = 0$. But T_ε and $T_\varepsilon^{-1/2}$ are both injective, so B is as well. Therefore $y = x$ and $\operatorname{Dom} B \subset \operatorname{Im} A$.

Let us verify that the domain is independent of $\varepsilon > 0$. We write $B_\varepsilon = T_\varepsilon T_\varepsilon^{-1/2}$ and show that $B_\varepsilon - B_\eta \in \mathcal{L}(H,H)$, when $0 < \varepsilon \le \eta < \infty$.

To simplify the proof, we observe that, by an obvious deformation of the contour, we can reduce (2.7) to

$$(2.8) \qquad T_\varepsilon^{-1/2} = \frac{1}{\pi} \int_0^\infty (T_\varepsilon + \lambda I)^{-1}\lambda^{-1/2}\, d\lambda\,.$$

Next, we note that the $\mathcal{L}(H,H)$ norms of the operators $T(T + \lambda I)^{-1}$ are uniformly bounded in $\lambda > 0$. To see this, it is enough to write $T(T + \lambda I)^{-1} = I - \lambda(T + \lambda I)^{-1}$ and apply (2.6). From this, it follows that $T(T_\varepsilon^{-1/2} - T_\eta^{-1/2}) \in \mathcal{L}(H,H)$, which implies $B_\varepsilon - B_\eta \in \mathcal{L}(H,H)$.

We have thus established the first part of the following lemma.

Lemma 1. *The image W of $(T + \varepsilon I)^{-1/2} : H \to H$ is independent of $\varepsilon > 0$ and this image is the domain of $(T + \varepsilon I)(T + \varepsilon I)^{-1/2}$ which, from now on, we shall denote by $(T + \varepsilon I)^{1/2}$. On V it is also true that $(T + \varepsilon I)^{1/2} = (T + \varepsilon I)^{-1/2}(T + \varepsilon I)$.*

To prove the second assertion of the lemma, we show that $T_\varepsilon^{1/2} T_\varepsilon^{1/2} = T_\varepsilon$ on V. Indeed, every $y \in V$ may be written as $y = T_\varepsilon^{-1}(x)$, for some $x \in H$. As a result,

$$T_\varepsilon^{1/2} T_\varepsilon^{1/2}(y) = T_\varepsilon^{1/2} T_\varepsilon^{1/2} T_\varepsilon^{-1}(x) = T_\varepsilon^{1/2}(T_\varepsilon T_\varepsilon^{-1/2})(T_\varepsilon^{-1/2} T_\varepsilon^{-1/2})(x)$$
$$= T_\varepsilon^{1/2}(T_\varepsilon T_\varepsilon^{-1}) T_\varepsilon^{-1/2} = T_\varepsilon^{1/2} T_\varepsilon^{-1/2}(x)$$
$$= T_\varepsilon T_\varepsilon^{-1/2} T_\varepsilon^{-1/2}(x) = x\,,$$

thus concluding the proof of the lemma.

We have already indicated that $T_\varepsilon^{1/2} - T_\eta^{1/2}$ is continuous on H, for $0 < \varepsilon \le \eta < \infty$. More exactly,

(2.9)
$$\|T_\varepsilon^{1/2} - T_\eta^{1/2}\| \le C\sqrt{\eta}\,,$$

where C is a constant. The proof, which we leave to the reader, depends on (2.8).

From (2.9), we deduce that, if x lies in the common domain W of the operators $T_\varepsilon^{1/2}$, then $\lim_{\varepsilon \downarrow 0} T_\varepsilon^{1/2}(x)$ exists. The convergence is to be understood as convergence in H and the limit is denoted by $T^{1/2}(x)$.

In what follows, we shall need an integral representation formula for calculating $T^{1/2}$ directly. This is given by the following proposition, which also summarizes the preceding discussion.

Proposition 1. *With the hypotheses* (2.4), (2.5) *and* (2.6), *we define the operators* $T_\varepsilon^{-1/2}$, *for* $\varepsilon > 0$, *by* (2.8). *The image* $W = T_\varepsilon^{-1/2}(H)$ *is independent of* $\varepsilon > 0$ *and contains* V.

For all $x \in V$, *the integral* $\int_0^\infty T(T+\lambda I)^{-1}(x)\lambda^{-1/2}\,d\lambda$ *is an H-valued, Bochner integral, equal to* $\int_0^\infty (T + \lambda I)^{-1} T(x)\lambda^{-1/2}\,d\lambda$. *For* $x \in V$, *we may define*

(2.10)
$$T^{1/2}(x) = \frac{1}{\pi} \int_0^\infty T(T + \lambda I)^{-1}(x)\lambda^{-1/2}\,d\lambda\,.$$

The operator $T^{1/2} : V \to H$ *then extends to* W *as an operator, still denoted by* $T^{1/2}$, *whose square on* V *is* T.

To show that the integral is a Bochner integral, we split $\int_0^\infty$ into $\int_0^1 + \int_1^\infty$. To deal with the first integral, we observe that $T(T+\lambda I)^{-1} = 1 - \lambda(T + \lambda I)^{-1}$ which, by (2.6), gives $\|T(T + \lambda I)^{-1}\| \le C$. Thus the integral $\int_0^1 T(T + \lambda I)^{-1}(x)\lambda^{-1/2}\,d\lambda$ converges for all $x \in H$. As far as the second integral is concerned, we observe that, for all $x \in V$, we have $T(T+\lambda I)^{-1}(x) = (T+\lambda I)^{-1} T(x)$, and thus $\|T(T+\lambda I)^{-1}(x)\| = O(\lambda^{-1})$. This ensures the convergence at infinity.

To prove (2.10), we return to $T_\varepsilon = T + \varepsilon I$, $\varepsilon > 0$, $x \in V$, and to the integral

$$(2.11) \qquad T_\varepsilon^{1/2}(x) = \frac{1}{\pi} \int_0^\infty T_\varepsilon (T_\varepsilon + \lambda I)^{-1}(x) \lambda^{-1/2} \, d\lambda.$$

We shall apply Lebesgue's dominated convergence theorem. We again observe that, for $0 < \lambda \leq 1$, $\|T_\varepsilon(T_\varepsilon + \lambda I)^{-1}\| \leq C$ and, for $\lambda \geq 1$, $\|(T_\varepsilon + \lambda I)^{-1}\| \leq C\lambda^{-1}$, uniformly in $\varepsilon > 0$. Also, in the second case, if $x \in V$, then $\|T_\varepsilon(x)\| \leq C'$, for $0 < \varepsilon \leq 1$. Thus the right-hand side of (2.11) converges to that of (2.10).

To conclude these remarks, we should note that the set of x for which the right-hand side of (2.10) is a Bochner integral is, in general, strictly contained in the domain W of $T^{1/2}$. That is, for an arbitrary element $x \in W$, the integral (2.10) is not necessarily a Bochner integral. This difficulty will be apparent when T is an accretive differential operator.

3 Accretive square roots

In what follows, we shall restrict to the special case where $T : V \to H$ is what T. Kato ([151]) calls a maximal accretive operator. This means that, for all $x \in V$,

$$(3.1) \qquad \Re\langle T(x), x \rangle \geq 0$$

and

$$(3.2) \qquad I + T : V \to H \text{ is an isomorphism.}$$

Let us begin by verifying that properties (2.5) and (2.6) of the preceding section hold. We shall establish a slightly more precise result.

Lemma 2. An operator $T : V \to H$ is maximal accretive if and only if, for all $\lambda > 0$, $T + \lambda I : V \to H$ is an isomorphism and $\|(T + \lambda I)^{-1}\| \leq \lambda^{-1}$.

So what distinguishes the maximal accretive operators from those studied in the previous section is that the constant C of (2.6) is 1.

Suppose that (3.1) and (3.2) are satisfied: let us establish (2.5). For that, it is enough to check that $S = I - (1 - \lambda)(I + T)^{-1} : H \to H$ is an isomorphism, because $T + \lambda I = S(I + T)$ and then $T + \lambda I$ will be an isomorphism.

Clearly, $S : H \to H$ is continuous. To show that S is an isomorphism, it is enough to show that there exists a constant $\delta > 0$ such that, for all $x \in H$,

$$(3.3) \qquad \Re\langle S(x), x \rangle \geq \delta \|x\|^2.$$

To establish (3.3), we use (3.2) to write $x = (T + I)(y)$, for some

$y \in V$. This gives

$$\begin{aligned}
\Re \langle S(x), x \rangle &= \Re \langle (T + \lambda I)(y), (T + I)(y) \rangle \\
&= \lambda \|y\|^2 + (\lambda + 1)\Re \langle T(y), y \rangle + \|T(y)\|^2 \\
&\geq \delta(\|y\|^2 + 2\Re \langle T(y), y \rangle + \|T(y)\|^2) \\
&= \delta \|x\|^2,
\end{aligned}$$

where $\delta = \min(\lambda, 1)$.

We now show that $\|(T + \lambda I)^{-1}x\| \leq \lambda^{-1}\|x\|$. Since $T + \lambda I : V \to H$ is an isomorphism, we can put $x = (T + \lambda I)(y)$. We must show that $\|(T + \lambda I)(y)\| \geq \lambda\|y\|$, for all $\lambda > 0$. But

$$\begin{aligned}
\|(T + \lambda I)(y)\|^2 &= \|T(y)\|^2 + 2\lambda\Re \langle T(y), y \rangle + \lambda^2\|y\|^2 \\
&\geq \lambda^2\|y\|^2.
\end{aligned}$$

Conversely, suppose that $\|(T + \lambda I)^{-1}\| \leq \lambda^{-1}$, for all $\lambda > 0$. Then, for all $x \in H$, $\|T(x)\|^2 + 2\lambda\Re \langle T(x), x \rangle + \lambda^2\|x\|^2 \geq \lambda^2\|x\|^2$. If $\Re \langle T(x), x \rangle$ were strictly negative, for some $x \in H$, then the preceding inequality would not hold, for some small enough value of $\lambda > 0$.

Similarly, we can prove the following result.

Lemma 3. *An operator $T : V \to H$ is maximal accretive if and only if $I + T : V \to H$ is an isomorphism and $S = (I - T)(I + T)^{-1}$ is a contraction.*

An operator S is a contraction if $\|S\| \leq 1$. The salient point is to show that (3.1) is equivalent to $\|(I - T)x\| \leq \|(I + T)x\|$, when (3.2) holds. To do this, it is enough to square both sides of the inequality and expand.

Part of the symbolic calculus on maximal, accretive operators is given by a classical theorem of von Neumann, which we now recall.

Theorem 1. *Let H be a Hilbert space, $S : H \to H$ be a contraction, and $P(z) = c_0 + c_1 z + \cdots + c_m z^m$ be a polynomial. Then*

$$\|P(S)\| \leq \sup_{|z| \leq 1} |P(z)|.$$

The idea of an accretive operator plays an indirect part in the proof of Theorem 1, via the following remark.

Lemma 4. *If $\|S\| \leq 1$, then $T = (I - S)^{-1}$ is accretive and, more precisely, $\Re \langle T(x), x \rangle \geq \|x\|^2/2$, for all $x \in H$.*

Indeed, if we put $y = Tx$, then the lemma is a matter of verifying that

$$\Re \langle y, (I - S)y \rangle \geq \frac{1}{2}\langle y - Sy, y - Sy \rangle,$$

which is equivalent to $\|S(y)\| \leq \|y\|$.

We now prove the theorem. Let $0 \le |z| < 1$. We define the sesquilinear form $B_z : H \to H$ by

$$B_z(x, y) = \langle x, y \rangle + \sum_{1}^{\infty} (z^k \langle S^k x, y \rangle + \bar{z}^k \langle (S^*)^k x, y \rangle).$$

The series converges because $\|S\| \le 1$ and $|z| < 1$. Putting $y = x$ gives

$$B_z(x, x) = \|x\|^2 + 2\Re e \sum_{1}^{\infty} z^k \langle S^k x, x \rangle = 2\Re e \langle (I - zS)^{-1} x, x \rangle - \|x\|^2 \ge 0,$$

by Lemma 4.

The proof of the Cauchy-Schwarz inequality now gives

$$|B_z(x, y)| \le \sqrt{B_z(x, x)} \sqrt{B_z(y, y)}.$$

Putting $z = re^{i\theta}$, $0 \le r < 1$, and writing B_θ for B_z shows us that

$$\int_0^{2\pi} |B_\theta(x, y)| \, d\theta \le \left(\int_0^{2\pi} |B_\theta(x, x)| \, d\theta \right)^{1/2} \left(\int_0^{2\pi} |B_\theta(y, y)| \, d\theta \right)^{1/2}$$

$$= 2\pi \|x\| \|y\|.$$

To finish the proof of von Neumann's theorem, we look at the integral

$$I = \frac{1}{2\pi} \int_0^{2\pi} B_\theta(x, y) P(e^{-i\theta}) \, d\theta = \langle P(rS)x, y \rangle.$$

For $\sup_{|z| \le 1} |P(z)| \le 1$, we get $|I| \le \|x\| \|y\|$, so that $\|P(rS)\| \le 1$ and thus, on passing to the limit as $r \to 1$, $\|P(S)\| \le 1$, as required.

Let us return to the symbolic calculus of accretive operators. We let A denote the Banach algebra of functions which are holomorphic on $\Re e\, z > 0$, continuous on $\Re e\, z \ge 0$, and tend to a limit as $|z| \to \infty$. An obvious consequence of von Neumann's theorem is the following proposition.

Proposition 2. *Let $T : V \to H$ be a maximal accretive operator. Then there exists a unique algebra homomorphism $\chi : A \to \mathcal{L}(H, H)$ such that $\chi((\lambda + z)^{-1}) = (\lambda I + T)^{-1}$, for all $\lambda > 0$, and such that $\|\chi(f)\| \le \|f\|_\infty$, for $f \in A$.*

To see this, we use the transformation $S = (I - T)(I + T)^{-1}$ to reduce to the corresponding statement, in which A is the disk algebra and S is a contraction. The proposition then follows from von Neumann's theorem and the fact that the polynomials are dense in the disk algebra.

For certain applications, we have to enlarge the algebra A a little, replacing it by the algebra B of bounded holomorphic functions on $\Re e\, z > 0$ which are continuous on $\Re e\, z \ge 0$. We no longer require a limit at infinity.

If $f \in B$, then $f_\varepsilon(z) = (1 + \varepsilon z)^{-1} f(z)$ belongs to A, for all $\varepsilon > 0$. As a consequence, the operators $f_\varepsilon(T) = \chi(f_\varepsilon)$ form a bounded family.

Let us show that, for each $x \in H$, $f_\varepsilon(T)(x)$ converges to a limit which, by definition, will be $f(T)(x)$. We may restrict attention to $x \in V$, because V is dense in H. So we write $x = (I + T)^{-1}y$, for some $y \in H$. Everything works out as if $f(z)$ were replaced by $g(z) = f(z)/(1 + z)$. Since the functions $g(z)/(1 + \varepsilon z)$ converge to $g(z)$ in the uniform norm on A, the operators $f_\varepsilon(T)(I + T)^{-1}$ converge to $g(T)$ in operator norm. The strong convergence of the $f_\varepsilon(T)$ follows.

We leave the reader the task of verifying that the extension of $\chi : A \to \mathcal{L}(H, H)$ to B is still an algebra homomorphism.

As an application of these ideas, consider the function $f(z) = e^{-tz}$, for $t \geq 0$, and form $S_t = f(T)$. We have constructed the semigroup of contractions with infinitesimal generator $-T$. This semigroup has the property that S_t converges strongly to the identity as $t \to 0$. Conversely, every strongly continuous semigroup of contractions is generated by a maximal accretive operator. The interested reader is referred to [151].

We return to the problem of square roots.

If T is maximal accretive, then $T(T + \lambda I)^{-1}$ is itself accretive, for any $\lambda > 0$. Indeed, we can write this operator as $I - \lambda(T + \lambda I)^{-1}$ and $\lambda(T + \lambda I)^{-1}$ is a contraction.

Every linear combination with positive coefficients of the operators $T(T + \lambda I)^{-1}$ is still accretive and we thus have $\Re e \, \langle T^{1/2}x, x \rangle \geq 0$, for all $x \in V$. By continuity, this property extends to W, the domain of $T^{1/2}$. But we still must show that $T^{1/2} : W \to H$ is maximal accretive. In other words, we must show that $T^{1/2} + I : W \to H$ is an isomorphism. Recall that W is defined as the image of $(T + I)^{-1/2}$. So it all comes down to showing that $(T^{1/2} + I)(T + I)^{-1/2} : H \to H$ is an isomorphism.

To do this, we compare $(T^{1/2} + I)(T + I)^{-1/2}$ with $(T_\varepsilon^{1/2} + I)(T + I)^{-1/2}$.

Now $T_\varepsilon^{1/2} : W \to H$ is an isomorphism, as we established in the previous section. Further, $T_\varepsilon^{1/2}$ is accretive, as is $T_\varepsilon^{-1/2}$. The operator $I + T_\varepsilon^{-1/2} : H \to H$ is thus an isomorphism. This operator may also be written in the form $(1 + T_\varepsilon^{1/2})T_\varepsilon^{-1/2}$, so $I + T_\varepsilon^{1/2} : W \to H$ is an isomorphism. Thus $(T_\varepsilon^{1/2} + I)(T + I)^{-1/2}$ is an isomorphism.

We now show that there is a constant $\gamma > 0$ such that, for $0 < \varepsilon \leq \eta$, where $\eta > 0$ is sufficiently small, and for all $x \in H$,

$$(3.4) \qquad \|(T_\varepsilon^{1/2} + I)(T + I)^{-1/2}x\| \geq \gamma\|x\| .$$

To establish (3.4), we first remark that $\|(T_\varepsilon^{1/2} + I)y\|^2 \geq \|T_\varepsilon^{1/2}y\|^2 + \|y\|^2$, for all $y \in W$, since $T_\varepsilon^{1/2}$ is accretive. But we already know that $\|T_\varepsilon^{1/2} - T^{1/2}\| \leq C\sqrt{\varepsilon}$. Also, $T^{1/2}$ and $(T + I)^{1/2}$ differ by a bounded operator on H. Lastly, $\|x\| + \|T^{1/2}x\|$ and $\|(T + I)^{1/2}x\|$ are equivalent norms on W. Putting these facts together gives (3.4).

Obviously, $\|(T_\varepsilon^{1/2} + I)(T + I)^{-1/2}\| \leq C_0$, if $0 < \varepsilon \leq 1$. Finally, $(T^{1/2} + I)(T + I)^{-1/2}$ is an isomorphism of H with itself, because the norm of the difference between this operator and $(T_\varepsilon^{1/2} + I)(T + I)^{-1/2}$ does not exceed $C_1\sqrt{\varepsilon}$, whereas C_0 and $\gamma > 0$ do not depend on $\varepsilon > 0$.

We summarize the above:

Proposition 3. *Let $T : V \to H$ be a maximal accretive operator. Then there is a maximal accretive operator, which we write as $T^{1/2}$, whose domain W contains V and whose square is T. In particular, $W = (I + T)^{-1/2}H$ and*

$$(3.5) \qquad T^{1/2} = \frac{1}{\pi} \int_0^\infty T(T + \lambda I)^{-1} \lambda^{-1/2}\, d\lambda.$$

As we have already remarked, the integral on the right-hand side is a Bochner integral when applied to $x \in V$.

Proposition 3 may be sharpened. In fact, there exists a unique maximal accretive operator L whose square is T. The reader is referred to [151], where this assertion is proved.

4 Accretive sesquilinear forms

We should not give applications without first constructing a maximal accretive operator.

Throughout this section, we shall use H_0 to denote the Hilbert space that we have hitherto denoted by H. We shall write $H_1 \subset H_0$ for a dense linear subspace which is itself a Hilbert space for an inner-product $\langle \cdot, \cdot \rangle$ and norm $\| \cdot \|$. The inner product and norm on H_0 will be denoted by $(\cdot, \cdot)$ and $|\cdot|$.

We suppose that the injection $H_1 \hookrightarrow H_0$ is continuous: there is a constant C such that $|x| \leq C\|x\|$, for all $x \in H_1$.

We have in mind an application in which $H_0 = L^2(\mathbb{R}^n)$, H_1 is the Sobolev space $H^1(\mathbb{R}^n)$, and $\langle f, g \rangle = \int_{\mathbb{R}^n} f(x)\bar{g}(x)\, dx + \int_{\mathbb{R}^n} \nabla f \cdot \nabla \bar{g}\, dx$.

The Riesz representation theorem shows that, for every $y \in H_0$, there is an element $z \in H_1$ such that $(x, y) = \langle x, z \rangle$, for all $x \in H_1$. We put $z = J(y)$ and it is a triviality to show that $J : H_0 \to H_1$ is linear, continuous and injective. We let $H_2 \subset H_1$ denote the image $J(H_0)$ and we have

$$(4.1) \qquad \langle x, y \rangle = (x, J^{-1}(y)), \qquad x \in H_1, \quad y \in H_2.$$

In the concrete example, $J = (I - \Delta)^{-1}$ and H_2 is the Sobolev space $H^2(\mathbb{R}^n)$.

Finally, we let $T : H_2 \to H_0$ be the operator J^{-1}.

All this has been done so that T is a self-adjoint positive operator. This preliminary construction is a special case of what we shall now do. We replace $\langle x, y \rangle$ by a sesquilinear form $B : H_1 \times H_1 \to \mathbb{C}$, that is, a form satisfying the following properties:

(4.2) $$B(\alpha x + \beta y, z) = \alpha B(x, z) + \beta B(y, z)$$

and

(4.3) $$B(x, \alpha y + \beta z) = \bar{\alpha} B(x, y) + \bar{\beta} B(x, z)$$

for $x, y, z \in H_1$ and $\alpha, \beta \in \mathbb{C}$.

We shall further suppose that there exist two constants $C \geq \gamma > 0$ such that

(4.4) $$|B(x, y)| \leq C \|x\| \|y\|$$

and

(4.5) $$\Re e\, B(x, x) \geq \gamma \|x\|^2 \,.$$

Ignoring H_0 for a moment, we see that there exists an isomorphism $S : H_1 \to H_1$ such that $B(x, y) = \langle S(x), y \rangle$, for all $x, y \in H_1$.

Using the isomorphism $T : H_2 \to H_0$ that we constructed above, we put $T_B = TS$. The domain of T_B is the subspace $V \subset H_1$ defined by the condition $S(x) \in H_2$.

By our construction, $S : V \to H_2$ is an isomorphism (between these two subspaces of H_1) and $T_B : V \to H_0$ is an isomorphism.

By (4.1), for $x \in V$ and $y \in H_1$,

(4.6) $$B(x, y) = (T_B x, y)$$

and, for all $x \in V$,

(4.7) $$\Re e\,(T_B x, x) \geq \gamma |x|^2 \,.$$

It is worth remarking that T_B and V depend non-linearly on B. The operator T_B is maximal accretive, with domain V.

We have established the following result.

Proposition 4. *Let $B : H_1 \times H_1 \to \mathbb{C}$ be a sesquilinear form satisfying (4.2) to (4.5). Then there exist a subspace $V \subset H_1$ and a maximal accretive operator $T_B : V \to H_0$ such that, for all $x \in V$ and $y \in H_1$,*

(4.8) $$B(x, y) = (T_B(x), y) \,.$$

Word of mouth attributes the following question to T. Kato ([151]). Is it true that the domain W of the accretive square root $T_B^{1/2}$ of the operator T_B, which we have just constructed, coincides with the space H_1? In other words, is the domain of the square root the domain of the form?

McIntosh constructed the first counter-examples and thus showed that

we could not expect a positive answer in complete generality. We therefore need to restrict ourselves to the special case of square roots of operators $T = -\operatorname{div} A(x)\nabla$, where $A(x)$ satisfies the conditions given in the
introduction. It is this restricted problem which we call Kato's conjec-
ture and we shall describe it in detail.

5 Kato's conjecture

We start with an $n \times n$ matrix $A(x)$, whose entries $a_{jk}(x)$, $1 \leq j, k \leq n$,
are in $L^{\infty}(\mathbb{R}^n)$. We shall suppose that there exists a constant $\delta > 0$ such
that, for each vector $(\zeta_1, \ldots, \zeta_n) \in \mathbb{C}^n$,

$$(5.1) \qquad \Re e \sum_1^n \sum_1^n a_{jk}(x)\zeta_j\bar{\zeta}_k \geq \delta(|\zeta_1|^2 + \cdots + |\zeta_n|^2),$$

for almost all $x \in \mathbb{R}^n$.

We consider the sesquilinear form B, defined on $H^1(\mathbb{R}^n) \times H^1(\mathbb{R}^n)$ by

$$(5.2) \qquad B(f,g) = \sum_1^n \sum_1^n \int_{\mathbb{R}^n} a_{jk}(x)\frac{\partial f}{\partial x_j}\frac{\partial \bar{g}}{\partial x_k}\,dx.$$

We then have

$$(5.3) \qquad |B(f,g)| \leq C\|\nabla f\|_2 \|\nabla g\|_2$$

and

$$(5.4) \qquad \Re e\, B(f,f) \geq \delta\|\nabla f\|_2^2.$$

We are not in the situation of the preceding section, because $\|\nabla f\|_2$
is not the norm of f in $H^1(\mathbb{R}^n)$. We right ourselves by considering the
form $\tilde{B}(f,g) = B(f,g) + \int_{\mathbb{R}^n} f\bar{g}\,dx$. Using $\tilde{B}(f,g)$, we construct the
maximal accretive operator $\tilde{T}_B$, with domain V. This operator can be
written as $\tilde{T}_B = T_B + I$ and we have

$$(5.5) \qquad B(f,g) = (T_B(f),g), \qquad f \in V, \quad g \in H^1.$$

The operator is maximal accretive and its domain is V. Returning
to the construction of a maximal accretive operator via an accretive
form, we immediately verify that the domain of T_B is the vector space
$V \subset H^1(\mathbb{R}^n)$ of those f for which $\operatorname{div}(A(x)\nabla f)$ is in $L^2(\mathbb{R}^n)$. We note
that $A(x)\nabla f$ is a vector whose n co-ordinates belong to $L^2(\mathbb{R}^n)$. Its
divergence must therefore be calculated in the sense of distributions.

This characterization of the domain of T_B is compatible with the
general rules defining the domain of an operator obtained by the composition of unbounded operators: at each stage of the composition, all
the calculations must be well-defined. So we may write

$$T_B(f) = -\operatorname{div}(A(x)\nabla f).$$

Kato's conjecture is that the domain of the accretive square root of T_B is $H^1(\mathbb{R}^n)$.

We shall prove the following theorem.

Theorem 2. *For each integer $n \geq 1$, there exists a constant $\varepsilon(n) > 0$ such that, if $\|I - A(x)\|_\infty < \varepsilon(n)$, the accretive square root $T_B^{1/2}$ of T_B may be written in the form*

$$(5.6) \qquad T_B^{1/2} = \sum_1^n R_j(A)D_j, \qquad where \qquad D_j = -i\frac{\partial}{\partial x_j},$$

and where the operators $R_j(A)$ are functions of $A(x)$, which are holomorphic on the open set $\|I - A(x)\|_\infty < \varepsilon(n)$ in $(L^\infty(\mathbb{R}^n))^{n^2}$ and take values in the algebra $\mathcal{L}(L^2(\mathbb{R}^n), L^2(\mathbb{R}^n))$ of bounded operators on $L^2(\mathbb{R}^n)$.

Further, the operators $R_1(I), \ldots, R_n(I)$ are the usual Riesz transforms.

Finally, the operators $R_j(A)$, $1 \leq j \leq n$, extend as continuous operators from $L^p(\mathbb{R}^n)$ to $L^p(\mathbb{R}^n)$, for $2 \leq p < \infty$, and from $L^\infty(\mathbb{R}^n)$ to $BMO(\mathbb{R}^n)$.

The theorem evidently implies that $H^1(\mathbb{R}^n)$ is contained in the domain W of $T_B^{1/2}$. Let us show that the theorem, once we have established it for the matrix $A(x)$ and its adjoint $A^\star(x)$, implies that $W = H^1(\mathbb{R}^n)$. Let $B^\star$ denote the adjoint of the form B. $B^\star(f,g) = \overline{B(g,f)}$. Then the adjoint of T_B is $T_{B^\star}$ and the adjoint of $T_B^{1/2}$ is $T_{B^\star}^{1/2}$. To verify this statement, all we need do is go back to the integral representation formula (2.10).

So $B(f,g) = (T_Bf,g) = (T_B^{1/2}T_B^{1/2}f,g) = (T_B^{1/2}f,T_{B^\star}^{1/2}g)$, if f is in the domain V of T_B and $g \in H^1(\mathbb{R}^n)$. We can extend the identity $B(f,g) = (T_B^{1/2}f,T_{B^\star}^{1/2}g)$ to $H^1(\mathbb{R}^n) \times H^1(\mathbb{R}^n)$.

Putting $g = f$, we get

$$(5.7) \quad \delta\|\nabla f\|_2^2 \leq \Re e\, B(f,f) = \Re e\, (T_B^{1/2}f, T_{B^\star}^{1/2}f) \leq \|T_B^{1/2}f\|_2\|T_{B^\star}^{1/2}f\|_2.$$

Since we are supposing that $\|T_{B^\star}^{1/2}f\|_2 \leq C\|\nabla f\|_2$, we get $\|T_B^{1/2}f\|_2 \geq \delta C^{-1}\|\nabla f\|_2$, and the domain W of $T_B^{1/2}$ is thus $H^1(\mathbb{R}^n)$.

6 The multilinear operators of Kato's conjecture

Following the approach of Chapter 13, we intend to construct the operators R_j of Theorem 2 by expanding each of them as a series of multilinear operators and then establishing the convergence of the series.

To simplify the notation, we shall write T instead of T_B in all that follows. The maximal accretive operator T corresponding to B was defined in the preceding section and we intend to show that its accretive square root may be written as

$$(6.1) \qquad T^{1/2} = \sum_{m=0}^{\infty} \sum_{j=1}^{n} R_{jm}(A) D_j \,,$$

where $D_j = -i(\partial/\partial x_j)$ and the operators $R_{jm}(A)$ are bounded on $L^2(\mathbb{R}^n)$. Moreover, we shall show that there is a constant $C = C(n) > 0$, depending only on the dimension, such that the norms $\|R_{jm}(A)\|$ of the operators $R_{jm}(A) : L^2(\mathbb{R}^n) \to L^2(\mathbb{R}^n)$ satisfy

$$(6.2) \qquad \|R_{jm}(A)\| \le C^m \|A(x) - I\|_{\infty}^m \qquad \text{for all } m \ge 1.$$

The operators $R_{j0}(A)$, $1 \le j \le n$, do not depend on A and are the usual Riesz transforms.

Once we have established (6.2), it will follow immediately that $T^{1/2} : H^1(\mathbb{R}^n) \to L^2(\mathbb{R}^n)$ is continuous, for $C\|A(x) - I\|_{\infty} < 1$.

The operators $R_{jm}(A)$ are multilinear in $B = I - A^{-1}$ and will be analysed by McIntosh's formal calculus (Chapter 13, section 7). The analysis of their structure will enable us to deal with a more general case than that needed for Kato's conjecture.

We return to identity (3.5). We substitute for the variable of integration λ by putting $\lambda = t^{-2}$, where $t > 0$. The reason for this change of variable is that t has an important geometrical significance.

We shall obtain the multilinear expansion of $T^{1/2}$, from the expansion of the resolvent $(I + t^2 T)^{-1}$, by integrating with respect to t.

To compute this resolvent, we shall write $t^2 T$ as the following composition of unbounded operators:

$$t^2 T = UAV \,,$$

where

$$V = t\nabla : H^1(\mathbb{R}^n) \to (L^2(\mathbb{R}^n))^n,$$

A is the operator of pointwise multiplication by the matrix $A(x)$,

and

$$U = -t\,\text{div} : E \to L^2(\mathbb{R}^n), \text{ where } E \subset L^2(\mathbb{R}^n) \text{ is the domain of}$$
the divergence operator.

A few obvious remarks will be useful in what follows. With the notation we have introduced, $I + UV = I - t^2\Delta$, where $\Delta = \partial^2/\partial x_1^2 + \cdots + \partial^2/\partial x_n^2$, and $(I + UV)^{-1} U = Q_t = -(I - t^2\Delta)^{-1} t\,\text{div}$. The operator Q_t is a convolution operator with symbol

$$\left(\frac{-it\xi_1}{1 + t^2|\xi|^2}, \ldots, \frac{-it\xi_n}{1 + t^2|\xi|^2} \right).$$

Similarly, the operator $Z_t = I - t\nabla Q_t$ is a convolution operator. It is a classical pseudo-differential operator of order 0 whose matrix symbol is

$$\left(\left(\delta_{jk} - \frac{t^2\xi_j\xi_k}{1+t^2|\xi|^2}\right)\right)^n_{j,k=1},$$

where $\delta_{jk} = 0$, if $j \neq k$, and 1, if $j = k$.

The symbol of the Riesz transform R_j is $\xi_j/|\xi|$. If we put $P_t = (I - t^2\Delta)^{-1}$, we may write

$$Z_t = ((\delta_{jk}I - (I - P_t)R_jR_k))^n_{j,k=1}.$$

Let us clarify the meaning of t. The operator P_t is the operation of convolution by ϕ_t, where $\phi_t(x) = t^{-n}\phi(t^{-1}x)$ and $\phi \in L^1(\mathbb{R}^n)$. More precisely, if $|x| \geq 1$, then $|\phi(x)| + |\nabla\phi(x)| \leq C_m|x|^{-m}$, for all integers $m \geq 1$, but if $|x| \leq 1$, then $|\phi(x)| \leq C|x|^{-n+1}$ and $|\nabla\phi(x)| \leq C|x|^{-n}$. This means that $\phi(x)$ is essentially supported by the unit ball and that, in essence, the "radius of influence" of P_t is t. The same remark is valid for Q_t, apart from two things. $Q_t(f) = f \star \psi_t$, where $\psi_t = t^{-n}\psi(t^{-1}x)$. This time, ψ is vector-valued, belongs to $L^1(\mathbb{R}^n)$, and satisfies $\int \psi(x)\,dx = 0$, whereas before $\int \phi(x)\,dx = 1$. On the other hand, $\psi(x)$ has the same properties of regularity and localization as $\phi(x)$.

The case of the operator Z_t is similar, except that the singularity at the origin is more pronounced. Indeed, there is a tempered distribution $Z \in \mathcal{S}'(\mathbb{R}^n)$ such that the distribution-kernel of Z_t is given by $t^{-n}Z(t^{-1}(x - y))$. The restriction of Z to $|x| \geq 1$ coincides with that of a function in the Schwartz class $\mathcal{S}(\mathbb{R}^n)$, Z is infinitely differentiable on $\mathbb{R}^n \setminus \{0\}$, and the radius of influence of Z_t is t.

We shall use the fact that $Z_t = \rho_t + \pi_t$, where the distribution-kernel of the singular part ρ_t is supported by $|x - y| \leq t$ and where π_t is the operator of convolution with $w_t = t^{-n}w(t^{-1}x)$, where $w(x) \in \mathcal{S}(\mathbb{R}^n)$.

We return to $(I + t^2T)^{-1}$. Recall that A is the operation of multiplication by the matrix $A(x)$. We put $A^{-1}(x) = I - B(x)$ and let B denote the operation of multiplication by the matrix $B(x)$. We shall expand $(I + t^2T)^{-1}$ as a series of powers of the operator B, using the identity

(6.3) $$(I + UAV)^{-1}UA = Q(I - BZ)^{-1},$$

where U, A, and V are now arbitrary elements of an associative algebra, where $Q = (I + UV)^{-1}U$, $Z = I - VQ$, and we suppose that all the inverses we have written down actually exist.

The proof of (6.3) is an exercise in algebra which we leave to the reader. The particular choices of the operators U, A, and V that we need play no rôle in that proof.

In our situation, $Z = Z_t : (L^2(\mathbb{R}^n))^n \to (L^2(\mathbb{R}^n))^n$ is bounded and the operator norm of Z_t is independent of t. As a consequence, the norm of

$BZ = BZ_t$ is strictly less than 1, if $\|I - A(x)\|_\infty < \varepsilon(n)$, where $\varepsilon(n) > 0$ is sufficiently small. We can, therefore, expand the right-hand side of (6.3) as a Neumann series $\sum_0^\infty Q(BZ)^m$. Going back to (3.5), we get

$$(6.4) \qquad T^{1/2} = \sum_1^n R_j(A) D_j \, ,$$

where

$$(6.5) \qquad \begin{pmatrix} R_1(A) \\ \cdot \\ \cdot \\ \cdot \\ R_n(A) \end{pmatrix} = \frac{2}{\pi} \int_0^\infty Q_t (I - BZ_t)^{-1} \frac{dt}{t} = \frac{2}{\pi} \sum_0^\infty L_m \, ,$$

with

$$(6.6) \qquad L_m = \int_0^\infty Q_t (BZ_t)^m \frac{dt}{t} \, .$$

We shall show that there exists a constant $C = C(n)$ such that $\|L_m\| \leq C^m \|B\|^m$, for all $m \geq 1$. This will let us establish the continuity of the operators $R_j(A)$ acting on $L^2(\mathbb{R}^n)$, on condition that $C\|B(x)\|_\infty < 1$, which will be the case if $\|I_n - A(x)\|_\infty < \varepsilon(n)$, where $\varepsilon(n)$ is small enough.

Let us forget about part of this setup to explore the operators L_m in a form which is both simpler and more general.

For one thing, we shall consider only scalar operators, which will correspond to the entries of the preceding matrix operators.

We start with two convolution operators. Let P_t and Q_t be the operators of convolution with ϕ_t and ψ_t, respectively, where $\phi_t(x) = t^{-n}\phi(t^{-1}x)$ and $\psi_t(x) = t^{-n}\psi(t^{-1}x)$. The functions ϕ and ψ satisfy the same conditions as before: if $|x| \leq 1$, then $|\phi(x)| \leq C|x|^{-n+1}$, $|\nabla\phi(x)| \leq C|x|^{-n}$, and similarly for ψ, while, for $|x| \geq 1$, ψ, ϕ, and their gradients decrease rapidly at infinity. Lastly $\int \phi(x)\, dx = 1$ and $\int \psi(x)\, dx = 0$.

Next, let $m(\xi)$ be a function which is infinitely differentiable on $\mathbb{R}^n \setminus \{0\}$ and homogeneous of degree 0: $m(\lambda\xi) = m(\xi)$ for all $\lambda > 0$ and $\xi \neq 0$. In the special case of Kato's problem, $m(\xi)$ is one of the functions $\xi_j\xi_k/|\xi|^2$. We let $R : L^2(\mathbb{R}^n) \to L^2(\mathbb{R}^n)$ denote the convolution operator whose symbol (or multiplier) is $m(\xi)$. Then R is a Calderón-Zygmund operator and thus bounded on $L^p(\mathbb{R}^n)$, for $1 < p < \infty$.

The final hypothesis is that we can decompose the operator $R_t = (I - P_t)R$ into $R_t = \rho_t + \pi_t$, where the distribution-kernel of ρ_t has support in $|x - y| \leq t$ and π_t is the operator of convolution with $w_t(x) = t^{-n}w(t^{-1}x)$, where $w(x)$ is in the Schwartz class $\mathcal{S}(\mathbb{R}^n)$.

We note that the operator ρ_t is bounded on $L^p(\mathbb{R}^n)$, for $1 < p < \infty$, and that its norm is independent of T.

Let $B_1, \ldots, B_m$ be operators of pointwise multiplication by functions $b_1(x), \ldots, b_m(x)$, for some integer $m \geq 1$. We suppose that $\|b_1\|_\infty \leq 1, \ldots, \|b_m\|_\infty \leq 1$, so that $B_1, \ldots, B_m$ are contractions on $L^2(\mathbb{R}^n)$.

With this notation and these hypotheses, the following result provides a generalization of Theorem 2.

Theorem 3. *There exists a constant C, depending only on the functions ϕ, ψ, and the operator R, such that, for all integers $m \geq 1$ and every function $\mu(t) \in L^\infty(0, \infty)$, the integral*

$$\int_0^\infty Q_t B_1 R_t \ldots B_m R_t \mu(t) \, \frac{dt}{t}$$

converges strongly to an operator $L_m : L^2(\mathbb{R}^n) \to L^2(\mathbb{R}^n)$ whose norm does not exceed $C^m \|\mu\|_\infty$.

Further, there exists a constant C' such that the restriction $L_m(x, y)$ of the distribution-kernel of L_m to the complement of the diagonal satisfies

$$(6.7) \qquad \int_{|x-y| \geq 2|x'-x|} |L_m(x, y) - L_m(x', y)| \, dy \leq C'^m \|\mu\|_\infty .$$

The rest of this chapter will be devoted to the proof of Theorem 3. To do this, we introduce two variants of the operator L_m, namely

$$L_m^{(1)} = \int_0^\infty Q_t B_1 R_t \ldots B_{m-1} R_t B_m P_t \mu(t) \, \frac{dt}{t}$$

and

$$L_m^{(2)} = \int_0^\infty Q_t B_1 R_t \ldots B_{m-1} R_t B_m W_t \mu(t) \, \frac{dt}{t},$$

where W_t is the same as P_t, except that the function ϕ is replaced by the Gaussian $\pi^{-n/2} e^{-|x|^2}$.

The three operators L_m, $L_m^{(1)}$ and $L_m^{(2)}$ are related in the following ways. Firstly, $L_m = L_{m-1} B_m R - L_m^{(1)} R$, because $R_t = (I - P_t)R$. Thus the continuity of L_m on $L^2(\mathbb{R}^n)$ follows from that of L_{m-1} and $L_m^{(1)}$. Secondly, $L_m^{(1)} - L_m^{(2)}$ is a continuous operator on $L^2(\mathbb{R}^n)$. To see this, we consider the more general operators of the form $J = \int_0^\infty Q_t A_t \tilde{Q}_t t^{-1} \, dt$, where Q_t and $\tilde{Q}_t$ are defined in the same way as the operator Q_t of Theorem 3 and where the norm of the operator $A_t : L^2(\mathbb{R}^n) \to L^2(\mathbb{R}^n)$ is bounded uniformly in t. The operator J is bounded on $L^2(\mathbb{R}^n)$, as the next lemma shows.

Lemma 5. *With the above notation, the integral $\int_0^\infty Q_t A_t \tilde{Q}_t t^{-1} \, dt$ converges strongly to a bounded operator on $L^2(\mathbb{R}^n)$.*

To begin with, we shall suppose that $A_t = 0$ for $0 < t < \varepsilon$ and for $t > R$ and we shall give a uniform estimate for the corresponding operator $J = \int_0^\infty Q_t A_t \tilde{Q}_t t^{-1} \, dt$. From this we shall deduce the strong convergence of the integral with respect to t.

To estimate $\|J\|$, we evaluate (Jf, g), when $\|f\|_2 \leq 1$ and $\|g\|_2 \leq 1$.

$$(Jf, g) = \int_0^\infty (A_t \tilde{Q}_t f, Q_t^\star g) \frac{dt}{t}$$

and the Cauchy-Schwarz inequality gives

$$|(Jf, g)| \leq \left(\int_0^\infty \|A_t \tilde{Q}_t f\|_2^2 \frac{dt}{t} \right)^{1/2} \left(\int_0^\infty \|Q_t^\star g\|_2^2 \frac{dt}{t} \right)^{1/2}.$$

Because the operators A_t are continuous, we can replace $\|A_t \tilde{Q}_t f\|_2$ by $C_0 \|\tilde{Q}_t f\|_2$. The operators Q_t, $\tilde{Q}_t$, and $Q_t^\star$ have the same structure, so it is enough to show that

(6.8)
$$\left(\int_0^\infty \|Q_t f\|_2^2 \frac{dt}{t} \right)^{1/2} \leq C_1 \|f\|_2.$$

To verify (6.8), we use Plancherel's formula and this leads to estimating the double integral

(6.9)
$$\int_0^\infty \int_{\mathbb{R}^n} |\hat{\psi}(t\xi)|^2 |\hat{f}(\xi)|^2 \, d\xi \frac{dt}{t}.$$

The conditions on ψ give $|\hat{\psi}(\xi)| \leq C_2 |\xi|$, when $|\xi| \leq 1$, and $|\hat{\psi}(\xi)| \leq C_\delta |\xi|^{-\delta}$, when $|\xi| \geq 1$ and $\delta \in (0, 1)$. As a consequence,

(6.10)
$$\int_0^\infty |\hat{\psi}(t\xi)|^2 \frac{dt}{t} \leq C_3,$$

from which we reach the desired conclusion.

To establish the strong convergence of the integral defining J, we must show, for fixed $f \in L^2(\mathbb{R}^n)$, that $\| \int_T^{T'} Q_t A_t \tilde{Q}_t f t^{-1} \, dt \|_2 \to 0$ as $T \to \infty$ and as $T' \to 0$. (We suppose, of course, that $T' > T$.) Proceeding as before, we have to calculate the limit of $\int_{\mathbb{R}^n} \int_T^{T'} |\hat{\psi}(t\xi)|^2 |\hat{f}(\xi)|^2 t^{-1} \, d\xi \, dt$. By (6.10), we can use Lebesgue's dominated convergence theorem to conclude.

The study of the operators L_m thus reduces to that of the operators $L_m^{(2)}$. J.L. Journé showed that the kernel $L_m^{(2)}(x, y)$ of $L_m^{(2)}$ satisfied the conditions of a Calderón-Zygmund kernel. This led him to use the $T(1)$ theorem to establish the L^2 continuity of $L_m^{(2)}$. Since $W_t(1) = 1$, we have

(6.11)
$$L_m^{(2)}(1) = L_{m-1}(b_m).$$

This means that the induction argument we use must zigzag between the operators L_m and $L_m^{(2)}$ and will require (6.7). The proof of (6.7) will be direct, making no appeal to induction. Now (6.7) gives us the constant C'. Then the $T(1)$ theorem, applied to $L_m^{(2)}$, will let us establish that

(6.12) $$\|L_m^{(2)}\| \le C_0\|L_{m-1}\| + C'^{\,m-1},$$

which will, in turn, lead to

(6.13) $$\|L_m\| \le C_0\|L_{m-1}\| + C'^{\,m-1} + C_1.$$

The latter inequality gives the claimed growth of the operator norms of the L_m.

Before concluding this section, we explain how the induction starts and how the other hypotheses of the $T(1)$ theorem are verified.

The operator L_0 is given by $L_0 = \int_0^\infty Q_t \mu(t) t^{-1} dt$. This is a convolution operator. The corresponding symbol is $\int_0^\infty \hat\psi(t\xi)\mu(t)t^{-1}\,dt$ and it is not hard to see that this symbol is in $L^\infty(\mathbb{R}^n)$, using the estimates of $\hat\psi$ that we used earlier.

To prove the weak continuity property, we use Proposition 6 of Chapter 8. We put $\chi_\xi(x) = e^{ix\cdot\xi}$ and try to estimate the norm of $L_m^{(2)}(\chi_\xi)$ in the space BMO. We get

(6.14) $$L_m^{(2)}(\chi_\xi) = L_{m-1}^{(\xi)}(\chi_\xi b_m),$$

where

$$L_{m-1}^{(\xi)} = \int_0^\infty Q_t B_1 R_t \cdots B_{m-1} R_t e^{-t|\xi|^2/4}\mu(t)\,\frac{dt}{t}.$$

The same argument will give $L_m^{(1)}(1) \in$ BMO and $L_m^{(2)}(\chi_\xi) \in$ BMO, the only change being that the function $\mu(t)$ differs in each case.

Finally ${}^tL_m^{(2)}(1) = 0$, because $\int\psi(x)\,dx = 0$.

The next two sections will be devoted to proving our assertions about the properties of the kernels of L_m and $L_m^{(2)}$. Once this has been done, Theorems 2 and 3 will have been established.

7 Estimates of the kernels of the operators $L_m^{(2)}$

We explain the calculations which follow by considering a special case. Consider the operator

(7.1) $$L = \int_0^\infty P_t A_t \tilde{P}_t \,\frac{dt}{t},$$

where P_t and $\tilde{P}_t$ are the operators of convolution with functions ϕ_t and $\tilde{\phi}_t$, respectively. These are defined by $\phi_t(x) = t^{-n}\phi(x/t)$ and $\tilde{\phi}_t(x) = t^{-n}\tilde{\phi}(x/t)$, where ϕ and $\tilde{\phi}$, unlike the functions we shall have to consider in a moment, are C^1 functions with supports in the unit ball. We further suppose that the operator $A_t : L^2(\mathbb{R}^n) \to L^2(\mathbb{R}^n)$ satisfies $\|A_t\| \le C_0$ and that the distribution-kernel $A_t(x,y)$ of A_t is identically zero on

$|x - y| > t$. In the terminology we used earlier, the radius of influence of A_t is not greater than t. We shall show that, under these conditions, the distribution-kernel $K(x, y)$ of our operator L satisfies the estimates

(7.2) $$|K(x, y)| \leq C_1 |x - y|^{-n}$$

and

(7.3) $$\left| \frac{\partial K}{\partial x_j}(x, y) \right| + \left| \frac{\partial K}{\partial y_j}(x, y) \right| \leq C_1 |x - y|^{-n-1}.$$

To simplify the calculation, we suppose that ϕ and $\tilde{\phi}$ are real-valued even functions. We write $\phi_t^{(x)}(u) = \phi_t(u - x)$ and $\tilde{\phi}_t^{(y)}(v) = \tilde{\phi}_t(v - y)$. The kernel $K(x, y)$ is then given by the triple integral

$$\int_0^\infty \int_{\mathbb{R}^n} \int_{\mathbb{R}^n} \phi_t(x - u) A_t(u, v) \tilde{\phi}_t(v - y) \, du \, dv \, \frac{dt}{t} \, ,$$

where $A_t(u, v)$ is the distribution-kernel of the operator A_t. In terms of that operator,

(7.4) $$K(x, y) = \int_0^\infty \langle A_t \tilde{\phi}_t^{(y)}, \phi_t^{(x)} \rangle \, \frac{dt}{t} \, .$$

To get an upper bound for $|K(x, y)|$, we use the following trivial remarks. On the one hand, $|\langle A_t \tilde{\phi}_t^{(y)}, \phi_t^{(x)} \rangle| \leq C_0 \|\tilde{\phi}_t^{(y)}\|_2 \|\phi_t^{(x)}\|_2 = Ct^{-n}$, for all $t > 0$. On the other hand, $\langle A_t \tilde{\phi}_t^{(y)}, \phi_t^{(x)} \rangle = 0$, if $|x - y| > 3t$, because the supports of $\phi_t^{(x)}$ and $A_t \tilde{\phi}_t^{(y)}$ are then disjoint. So the integral on the right-hand side of (7.4) may be written as $\int_{|x-y|/3}^\infty \cdots t^{-1} dt$ and the estimate follows immediately.

The verification of (7.3) is similar and is left to the reader.

The treatment of the kernel $L_m^{(2)}(x, y)$ of the operator $L_m^{(2)}$ follows the lines we have just described. There are, however, two new difficulties. The first is to do with the localization of the functions ϕ_t and ψ_t and the operators R_t which are involved. The localization is no longer exact, but only approximate. The second problem is that the function ψ, which defines the operator Q_t, does not lie in L^2, but only in L^p, where $1 < p < n/(n-1)$. Instead of using the continuity of A_t on L^2, we shall need to use its continuity on such an L^p. But, at the same time, the second function $\tilde{\phi}$, which we use, must lie in L^q, where $1/p + 1/q = 1$, so that $\langle A_t \tilde{\phi}_t^{(y)}, \phi_t^{(x)} \rangle$ is well-defined. This is certainly not the case for the operator $L_m^{(1)}$, because the functions ψ and ϕ, from which Q_t and P_t are constructed, have the same type of singularity. However, all works out well for $L_m^{(2)}$, because ϕ is then replaced by a Gaussian function which belongs to all the L^q spaces.

It is time to get down to details. The following lemma is due to J.L. Journé. We consider functions $a(x)$ and $b(x)$ in $\mathcal{D}(\mathbb{R}^n)$, which are positive, radial, and such that $c(x) = (a \star b)(x) = 1$ on the ball $|x| \leq 1$. If $1 < p, q < \infty$ are conjugate exponents, we put $\alpha(x) = (a(x))^p$ and $\beta(x) = (b(x))^q$.

Lemma 6. *Let $T : L^p(\mathbb{R}^n) \to L^p(\mathbb{R}^n)$ be a continuous linear operator whose distribution-kernel has support in $|x - y| \leq r$. Then, for all $f \in L^p(\mathbb{R}^n)$ and $g \in L^q(\mathbb{R}^n)$,*

$$(7.5) \qquad |\langle Tf, g \rangle| \leq \|T\| \int_{\mathbb{R}^n} (|f|^p \star \alpha_r)^{1/p} (|g|^q \star \beta_r)^{1/q} \, dx \,,$$

where $\alpha_r(x) = r^{-n}\alpha(x/r)$ and $\beta_r(x) = r^{-n}\beta(x/r)$.

This inequality is an improvement of the obvious estimate $|\langle Tf, g \rangle| \leq \|T\|\|f\|_p\|g\|_q$, given by the continuity of T. We note that applying Hölder's inequality to the right-hand side of (7.5) would take us back to the obvious estimate. The inequality (7.5) is particularly interesting if, on the scale given by r, $|f|$ is small where $|g|$ is large, and vice versa.

To prove Lemma 6, we write

$$\langle Tf, g \rangle = \int\int T(x, y) f(y) g(x) \, dy \, dx = \int\int T(x, y) c\Big(\frac{x-y}{r}\Big) f(y) g(x) \, dy \, dx \,.$$

Now

$$c\left(\frac{x - y}{r}\right) = \frac{1}{r^n} \int a\Big(\frac{y - u}{r}\Big) b\Big(\frac{u - x}{r}\Big) \, du \,,$$

which leads to

$$c\Big(\frac{x - y}{r}\Big) f(y) g(x) = \int f_u(y) g_u(x) \, du \,,$$

where

$$f_u(y) = \frac{1}{r^{n/p}} a\Big(\frac{y - u}{r}\Big) f(y)$$

and similarly for $g_u(x)$. Finally, the continuity of T gives

$$|\langle Tf, g \rangle| = \left| \int \langle Tf_u, g_u \rangle \, du \right| \leq \|T\| \int \|f_u\|_p \|g_u\|_q \, du$$

$$= \|T\| \int_{\mathbb{R}^n} (|f|^p \star \alpha_r)^{1/p} (|g|^q \star \beta_r)^{1/q} \, du \,.$$

Here is our first application of Lemma 6. We let $p_t(x) = t^{-n}p(x/t)$ denote the Poisson kernel: $p(x) = c_n(1 + |x|^2)^{-(n+1)/2}$, where $c_n > 0$ is chosen so that $\int p(x) \, dx = 1$. Let $S_t^{(1)}$ and $S_t^{(2)}$ be operators defined by kernels $S_t^{(1)}(x, y)$ and $S_t^{(2)}(x, y)$, satisfying

$$|S_t^{(1)}(x, y)| \leq Cp_t(x - y), \qquad |S_t^{(2)}(x, y)| \leq Cp_t(x - y) \,,$$

and

$$\left|\frac{\partial S_t^{(1)}}{\partial x_j}(x, y)\right| + \left|\frac{\partial S_t^{(2)}}{\partial y_j}(x, y)\right| \leq Ct^{-1}p_t(x - y), \qquad 1 \leq j \leq n \,.$$

We consider the operator $L = \int_0^\infty S_t^{(1)} A_t S_t^{(2)} t^{-1} \, dt$, where A_t satisfies the same hypotheses as in the special case at the start of this section. Then the distribution-kernel $K(x, y)$ of L, restricted to $x \neq y$, will satisfy (7.2) and (7.3). To see this, we put

$$f_{(t,y)}(u) = S_t^{(2)}(u, y), \qquad g_{(t,x)}(u) = \bar{S}_t^{(1)}(u, x),$$

followed by

$$K_t(x, y) = \langle A_t f_{(t,y)}, g_{(t,x)} \rangle,$$

and we have

$$K(x, y) = \int_0^\infty K_t(x, y) \frac{dt}{t}.$$

Journé's lemma applies to $K_t(x, y)$ and gives

$$|K_t(x, y)| \leq C_0 \int_{\mathbb{R}^n} (|f_{(t,y)}|^2 \star \alpha_t)^{1/2} (|g_{(t,x)}|^2 \star \beta_t)^{1/2} \, du.$$

By our hypotheses,

$$(|f_{(t,y)}|^2 \star \alpha_t)^{1/2} \leq C_1 p_t(u - y)$$

and

$$(|g_{(t,x)}|^2 \star \beta_t)^{1/2} \leq C_1 p_t(u - x).$$

The calculation ends by observing that

$$\int_{\mathbb{R}^n} p_t(u - y) p_t(u - x) \, du = (p_t \star p_t)(x - y) = p_{2t}(x - y) \leq 2 p_t(x - y)$$

and that

$$\int_0^\infty p_t(x - y) \frac{dt}{t} = c_n |x - y|^{-n}.$$

The calculation of $(\partial/\partial x_j) K(x, y)$ and of $(\partial/\partial y_j) K(x, y)$ is similar and is left to the reader.

Armed with Lemma 6, we are ready to describe the regularity of the kernel $L_m^{(2)}(x, y)$ of the operator $L_m^{(2)}$. The hypotheses and the notation are those of Theorem 3.

Proposition 5. *The kernels $L_m^{(2)}(x, y)$ of the operators $L_m^{(2)}$ satisfy the estimates*

(7.6) $$|L_m^{(2)}(x, y)| \leq C_0^m |x - y|^{-n},$$

(7.7) $$\left| \frac{\partial L_m^{(2)}}{\partial y_j}(x, y) \right| \leq C_0^m |x - y|^{-n-1}, \qquad 1 \leq j \leq n,$$

and, if $|x' - x| \leq |x - y|/2$ and $\varepsilon \in (0, 1)$,

(7.8) $$|L_m^{(2)}(x', y) - L_m^{(2)}(x, y)| \leq C_0^m C_\varepsilon |x' - x|^\varepsilon |x - y|^{-n-\varepsilon}.$$

We recall that $L_m^{(2)} = \int_0^\infty Q_t B_1 R_t \ldots B_{m-1} R_t B_m W_t \mu(t) t^{-1} \, dt$, where $R_t = \rho_t + \pi_t$, the singular part ρ_t having a radius of influence not greater than t (which allows us to apply Journé's lemma.) The regular part π_t is the operator of convolution with a function $w_t(x) = t^{-n}w(x/t)$, with w in the Schwartz class $\mathcal{S}(\mathbb{R}^n)$. For the purposes of the proof, the operators π_t and W_t will be included in a more general class, whose elements are denoted by S_t. An operator will be written as S_t if it is defined by a kernel $S_t(x, y)$ which behaves like the Poisson kernel $p_t(x-y) = c_n t(t^2 + |x-y|^2)^{-(n+1)/2}$, in the sense that, for some constant C,

$$(7.9) \quad |S_t(x, y)| \le C p_t(x - y) \quad \text{and} \quad \left|\frac{\partial S_t}{\partial y_j}(x, y)\right| \le C t^{-1} p_t(x - y).$$

We begin by replacing all the operators R_t which appear in $L_m^{(2)}$ by the operators ρ_t. Put $\Lambda_t = B_1 \rho_t B_2 \cdots B_{m-1} \rho_t B_m$. We study $\mathcal{T}_m = \int_0^\infty Q_t \Lambda_t W_t t^{-1}\, dt$, or, more generally, $\mathcal{T}_m = \int_0^\infty Q_t \Lambda_t S_t t^{-1}\, dt$.

The distribution-kernel $\Lambda_t(u, v)$ of Λ_t vanishes when $|u - v| > (m-1)t$ and the distribution-kernel $\mathcal{T}_m(x, y)$ of $\mathcal{T}_m$ is given by

$$(7.10) \qquad \mathcal{T}_m(x, y) = \iint \psi_t(x - u)\Lambda_t(u, v)S_t(v, y)\, du\, dv.$$

We shall use Journé's lemma to estimate this integral. To that end, we fix $p \in (1, n/(n-1))$ and let c_0 denote the norm of $\rho_t : L^p(\mathbb{R}^n) \to L^p(\mathbb{R}^n)$, which is independent of t. Then the norm of $\Lambda_t : L^p \to L^p$ is not greater than c_0^{m-1}, and Journé's lemma gives

$$(7.11) \quad |\mathcal{T}_m(x, y)| \le C c_0^{m-1} \int_{\mathbb{R}^n} (|\psi_t|^p \star \alpha_r)^{1/p}(u - x)$$

$$\times (|p_t|^q \star \beta_r)^{1/q}(u - y)\, du,$$

where $r = (m-1)t$. The constant C comes from $|S_t(x, y)| \le C p_t(x-y)$, where p_t is the Poisson kernel.

To evaluate $|\psi_t|^p \star \alpha_r$, when $0 < t \le r$, it is useful to note that, up to a constant factor which is irrelevant to our needs, $t^{n(p-1)}|\psi_t|^p$ is an approximation, scaled by t, to the Dirac measure at 0. Taking account of the normalizations of α_r and of the Poisson kernel p_r, we thus have

$$(7.12) \qquad (t^{n(p-1)}|\psi_t|^p \star \alpha_r)(u) \le C r^{n(p-1)}(p_r(u))^p,$$

for $0 < t \le r$. Since $r = (m-1)t$, we get the simpler form

$$(7.13) \qquad (|\psi_t|^p \star \alpha_r)^{1/p} \le C m^{n/q} p_r$$

and, similarly,

$$(7.14) \qquad (|p_t|^q \star \beta_r)^{1/q} \le C m^{n/p} p_r.$$

Thus, to estimate the right-hand side of (7.11), we calculate

$$\int p_r(u - x)p_r(y - u)\, du = (p_r \star p_r)(x - y) = p_{2r}(x - y)$$

$$\le 2m p_t(x - y).$$

Finally, we observe that $\int_0^\infty p_t(x-y)t^{-1}\,dt = c_n|x-y|^{-n}$, which proves (7.6), in the case of the operators $\mathcal{T}_m$.

The geometric growth appearing in the right-hand side of (7.6) thus comes from the operator norm of $\rho_t : L^p \to L^p$, this operator being applied $m-1$ times.

The proof of (7.7) is virtually identical. The only difference comes from $S_t(x,y)$ being replaced by $(\partial/\partial y_j)S_t(x,y)$, which is bounded by $Ct^{-1}p_t(x-y)$.

We come to the regularity in x of the kernel $\mathcal{T}_m(x,y)$. We put $d = |x'-x|$ and $f_t(u) = \psi_t(x-u) - \psi_t(x'-u)$, which leads to estimating $|I(x,x',y)|$, where

$$I(x,x',y) = \int_0^\infty \int\int f_t(u)\Lambda_t(u,v)S_t(v,y)\,du\,dv\,\frac{dt}{t}.$$

We split the outer integral into $\int_0^{2d} + \int_{2d}^\infty = I_1 + I_2$.

To estimate $|I_1|$, we deal separately with the two terms of which $f_t(u)$ is the difference and get, using the previous calculations,

$$c_0^m \int_0^{2d} p_t(x-y)\,\frac{dt}{t} + c_0^m \int_0^{2d} p_t(x'-y)\,\frac{dt}{t} \le C'c_0^m d|x-y|^{-(n+1)},$$

because $d = |x'-x| \le |x-y|/2$.

On the other hand, if $t \ge 2d$, we apply Lemma 6 to get an upper bound for $|\int\int f_t(u)\Lambda_t(u,v)S_t(v,y)\,du\,dv|$. An explicit calculation, which we leave to the reader, gives

$$(7.15) \qquad (|f_t|^p \star \alpha_2)^{1/p}(u) \le C \left(\frac{d}{t}\right)^\varepsilon m^{n/q}p_r(u-x),$$

for $\varepsilon = n/p - (n-1)$. This estimate replaces (7.13) and, together with (7.14), gives (7.8).

To finish, we show how to pass from the operators $\mathcal{T}_m$ to $L_m^{(2)}$. We shall split $L_m^{(2)}$ into m operators, each with the same structure as $\mathcal{T}_m$. Here is how to do it.

By our first application of Lemma 6, we know that the kernel $S_t(x,y)$ of $S_t = \pi_t B_l \rho_t \cdots \rho_t B_m W_t$ satisfies the estimate (7.9), for $1 \le l \le m$.

So we start with a "word" $Q_t B_1 R_t \cdots B_{m-1} R_t B_m W_t$ and decompose it using the following algorithm. We start on the right and split the last of the occurrences of R_t as $\rho_t + \pi_t$. This gives two terms, of which that containing π_t is not split any further. The term containing ρ_t is again split into two by writing the (originally) penultimate occurrence of R_t in the form $R_t = \rho_t + \pi_t$, and so on.

This algorithm produces m terms. $Q_t B_1 \rho_t B_2 \cdots B_{m-1}\rho_t B_m W_t$ is the last and, after integration with respect to t, leads to the operator $\mathcal{T}_m$. The others are of the form $Q_t B_1 R_t \cdots B_{l-1}S_t$, where $l \le m$. They lead to operators with the same structure as $L_m^{(2)}$, except that W_t is replaced by S_t and m by $l - 1 < m$.

Finally, we are led to enlarge the collection of operators $L_m^{(2)}$ by replacing W_t by S_t and to conclude the proof by using induction on m. Attention to the details shows that the growth of the constants is, indeed, geometric in m.

8 The kernels of the operators L_m

We start with a simpler situation, which serves as a model for what we want to do. We consider an operator $L = \int_0^\infty Q_t A_t t^{-1} \, dt$, defined by operators Q_t and A_t with the following properties.

The kernel $q_t(x, y)$ of Q_t satisfies the conditions $|q_t(x, y)| \le t^{-n}$, $|(\partial/\partial x_j)q_t(x, y)| \le t^{-n-1}$, for $1 \le j \le n$, and $q_t(x, y) = 0$, for $|x - y| > t$.

As far as the operators A_t are concerned, we suppose that there is a constant C_0, such that the norm of the operator $A_t : L^2(\mathbb{R}^n) \to L^2(\mathbb{R}^n)$ does not exceed C_0, and that $A_t(x, y) = 0$ if $|x - y| > t$.

With these conditions, the distribution-kernel $K(x, y)$ of L satisfies

$$(8.1) \qquad \int_{|x-y| \ge 2|x'-x|} |K(x', y) - K(x, y)| \, dy \le C_1 \, .$$

Here's why.

We put $d = |x' - x|$ and let E_j denote the dyadic shell defined by $2^j d \le |x - y| \le 2^{j+1} d$. The distribution-kernel $K_t(x, y)$ of $Q_t A_t$ is

$$(8.2) \qquad K_t(x, y) = \int q_t(x, u) A_t(u, y) \, du \, .$$

We shall verify (8.1) by evaluating

$$(8.3) \qquad \int_0^\infty \sum_{j \ge 1} \int_{E_j} |K_t(x', y) - K_t(x, y)| \, dy \, \frac{dt}{t} \, .$$

Since $K_t(x, y) = 0$, if $|x - y| > 2t$, we get

$$\int_{E_j} |K_t(x', y) - K_t(x, y)| \, dy = 0$$

unless $t \ge 2^{j-1} d$. In the latter case,

$$\int_{E_j} |K_t(x', y) - K_t(x, y)| \, dy \le |E_j|^{1/2} \Big(\int_{E_j} |K_t(x', y) - K_t(x, y)|^2 \, dy \Big)^{1/2} \, .$$

To evaluate the right-hand side, we linearize the problem, by considering $I_j = \int [K_t(x', y) - K_t(x, y)] f_j(y) \, dy$, where f_j has support in E_j and satisfies $\|f_j\|_2 \le 1$.

Now $I_j = \int\int \Delta_t(u) A_t(u, y) f_j(y) \, dy \, du$, where $\Delta_t(u) = q_t(x', u) - $

$q_t(x, u)$. Returning to the operators, we get $I_j = \langle A_t f_j, \Delta_t \rangle$ and we can use the fact that the A_t are continuous. So

$$|I_j| \le C_0 \|f_j\|_2 \|\Delta_t\|_2 \le Ct^{-n/2-1} d.$$

Then (8.3) is bounded above by a constant times

$$d^{n/2+1} \sum_1^\infty 2^{nj/2} \int_{2^{j-1}d}^\infty t^{-n/2-1} \frac{dt}{t} = C_1.$$

Having treated the simplified situation, we come to the case we really want to deal with, in which the operators are no longer supported by simple geometrical sets but, instead, decrease rapidly at infinity. This introduces error terms. A further complication is that the function ψ, which defines Q_t, is singular at 0, so that we have to use L^p norms rather than L^2 norms.

Despite these differences, the structure of the proof remains the same. As in the previous section, we consider the distribution-kernel $R_m(x, y)$ of the simplified operator

$$(8.4) \qquad \int_0^\infty Q_t B_1 \rho_t \cdots B_m \rho_t \mu(t) \frac{dt}{t} = \int_0^\infty Q_t \Lambda_t \frac{dt}{t}.$$

Keeping the notation of the beginning of the section, we intend to show that there are an exponent $\varepsilon > 0$ and a constant C_0 such that

$$(8.5) \qquad \int_{E_j} |R_m(x', y) - R(x, y)| \, dy \le C_0^m 2^{-\varepsilon j}.$$

This estimate will follow from the inequality

$$(8.6) \qquad \left(\int_{E_j} |R_m(x', y) - R(x, y)|^p \, dy \right)^{1/p} \le C_0^m (2^j d)^{-n/q} 2^{-\varepsilon j}$$

and Hölder's inequality.

To establish (8.6), we again linearize the problem by considering the integral

$$I_j = \int_{E_j} (R_m(x', y) - R(x, y)) f_j(y) \, dy,$$

where the support of f_j is contained in E_j and $\|f_j\|_q \le 1$.

Using the definition of R_m gives

$$I_j = \int_0^\infty \int \int (\psi_t(x' - u) - \psi_t(x - u)) \Lambda_t(u, y) f_j(y) \, du \, dy \frac{dt}{t}.$$

As in the case at the start of this section, we introduce a threshold value into the range of integration with respect to t. Let $\gamma \in (n/q, 1)$ be an exponent and let I_j' and I_j'' be the two integrals corresponding to the ranges $0 < t < 2^{\gamma j} d$ and $2^{\gamma j} d \le t < \infty$.

To estimate $|I_j''|$, we use only the continuity of $\rho_t : L^p \to L^p$ and get

$$|I_j''| \le C_0^m \int_{d2^{\gamma j}}^\infty \|\psi_t(x'-u) - \psi_t(x-u)\|_p \frac{dt}{t} \, .$$

We again put $\varepsilon = n/p - (n-1)$ and an explicit calculation gives

(8.7) $$\|\psi_t(x'-u) - \psi_t(x-u)\|_p \le Cd^\varepsilon t^{-1} \, ,$$

which gives $|I_j''| \le Cd^\varepsilon (2^{\gamma j} d)^{-1}$ and

$$\sum_1^\infty (2^j d)^{n/q} |I_j''| \le C \sum_1^\infty 2^{j(n/q - \gamma)} < \infty \, .$$

To estimate $|I_j'|$, we ignore the difference between $\psi_t(x-u)$ and $\psi_t(x'-u)$ and consider the integral

$$A_j = \int_0^{d2^{\gamma j}} \int\int \psi_t(x-u) \Lambda_t(u,y) f_j(y) \, du \, dy \, \frac{dt}{t} \, .$$

The integral B_j (where x is replaced by x') is of the same type.

Once again, we use Journé's lemma to evaluate the double integral with respect to u and y. This leads to

(8.8) $$C_0^m \int_{\mathbf{R}^n} (|\psi_t|^p \star \alpha_r)^{1/p} (u-x)(|f_j|^q \star \beta_r)^{1/q}(u) \, du \, .$$

Since $r = (m-1)t$, we have $(|\psi_t|^p \star \alpha_r)^{1/p}(u) \le Cm^{n/q} p_r(u)$. Then, putting $g = (|f_j|^q \star \beta_r)^{1/q}$, we observe that, by convexity,

$$\int p_r(u-x) g(u) \, du \le \left(\int p_r(u-x) g^q(u) \, du \right)^{1/q} \, .$$

We thus have

(8.9) $$\left| \int\int \psi_t(x-u) \Lambda_t(u,y) f_j(y) \, du \, dy \right|$$

$$\le C_0^m Cm^{n/q} (p_r \star \beta_r \star |f_j|^q)^{1/q}(x) \, .$$

We then remark that $p_r \star \beta_r \le Cp_r \le Cmp_t$ and it remains to estimate $p_t \star |f_j|^q(x)$.

To do this, we must take account of the different geometric scales. On the one hand, $0 < t < 2^{\gamma j} d$, while, on the other, $|f_j|^q$ has support in the spherical shell E_j. But E_j is very far from x on the scale given by t. Thus

$$(|f_j|^q \star p_t)(x) = \int_{E_j} p_t(x-y) |f_j(y)|^q \, dy \le C' t(2^j d)^{-n-1} \, .$$

All we now have to do is remark that $\int_0^{2^{\gamma j} d} t^{1/q} t^{-1} dt = C(d2^{\gamma j})^{1/q}$, so that the series converges for $\gamma < 1$.

9 Additional remarks

On one of his visits to Paris (1980-81), McIntosh drew our attention to the multilinear aspect of Kato's problem and to the deep connections between the multilinear operators arising from Kato's problem in dimension 1 and those coming from the Cauchy operator on Lipschitz curves. At that time (autumn 1980) neither of those problems had been solved.

McIntosh suggested a new way of attacking the higher order Calderón commutators (defined by the kernels $\mathrm{PV}\big((A(x) - A(y))^m/(x - y)^{m+1}\big)$, with $A' \in L^\infty(\mathbb{R})$) and that method proved much easier than those we had followed until then. In collaboration with McIntosh, we obtained the L^2 continuity of the higher order commutators with an upper bound of $C\|A'\|^m(1+m)^4$ for the operator norms, where $A(x)$ is real- or complex-valued. A renormalization technique, due to McIntosh, finally makes it possible to prove that the operator defined by the Cauchy kernel is L^2 continuous, for all Lipschitz curves ([65]). The same article contains the full proof of the one-dimensional Kato conjecture.

We now have a better understanding of the connection between these problems.

Keeping to the one-dimensional case, put $D = -id/dx$. Let A and B be the operators of pointwise multiplication by $a(x) \in L^\infty(\mathbb{R})$ and $b(x) \in L^\infty(\mathbb{R})$. These functions are complex-valued and satisfy the following accretive condition: there exists a constant $\delta > 0$ such that $\Re e\, a(x) \geq \delta > 0$ and $\Re e\, b(x) \geq \delta > 0$. With these hypotheses, the operator $T = BDAD$ satisfies (2.6) and we can define $T^{1/2}$ by applying Proposition 1. In collaboration with C. Kenig ([160]), we showed that $T^{1/2} = J(A, B)D$, where $J(A, B)$ is an isomorphism of $L^2(\mathbb{R})$ with itself.

Now, if $A = B$, then $J(A, B)$ is the Cauchy operator on the Lipschitz graph whose parametric representation is given by the primitive of $1/a(t)$, $t \in \mathbb{R}$.

On the other hand, if $B = I$, $T^{1/2}$ is the Kato operator.

The square roots of accretive differential operators and the Cauchy operators on Lipschitz curves thus belong to the same family. In a sense, this has been confirmed by the approach we have followed in this chapter: J.L. Journé's method, which consists of relating Kato's operator to a singular integral operator.

Finally, we draw the reader's attention to two other solutions of Kato's problem ([62] and [105]), obtained under the same restrictions as Theorem 2.

15

Potential theory in Lipschitz domains

1 Introduction

Calderón's research programme was motivated by the study of elliptic partial differential equations in domains with irregular boundary. Calderón's method consists of replacing the partial differential equation on the interior by a pseudo-differential equation on the boundary. But if the boundary is only Lipschitz, the nature of this equation changes: the operators which appear are no longer pseudo-differential, but are of the new type discussed in Chapter 9.

When Calderón inaugurated his programme, there were two difficulties. The very existence of the operators, needed for the method, was problematic. And, supposing that, in the fullness of time, such operators could be constructed, it would be necessary to solve the equations on the boundary that this process led to. For the solution, one would have to invert some operator and so would need a symbolic calculus. The regrettable absence of such a symbolic calculus was signalled in Chapter 9, and is the second problem of Calderón's programme.

We shall illustrate these remarks by examining a classical problem, which goes back to Poincaré, Neumann, and Hilbert. This is the solution, by the double-layer potential method, of the Dirichlet and Neumann problems for a domain Ω, in $\mathbb{R}^{n+1}$.

When Ω is a bounded, regular, open set, the operators of Calderón's method are classical pseudo-differential operators; they are also singular integral operators whose kernels are given by a "double-layer potential".

After the reduction given by Calderón's method, resolving a Dirichlet or Neumann problem just amounts to inverting an operator of the form $I/2 + K$ acting on the boundary. The ambient Banach space will be $L^2(\partial\Omega, d\sigma)$, where $d\sigma$ is the surface measure on the boundary $\partial\Omega$ of Ω.

When Ω is a bounded, regular, open set (of class $C^{1+\alpha}$, for some $\alpha > 0$), the operator $K : L^2(\partial\Omega, d\sigma) \to L^2(\partial\Omega, d\sigma)$ is compact, so Fredholm theory (which was invented for this purpose) allows us to invert $I/2 + K$.

When Ω is just of class C^1, the principal difficulty is to prove that K is continuous on the space $L^2(\partial\Omega, d\sigma)$. Once this continuity had been established (Calderón, 1977), Fabes, Jodeit, and Rivière ([104]) observed that the operator K was still compact, so that the rest could be done just as in the regular case. However, K no longer was a classical pseudo-differential operator.

Finally, for Ω a bounded, Lipschitz, open set, the continuity of K was established in 1981 in [65]. It follows from the results of Chapter 9. However, the operator K is no longer compact. This essential difficulty was overcome by G. Verchota. He showed that $I/2 + K$ was invertible, using the Jerison and Kenig energy inequalities ([136]).

After carefully stating Verchota's results, we shall examine only the special case where Ω is the open set above a Lipschitz graph. We shall give a complete proof for that case.

The modifications needed for the general case are ingenious, but do not need any new ideas. The reader who is interested may consult Verchota's thesis ([232]).

2 Statement of the results

In all that follows, $\Omega \subset \mathbb{R}^{n+1}$ is a bounded, connected, open set. We say that Ω is a Lipschitz open set, if the following conditions are satisfied. Firstly, the boundary $\partial\Omega$ of Ω has only a finite number of connected components. Secondly, for each $x_0 \in \partial\Omega$, we can find an orthonormal co-ordinate frame $\mathcal{R}(x_0)$, with origin x_0, two numbers $\varepsilon, \eta > 0$, and a Lipschitz function $\phi : \mathbb{R}^n \to \mathbb{R}$ such that, if $\mathcal{C} = \mathcal{C}(\mathcal{R}(x_0), \varepsilon, \eta)$ denotes the solid cylinder given by $\sqrt{x_1^2 + \cdots + x_n^2} \le \varepsilon$ and $-\eta \le x_{n+1} \le \eta$, then

(2.1) $\mathcal{C} \cap \Omega = \{x = (x', x_{n+1}) : |x'| \le \varepsilon \text{ and } \phi(x') < x_{n+1} \le \eta\}$

and

(2.2) $\mathcal{C} \cap \partial\Omega = \{x = (x', x_{n+1}) : |x'| \le \varepsilon \text{ and } x_{n+1} = \phi(x')\}.$

Since $\partial\Omega$ is a compact set, we need only a finite number N of cylinders $C_1,\ldots,C_N$ of the above type, to get a cover of $\partial\Omega$ by the corresponding open cylinders.

To deal with local problems on the boundary $\partial\Omega$ of Ω, it therefore is enough to work in the interior of one of these cylinders. We may then fix the orthonormal frame $\mathcal{R}$, ignore the limitation corresponding to ε and suppose that Ω is defined globally by $x_{n+1} > \phi(x')$, $x' \in \mathbb{R}^n$, $x_{n+1} \in \mathbb{R}$ and that ϕ is globally Lipschitz, that is, that $\|\nabla\phi\|_\infty \leq M$, where M is a (finite) constant. With this simplification, we fix another constant $M' > M$ and we attach a cone $\Gamma(x)$ of non-tangential approach to each point $x \in \partial\Omega$. The cone $\Gamma(x)$, with the exception of the point x itself, is contained entirely in Ω and is defined by $y_{n+1} - x_{n+1} \geq M'|y' - x'|$. The geometric significance of this cone is that there is a constant $\delta > 0$ such that the distance from $y \in \Gamma(x)$ to $\partial\Omega$ is not less than $\delta|y - x|$. When y tends to x, while remaining in $\Gamma(x)$, y is "relatively far" from the other points of the boundary.

Our simplification has created an unbounded Lipschitz open set. In the case of a bounded Lipschitz open set, the approach cones have to be cut off. We still use $\Gamma(x)$, $x \in \partial\Omega$, to denote them. They have a uniform angle $\alpha > 0$, a uniform height $\tau > 0$ and their geometric significance is the same as in the unbounded case.

The construction of these cut-off cones is very carefully described in ([232]), to which we refer the reader. The details do not have any part to play in what follows.

We now look at the Dirichlet problem, which is to find a function u which is harmonic in Ω, with given boundary values. Recall that u is harmonic if $\Delta u = 0$, where $\Delta = \partial^2/\partial x_1^2 + \cdots + \partial^2/\partial x_{n+1}^2$.

We intend to resolve Dirichlet's problem with boundary values $g(x)$ in $L^2(\partial\Omega, d\sigma)$, where $d\sigma$ is surface measure on the boundary. One might think that this problem is not well-defined because, on the one hand, $g(x)$ is only defined up to a set of measure zero on $\partial\Omega$ while, on the other, it is not sufficient to know the non-tangential limits, up to a set of measure zero, of a function u which is harmonic in Ω, in order to determine u. A counter-example is given in $\Omega = \{(x', x_{n+1}) \in \mathbb{R}^{n+1} : x_{n+1} > 0\}$ by $u(x) = u(x', x_{n+1}) = x_{n+1}(|x'|^2 + x_{n+1}^2)^{-(n+1)/2}$. Then u is harmonic in Ω and $u(x', 0) = 0$ everywhere except when $x' = 0$.

To exclude such pathological solutions of Dirichlet's problem, it is enough to introduce a condition like that of Lebesgue's dominated convergence theorem. This consists of restricting attention to those functions u which are harmonic in Ω and such that the maximal function

$u^*(x)$, corresponding to non-tangential approach to the boundary, is in $L^2(\partial\Omega, d\sigma)$. To do this, we make the following definition.

Definition 1. *Let u be a function which is harmonic in Ω. We write $u \in \mathcal{H}^2(\Omega)$ if $u^*(x) \in L^2(\partial\Omega, d\sigma)$, where*

$$(2.3) \qquad u^*(x) = \sup_{y \in \Gamma(x)} |u(y)|.$$

The space $\mathcal{H}^2(\Omega)$ is independent of the particular choice of truncated cones $\Gamma(x)$.

Now to describe the trace operator $\theta : \mathcal{H}^2(\Omega) \to L^2(\partial\Omega, d\sigma)$. It is clearly defined as the usual restriction to $\partial\Omega$ of $u \in \mathcal{H}^2(\Omega)$ when the harmonic function u extends to the boundary by continuity. In the general case, we say that u has a non-tangential limit at $x \in \partial\Omega$ if $\lim u(y)$ exists, as $y \to x$ with $y \in \Gamma(x)$. If $u \in \mathcal{H}^2(\Omega)$, then, for all $x \in \partial\Omega$, except for a set of $d\sigma$ measure zero, $u(y)$ has a non-tangential limit, which we denote by $u(x)$. The function defined in this way belongs to $L^2(\partial\Omega, d\sigma)$ and a different choice of non-tangential approach cones would give the same trace for almost all $x \in \partial\Omega$.

The trace operator $\theta : \mathcal{H}^2(\Omega) \to L^2(\partial\Omega, d\sigma)$ is, in fact, an isomorphism between these two spaces. This theorem, due to B. Dahlberg, depends on the equivalence of the harmonic measure to the surface measure on $\partial\Omega$ ([81] and [82]).

Dahlberg's theorem gives a theoretical solution of the Dirichlet problem, which consists of finding $u \in \mathcal{H}^2(\Omega)$ which is a solution of the equation $\theta(u) = g$, for a given $g \in L^2(\partial\Omega, d\sigma)$.

We shall deal with this same Dirichlet problem by a different route, which will give an algorithm for the solution.

To state the Neumann problem, we introduce the space $\mathcal{H}_1^2(\Omega)$ of functions u which are harmonic in Ω and such that $v(x) = \sup_{y \in \Gamma(x)} |\nabla u(y)|$ belongs to $L^2(\partial\Omega, d\sigma)$. Then the non-tangential limit of the gradient $\nabla u(y)$ exists for almost all $x \in \partial\Omega$ (with respect to the surface measure $d\sigma$), as $y \in \Gamma(x)$ tends to x.

Solving Neumann's problem is to find, for a given $g \in L^2(\partial\Omega, d\sigma)$, a function $u \in \mathcal{H}_1^2(\Omega)$ such that $\partial u/\partial n = g$, almost everywhere on $\partial\Omega$ (with respect to $d\sigma$). Here, $n(x)$, $x \in \partial\Omega$, denotes the vector at x which is normal to $\partial\Omega$. This vector exists for almost all $x \in \partial\Omega$ and we have put

$$(2.4) \qquad \frac{\partial u}{\partial n}(x) = \lim_{y \to x, y \in \Gamma(x)} \nabla u(y) \cdot n(x).$$

To state the main results of this chapter, we must now introduce the double-layer potential. It is constructed in the following way. We start with the fundamental solution $-\omega_n^{-1}(n-1)^{-1}|x|^{-n+1}$ of the Laplace equation in $\mathbb{R}^{n+1}$ (ω_n is the surface area of the unit sphere $S^n \subset \mathbb{R}^{n+1}$). The fundamental solution enables us to calculate the potential generated by a charge distribution. A double-layer distribution is a distribution S, with support in $\partial\Omega$, defined on the test functions $g \in \mathcal{D}(\mathbb{R}^{n+1})$ by

$$\langle S, g \rangle = -\int_{\partial\Omega} f(y) \frac{\partial g}{\partial n}(y) \, d\sigma(y),$$

where $f(y) \in L^1(\partial\Omega, d\sigma)$ is a given density.

The potential at $x \in \Omega$ created by the double-layer distribution of density f on $\partial\Omega$ is thus

$$(2.5) \qquad \mathcal{K}f(x) = \frac{1}{\omega_n} \int_{\partial\Omega} \frac{(y-x) \cdot n(y)}{|y-x|^{n+1}} f(y) \, d\sigma(y).$$

If u and v are functions which are harmonic on a bounded regular open set Ω, and if u and v are C^1 in a neighbourhood of the closure $\overline{\Omega}$ of Ω, then Green's theorem (whose use we shall justify in section 5) gives

$$\int_{\partial\Omega} \left(u \frac{\partial v}{\partial n} - v \frac{\partial u}{\partial n} \right) d\sigma = 0.$$

Applying this remark to $v(y) = -\omega_n^{-1}(n-1)^{-1}|x-y|^{-n+1}$, excising a ball centre x, radius ε, from Ω, and passing to the limit as $\varepsilon \to 0$ gives

$$(2.6) \qquad u(x) = \mathcal{K}u(x) + \frac{1}{(n-1)\omega_n} \int_{\partial\Omega} |x-y|^{-n+1} \frac{\partial u}{\partial n}(y) \, d\sigma(y).$$

This identity means that $\mathcal{K}u(x)$ is a reasonable approximation to u. For example, if u is identically 1, we get $\mathcal{K}(u) = 1$.

By our construction, $\mathcal{K}f(x)$ is a harmonic function on Ω, when f lies in $L^1(\partial\Omega, d\sigma)$. From now on, we shall restrict attention to the case where $f \in L^2(\partial\Omega, d\sigma)$, a choice which we shall justify after the event. It is not at all obvious that $\mathcal{K}f(x)$ belongs to $\mathcal{H}^2(\Omega)$, when f lies in $L^2(\partial\Omega, d\sigma)$. It was to prove precisely this result that Calderón engaged on the closer investigation of operators defined by generalized singular integrals.

The following theorem describes the properties of the operator $\mathcal{K}$.

Theorem 1. *Let Ω be a Lipschitz domain. Then the operator $\mathcal{K}$, defined by the double-layer potential, is continuous, as an operator from $L^2(\partial\Omega, d\sigma)$ to $\mathcal{H}^2(\Omega)$. Moreover, for each function $f \in L^2(\partial\Omega, d\sigma)$, we have, for almost all $x \in \partial\Omega$ (with respect to surface measure $d\sigma$),*

$$(2.7) \qquad \lim_{y \to x, y \in \Gamma(x)} \mathcal{K}f(y) = (\frac{1}{2}I + K)f(x),$$

where K is defined on $L^2(\Omega, d\sigma)$ by

$$(2.8) \qquad Kf(x) = \frac{1}{\omega_n} \lim_{\varepsilon \downarrow 0} \int_{|y-x| \geq \varepsilon} \frac{(y-x) \cdot n(y)}{|y-x|^{n+1}} f(y) \, d\sigma(y).$$

The operator K is thus continuous on $L^2(\partial\Omega, d\sigma)$ and its definition is obtained from a singular integral calculated entirely on $\partial\Omega$. The singular integral exists almost everywhere on $\partial\Omega$ (equipped, as always, with the surface measure $d\sigma$).

The solution to the Dirichlet problem is then given by the following statement.

Theorem 2. *Let Ω again be a Lipschitz domain. Then the "boundary" operator*

$$\frac{1}{2}I + K : L^2(\partial\Omega, d\sigma) \to L^2(\partial\Omega, d\sigma)$$

is an isomorphism. If g lies in $L^2(\partial\Omega, d\sigma)$, then the harmonic function $u = \mathcal{K}(I/2 + K)^{-1}g$ is the unique solution of the following Dirichlet problem:

(2.9) $$\Delta u = 0 \quad \text{in } \Omega \qquad \text{when} \quad u^\star \in L^2(\partial\Omega, d\sigma)$$

and

(2.10) $$\lim_{y \to x, y \in \Gamma(x)} u(y) = g(x) \qquad \text{for almost all } x \in \partial\Omega.$$

To deal with the Neumann problem, let $K^\star : L^2(\partial\Omega, d\sigma) \to L^2(\partial\Omega, d\sigma)$ denote the adjoint of the operator K. We write f_0 for the unique function in $L^2(\partial\Omega, d\sigma)$ satisfying

$$(\frac{1}{2}I - K^\star)f_0 = 0 \qquad \text{and} \qquad \int_{\partial\Omega} f_0 \, d\sigma = 1,$$

and let $L_0^2(\partial\Omega, d\sigma)$ denote the subspace of functions of zero mean in $L^2(\partial\Omega, d\sigma)$.

The single-layer operator $S : L^2(\partial\Omega, d\sigma) \to \mathcal{H}^2(\Omega)$ is defined by

(2.11) $$Sf(x) = -\frac{1}{(n-1)\omega_n} \int_{\partial\Omega} |x - y|^{-n+1} f(y) \, d\sigma(y),$$

for $x \in \Omega$. With this notation, we can state Theorem 3.

Theorem 3. *Let Ω be a Lipschitz domain. Then, for each function $f \in L^2(\partial\Omega, d\sigma)$, $u = S(f)$ lies in $\mathcal{H}_1^2(\Omega)$.*

Secondly, the operator $I/2 - K^\star : L_0^2(\partial\Omega, d\sigma) \to L_0^2(\partial\Omega, d\sigma)$ is an isomorphism.

Lastly, for each $g \in L_0^2(\partial\Omega, d\sigma)$, the function $u = -S(I/2 - K^\star)^{-1}g$ is the unique solution of the following Neumann problem:

(2.12) $$\Delta u = 0 \text{ in } \Omega \qquad \text{and} \qquad |\nabla u|^\star \in L^2(\partial\Omega, d\sigma);$$

(2.13) $$\frac{\partial u}{\partial n} = g \text{ almost everywhere } d\sigma \qquad \text{and} \qquad \int_{\partial\Omega} u f_0 \, d\sigma = 0.$$

Some explanantions are called for. We already know that $\mathcal{K}(1) = 1$. As a consequence, for every $f \in L^2(\partial\Omega, d\sigma)$, the function $h = (I/2 -$

$K^*)f$ necessarily has zero mean. Thus $I/2 - K^*$ could not possibly be an automorphism of $L^2(\partial\Omega, d\sigma)$, but could, at best, be an isomorphism of $L_0^2(\partial\Omega, d\sigma)$ with itself.

In (2.13), $\partial u/\partial n = g$, almost everywhere $d\sigma$, clearly means that $g(x)$ is, for almost all $x \in \partial\Omega$, the non-tangential limit of $\nabla u(y) \cdot n(x)$, as $y \in \Gamma(x)$ tends to x.

Next, the condition $\int_{\partial\Omega} uf_0 \, d\sigma = 0$ arises because $u = S(h)$ with $\int_{\partial\Omega} h \, d\sigma = 0$. Indeed, $S(f_0)$ is a constant and, since $S : L^2(\partial\Omega, d\sigma) \to L^2(\partial\Omega, d\sigma)$ is self-adjoint, $\int_{\partial\Omega} uf_0 \, d\sigma = \int_{\partial\Omega} S(h)f_0 \, d\sigma = \int_{\partial\Omega} hS(f_0) \, d\sigma = c \int_{\partial\Omega} h \, d\sigma = 0$.

Finally, if $u, v \in \mathcal{H}_1^2(\Omega)$, by Green's theorem, we have $\int_{\partial\Omega} (u(\partial v/\partial n) - v(\partial u/\partial n)) \, d\sigma = 0$. The details of passing to the limit, needed in the Lipschitz case, will be given later. Taking $v = 1$, we do get $\int_{\partial\Omega} (\partial u/\partial n) \, d\sigma = 0$, for $u \in \mathcal{H}_1^2(\Omega)$. The condition $g \in L^2(\partial\Omega, d\sigma)$ is, of course, necessary for the Neumann problem to have a solution.

3 Almost everywhere existence of the double-layer potential

We now turn our backs on the general case and study the situation where Ω is defined globally by $t > \phi(x)$, $t \in \mathbb{R}$, $x \in \mathbb{R}^n$, and where $\phi(x)$ is a real-valued Lipschitz function on the whole of $\mathbb{R}^n$, satisfying $\|\nabla\phi\|_\infty < M < \infty$. To ease the burden of notation, we denote the points of $\mathbb{R}^{n+1}$ by $X = (x, t)$ and $Y = (y, s)$, where $x, y \in \mathbb{R}^n$ and $s, t \in \mathbb{R}$. We shall also write V for the graph of $\phi : \mathbb{R}^n \to \mathbb{R}$.

Theorems 1, 2, and 3 will be proved in this setting. $L_0^2(V, d\sigma)$ has to be replaced by $L^2(V, d\sigma)$ in Theorem 3, because the constant function 1 is no longer square-integrable. Nor does the function f_0 exist any more.

The changes necessary for the bounded case are ingenious, but of a more technical nature. So it seems sensible to refer the interested reader to [232].

The first remark is about the form of the double-layer potential. If $X, Y \in V$, then

(3.1) $$\frac{(Y - X) \cdot N(Y)}{\omega_n |Y - X|^{n+1}} \, d\sigma(Y) = K(\phi; x, y) \, dy \,,$$

where

(3.2) $$K(\phi; x, y) = \frac{\phi(x) - \phi(y) - (x - y) \cdot \nabla\phi(y)}{\omega_n \big(|x - y|^2 + (\phi(x) - \phi(y))^2\big)^{(n+1)/2}} \,.$$

We have written $N(Y)$ for the unit vector normal to V at Y, oriented downwards, that is, away from Ω.

With this notation, the analogue of Theorem 1 is

Theorem 4. *For each function* $f \in L^2(\mathbb{R}^n, dx)$,

$$(3.3) \qquad T_\phi(f)(x) = \lim_{\varepsilon \downarrow 0} \int_{|x-y| \geq \varepsilon} K(\phi; x, y) f(y)\, dy$$

exists almost everywhere and the operator T_ϕ *defined by (3.3) is bounded on* $L^2(\mathbb{R}^n)$.

The operator T_ϕ will be called the singular part of the double-layer potential.

From Chapter 8, we know that, to prove such a result, it is sufficient to establish the existence of the limit when f lies in the Schwartz class $\mathcal{S}(\mathbb{R}^n)$ and to show that there is a constant C such that, for

$$(3.4) \qquad T_\phi^\star(f)(x) = \sup_{\varepsilon > 0} \left| \int_{|x-y| \geq \varepsilon} K(\phi; x, y) f(y)\, dy \right|,$$

we have

$$(3.5) \qquad \|T_\phi^\star(f)\|_2 \leq C \|f\|_2 .$$

Let us start by establishing that the limit (3.3) exists, when $f \in \mathcal{S}(\mathbb{R}^n)$. To do this, we use the following lemmas.

Lemma 1. *A Lipschitz function* $F : \mathbb{R}^n \to \mathbb{R}$ *is almost everywhere differentiable, in the sense that, for almost all* $x \in \mathbb{R}^n$,

$$\lim_{|y| \to 0} |y|^{-1} \big(F(x + y) - F(x) - y \cdot \nabla F(x) \big) = 0 .$$

We shall recall the proof of this classical result in the appendix to this chapter.

The second lemma justifies our use of integration by parts to dispose of the singularity of $K(\phi; x, y)$. Let $\lambda(t)$, $t \in \mathbb{R}$, denote the odd primitive of the function $(1 + t^2)^{-(n+1)/2}$.

Lemma 2. *At every point* $x \in \mathbb{R}^n$ *where the Lipschitz function* ϕ *is differentiable, and for every function* f *in the Schwartz class* $\mathcal{S}(\mathbb{R}^n)$,

$$(3.6) \quad \lim_{\varepsilon \downarrow 0} \int_{|x-y| \geq \varepsilon} K(\phi; x, y) f(y)\, dy = -\int \frac{(x - y) \cdot \nabla f(y)}{|x - y|^n} \lambda\Big(\frac{\phi(x) - \phi(y)}{|x - y|}\Big)\, dy.$$

The right-hand side of (3.6) is an absolutely convergent integral.

To prove (3.6), we fix x and apply the divergence theorem to

$$\int_{|x-y| \geq \varepsilon} \operatorname{div}\left(\frac{x - y}{|x - y|^n} \lambda\Big(\frac{\phi(x) - \phi(y)}{|x - y|}\Big) f(y) \right) dy .$$

The divergence is taken with respect to the variable y. Now

$$\frac{1}{\omega_n} \operatorname{div}\left(\frac{x - y}{|x - y|^n} \lambda\Big(\frac{\phi(x) - \phi(y)}{|x - y|}\Big) \right) = K(\phi; x, y) .$$

The proof of (3.6) thus depends on the resulting surface integral's tending to 0 with ε. Indeed, this surface integral is, up to sign,

$$I(\varepsilon) = \int_{S^{n-1}} f(x + \varepsilon\nu)\lambda\Big(\frac{\phi(x + \varepsilon\nu) - \phi(x)}{\varepsilon}\Big)\, d\sigma(\nu)\,,$$

where $d\sigma(\nu)$ is the surface measure on the unit sphere S^{n-1}.

Since ϕ is differentiable at x, the dominated convergence theorem applies and

$$\lim_{\varepsilon\downarrow 0} I(\varepsilon) = f(x) \int_{S^{n-1}} \lambda(\nu \cdot \nabla\phi(x))\, d\sigma(\nu)\,.$$

To conclude, we observe that λ is odd. We pair up the antipodal points and get $\lim_{\varepsilon\downarrow 0} I(\varepsilon) = 0$.

The estimate (3.5) of the maximal operator T_ϕ^* basically comes from the general theory of Calderón-Zygmund operators. We must, however, make some adjustments, because the kernel $K(\phi; x, y)$ does not have the regularity with respect to y that is part of the hypotheses of Chapter 7.

We start by splitting up the kernel $K(\phi; x, y)$ as

$$\frac{1}{\omega_n}\Big(K_0(x,y) - \sum_1^n K_j(x,y)\frac{\partial\phi}{\partial y_j}(y)\Big)\,,$$

where

$$K_0(x,y) = \frac{\phi(x) - \phi(y)}{\big(|x-y|^2 + (\phi(x) - \phi(y))^2\big)^{(n+1)/2}}$$

and

$$K_j(x,y) = \frac{x_j - y_j}{\big(|x-y|^2 + (\phi(x) - \phi(y))^2\big)^{(n+1)/2}}\,.$$

Since the operators act on functions $f \in L^2(\mathbb{R}^n)$, we may absorb L^∞ functions into such f, which lets us reduce to the operators T_0 and T_j, defined by the kernels PV K_0 and PV K_j.

We can thus use the results of previous chapters. On the one hand, Theorem 11 of Chapter 9 gives the basic L^2 estimate. On the other, the theory of Chapter 7 gives the continuity of the corresponding maximal operators.

In order to deduce the results of Theorem 1, we must compare the principal value with the limit defined by non-tangential convergence.

To do this, we use the following remark.

Lemma 3. *With the notation above and for $t \neq \phi(x)$,*

$$(3.7) \quad \int_{\mathbb{R}^n} \frac{t - \phi(y) - (x - y) \cdot \nabla\phi(y)}{\big(|y - x|^2 + (\phi(y) - t)^2\big)^{(n+1)/2}} f(y)\, dy$$

$$= \frac{\omega_n}{2}\, \mathrm{sgn}(t - \phi(x)) f(x) - \int_{\mathbb{R}^n} \frac{(y - x) \cdot \nabla f(y)}{|y - x|^n}\lambda\Big(\frac{\phi(y) - t}{|y - x|}\Big)\, dy\,.$$

To prove this result, we repeat the method of Lemma 2. That is, we start with the identity

$$\operatorname{div}\left(\frac{x-y}{|x-y|^n}\lambda\left(\frac{\phi(y)-t}{|x-y|}\right)\right) = \frac{\phi(y)-t-(y-x)\cdot\nabla\phi(y)}{(|y-x|^2+(\phi(y)-t)^2)^{(n+1)/2}}.$$

Regrouping the integrals of Lemma 3 gives

$$I(\varepsilon) = \int_{|x-y|\geq\varepsilon}\operatorname{div}\left(\frac{x-y}{|x-y|^n}\lambda\left(\frac{\phi(x)-t}{|x-y|}\right)\right)f(y)\,dy,$$

which becomes, by the divergence theorem,

$$(3.8)\qquad I(\varepsilon) = \varepsilon^{-n+1}\int_{|y-x|=\varepsilon}f(y)\lambda\left(\frac{\phi(y)-t}{|y-x|}\right)d\sigma(y).$$

Once again, we use Lebesgue's dominated convergence theorem to calculate $\lim_{\varepsilon\to 0}I(\varepsilon)$. It is convenient to change the surface integral so that it is taken over the unit sphere S^{n-1}. We observe that $\lambda(\pm\infty) = \pm\int_0^\infty(1+t^2)^{-(n+1)/2}\,dt$, that $\omega_{n-1}\int_0^\infty(1+t^2)^{-(n+1)/2}\,dt = \omega_n/2$ and that this gives $\lim_{\varepsilon\to 0}I(\varepsilon) = (\omega_n/2)\operatorname{sgn}(\phi(x)-t)f(x)$.

Let us go back to Theorem 1. We intend, for the first stage, to use the definition of the operator K given by (3.3) and to prove (2.7). After that, we shall indicate why the operator K may just as well be given by (2.8). We should observe that, in (2.8), $|y-x|\geq\varepsilon$ has become $|y-x|^2+(\phi(y)-\phi(x))^2\geq\varepsilon$, because of our change of notation.

To establish (2.7), we follow our accustomed route. We verify that this identity holds when f belongs to the Schwartz class $\mathcal{S}(\mathbb{R}^n)$ and we then prove an estimate for the corresponding maximal operator.

The non-tangential approach cone is given, via a parameter $\alpha > M \geq \|\nabla\phi\|_\infty$, by $\Gamma(x_0) = \{(x,t): t-\phi(x_0)\geq\alpha|x-x_0|\}$.

Let $F(x,t)$ denote the left-hand side of (3.7). This satisfies $F(x,t) = -\omega_n\mathcal{K}f(x,t)$. Let us calculate the limit of $F(x,t)$ as $(x,t)\in\Gamma(x_0)$ tends to $(x_0,\phi(x_0))$.

We make the change of variable $y-x = u-x_0$ in the integral on the right-hand side of (3.7), which then becomes

$$(3.9)\qquad \int_{\mathbb{R}^n}\frac{(u-x_0)\cdot\nabla f(u+x-x_0)}{|u-x_0|^n}\lambda\left(\frac{\phi(u+x-x_0)-t}{|u-x_0|}\right)du.$$

Since $f\in\mathcal{S}(\mathbb{R}^n)$ and we can restrict to the ball $|x-x_0|\leq 1$, Lebesgue's dominated convergence theorem applies when calculating the limit of (3.9), as $x\to x_0$ and $t\to t_0 = \phi(x_0)$. The non-tangential convergence plays no part in this, and the limit is

$$\int_{\mathbb{R}^n}\frac{(u-x_0)\cdot\nabla f(u)}{|u-x_0|^n}\lambda\left(\frac{\phi(u)-t_0}{|u-x_0|}\right)du.$$

However, the geometry of non-tangential convergence is crucial in proving the maximal estimate.

We put $F_*(x_0) = \sup_{(x,t) \in \Gamma(x_0)} |F(x,t)|$ and we intend to show that there is a constant C such that, for all functions $f \in L^2(V, d\sigma)$,

$$(3.10) \qquad \left(\int_{\mathbb{R}^n} |F_*(x)|^2 \, dx \right)^{1/2} \leq C\|f\|_2 \,.$$

Define the truncated kernel $K_\varepsilon(\phi; x, y)$ by $K_\varepsilon(\phi; x, y) = K(\phi; x, y)$ if $|x - y| \geq \varepsilon$ and by 0 otherwise. Then, if $\alpha^{-1}(t - \phi(x_0)) = \varepsilon$ and $(x, t) \in \Gamma(x_0)$, we can write

$$(3.11) \qquad \frac{t - \phi(y) - (x - y) \cdot \nabla \phi(y)}{\left(|y - x|^2 + (\phi(y) - t)^2\right)^{(n+1)/2}} = K_\varepsilon(\phi; x_0, y) + R_\varepsilon(x_0, x, y) \,,$$

where, for a certain constant $C = C(M, \alpha)$, we have

$$|R_\varepsilon(x_0, x, y)| \leq C\varepsilon^{-n} \,, \qquad \text{if } |x_0 - y| \leq \varepsilon,$$

and

$$|R_\varepsilon(x_0, x, y)| \leq C\varepsilon|x_0 - y|^{-n-1} \,, \qquad \text{if } |x_0 - y| \geq \varepsilon.$$

Getting the estimates for $R_\varepsilon(x_0, x, y)$ is a simple geometrical exercise, using the fact that $t - \phi(x_0) > \alpha|x - x_0|$ and that $|\phi(y) - \phi(x_0)| \leq M|y - x_0|$, with $\alpha > M \geq 0$.

Furthermore,

$$\varepsilon \int_{|x-y| \geq \varepsilon} |x_0 - y|^{-n-1} |f(y)| \, dy \leq C_n f^*(x_0) \,,$$

where $f^*(x_0)$ is the Hardy-Littlewood maximal function of f.

Thus

$$(3.12) \qquad F_*(x) \leq K_* f(x) + C f^*(x) \,,$$

where

$$K_* f(x) = \sup_{\varepsilon > 0} \left| \int K_\varepsilon(\phi; x, y) f(y) \, dy \right| \,.$$

Inequality (3.12) is consistent with the results of Chapter 7.

To conclude, the operator defined by the kernel $K(\phi; x, y)$ can be split into $n + 1$ Calderón-Zygmund operators preceded by n operators of multiplication by $L^\infty(\mathbb{R}^n)$ functions. Thus $K_* f \in L^2$ whenever $f \in L^2$ (Chapter 7, Theorem 4). Finally F_*, the non-tangential maximal function corresponding to the double-layer potential, also belongs to L^2 when f does.

To finish the proof of Theorem 1, let us see how the principal-value operator defined by (2.8) differs from that given by (3.3). We put $X =$

$(x, \phi(x))$, $Y = (y, \phi(y))$, write $N(Y)$ for $n(y)$ in (2.8) and consider

$$\Delta_\varepsilon f(x) = \int_{|X-Y| \geq \varepsilon} \frac{(Y-X) \cdot N(Y)}{|Y-X|^{n+1}} f(y) \, d\sigma(Y)$$

$$- \int_{|x-y| \geq \varepsilon} \omega_n K(\phi; x, y) f(y) \, dy$$

$$= \int_{R(\varepsilon)} \frac{(Y-X) \cdot N(Y)}{|Y-X|^{n+1}} f(y) \, d\sigma(Y),$$

where $R(\varepsilon)$ is defined by $|x-y| \leq \varepsilon$ and $\left(|x-y|^2 + (\phi(x) - \phi(y))^2\right)^{1/2} \geq \varepsilon$. In particular,

$$|x - y| \geq (1 + M^2)^{-1/2} \varepsilon,$$

if $y \in R(\varepsilon)$, and $|(Y-X) \cdot N(Y)||Y-X|^{-n-1} \leq C\varepsilon^{-n}$. The volume of $R(\varepsilon)$ is exactly $C\varepsilon^n$, and we deduce from this that $|\Delta_\varepsilon f(x)| \leq Cf^*(x)$, where $f^*(x)$ is the Hardy-Littlewood maximal function of f.

To show that $\lim_{\varepsilon \downarrow 0} \Delta_\varepsilon f(x) = 0$ almost everywhere, for $f \in L^2(\mathbb{R}^n)$, it is enough to consider the case where $f \in \mathcal{S}(\mathbb{R}^n)$. In the end, this amounts to showing that

$$\lim_{\varepsilon \downarrow 0} \int_{R(\varepsilon)} \frac{(Y-X) \cdot N(Y)}{|Y-X|^{n+1}} \, d\sigma(Y) = 0.$$

We start by observing that the function $N(Y) = N(y, \phi(y))$ is in $L^\infty(\mathbb{R}^n)$. Thus, almost every point $x \in \mathbb{R}^n$ is a Lebesgue point for this function. That is, $\varepsilon^{-n} \int_{|x-y| \leq \varepsilon} |N(Y) - N(X)| \, dy$ tends to 0 with ε. Since $\varepsilon \leq |Y - X| \leq (1 + M^2)^{1/2}\varepsilon$ on the shell $R(\varepsilon)$, we can replace $N(Y)$ by $N(X)$, whenever y is a Lebesgue point of $N(Y)$.

Now observe that

$$\lim_{\varepsilon \downarrow 0} \int_{R(\varepsilon)} \frac{(Y-X)}{|Y-X|^{n+1}} \, d\sigma(Y) = 0,$$

for each point x where ϕ is differentiable. This is because the function $Y|Y|^{-n-1}$ is odd and the range of integration $R(\varepsilon)$ is essentially symmetric in X, for such x (the volume of the symmetric difference of $R(\varepsilon)$ and its symmetric hull is $O(\varepsilon^n)$). The details are left to the reader.

We have finished the proof of Theorem 1.

4 The single-layer potential and its gradient

Before proving Theorem 2, we establish the analogue of Theorem 1 for the gradient of the single-layer potential. The methods used are the same as those of the previous section and we use the same notation.

The points of $\mathbb{R}^{n+1}$ are denoted by $X = (x, t)$, where $x \in \mathbb{R}^n$ and

$t \in \mathbb{R}$. The set $V \subset \mathbb{R}^{n+1}$ is the graph of the global Lipschitz function $\phi : \mathbb{R}^n \to \mathbb{R}$, with $\|\nabla \phi\|_\infty \leq M < \infty$. Finally, $d\sigma$ denotes surface measure on V.

If $n \geq 3$, the single-layer potential arising from $f \in L^2(V, d\sigma)$ is defined in $\mathbb{R}^{n+1} \setminus V$ by

$$(4.1) \qquad Sf(X) = \frac{1}{(n-1)\omega_n} \int_V |X - Y|^{-n+1} f(Y) \, d\sigma(Y).$$

This is just the convolution of the function $-(n-1)^{-1}\omega_n^{-1} \int_V |X|^{-n+1}$, a fundamental solution of the Laplace operator, with the surface measure $f \, d\sigma$, which is the single-layer distribution in question.

In dimension 3 $(n = 2)$, this integral may diverge at infinity, for an arbitrary $f \in L^2(V, d\sigma)$. To make it converge, it is enough to fix a point X_0 off V and to replace the kernel $|X - Y|^{-n+1}$, throughout, by $|X - Y|^{-n+1} - |X_0 - Y|^{-n+1}$. Indeed, the only things that matter are the partial derivatives of Sf: the choice of X_0 is of no importance.

If $X = (x, t)$ lies off V, we calculate the gradient of Sf at X by differentiating under the integral sign to get

$$(4.2) \qquad \nabla Sf(x) = \frac{1}{\omega_n} \int_V \frac{X - Y}{|X - Y|^{n+1}} f(Y) \, d\sigma(Y).$$

What happens as X approaches V?

In the following theorem, we no longer choose the open set Ω above V. For each $(x_0, \phi(x_0)) \in V$, we let $\Gamma_\pm(x_0)$ denote the two cones defined, respectively, by $t - \phi(x_0) > \alpha|x - x_0|$ and $t - \phi(x_0) < -\alpha|x - x_0|$ (where $\alpha > M$ is a fixed constant). We shall attempt to determine the limit of $\nabla Sf(X)$ as X tends to $X_0 = (x_0, \phi(x_0))$ while staying within $\Gamma_+(x_0)$ or $\Gamma_-(x_0)$.

Theorem 5. *If $f \in L^2(V, d\sigma)$, then $\nabla Sf(x)$ has a limit as $X \in \Gamma_\pm(x_0)$ tends to $X_0 = (x_0, \phi(x_0))$. The limit is given by*

$$(4.3) \qquad \pm\frac{1}{2} f(X_0) N(X_0) + \frac{1}{\omega_n} \mathrm{PV} \int_V \frac{(X_0 - Y)}{|X_0 - Y|^{n+1}} f(Y) \, d\sigma(Y),$$

where $N(X_0)$ is the unit normal vector at X_0, pointing downwards.

Further, the maximal operators $\sup_{x \in \Gamma_\pm(x_0)} |\nabla Sf(X)|$ are bounded on $L^2(V, d\sigma)$.

As in the previous section, we split the proof of Theorem 5 into two parts. Firstly, we establish that the non-tangential limit and the principal value exist when f belongs to a suitable dense subspace E of $L^2(\mathbb{R}^n)$. To pass from the special functions $f \in E$ to general functions, we need, secondly, to prove an L^2 estimate for the maximal operator corresponding to this problem. But the second stage is identical to that of the previous section. We shall concentrate on the first part. Let us define E.

As in the previous section, it is a matter of integrating by parts to make the singularity of the kernel $(X - Y)|X - Y|^{-n-1}$ disappear. But, first of all, we must undertake certain transformations of the kernel.

We put $\omega(x) = (1 + |\nabla\phi(x)|^2)^{-1/2}$ and consider the tangent vectors at $X = (x, \phi(x)) \in V$, defined by

$$T_1(X) = (\omega(x), 0, 0, \ldots, 0, \omega(x)\partial\phi/\partial x_1),$$

$$T_2(X) = (0, \omega(x), 0, \ldots, 0, \omega(x)\partial\phi/\partial x_2),$$

$$\cdots$$

$$T_n(X) = (0, 0, 0, \ldots, \omega(x), \omega(x)\partial\phi/\partial x_n),$$

together with the unit normal vector $N(X)$, given by

$$N(X) = \left(\omega(x)\frac{\partial\phi}{\partial x_1}, \ldots, \omega(x)\frac{\partial\phi}{\partial x_n}, -\omega(x)\right).$$

The lengths of these vectors lie between $(1 + M^2)^{-1/2}$ and 1. The determinant $\det(T_1, \ldots, T_n, N)$ has value $-(\omega(x))^{n-1}$. The absolute value of the determinant is thus greater than $(1 + M^2)^{-(n-1)/2}$, so the inverse matrix of $(T_1, \ldots, T_n, N)$ belongs to $L^\infty(\mathbb{R}^n)$.

Let $(T_1^\star, \ldots, T_n^\star, N)$ denote the dual basis of $(T_1, \ldots, T_n, N)$. Then every vector $Z \in \mathbb{R}^{n+1}$ may be written

$$(4.4) \qquad Z = (Z \cdot T)T_1^\star + \cdots + (Z \cdot T_n)T_n^\star + (Z \cdot N)N.$$

The vectors $T_1^\star, \ldots, T_n^\star$ belong to $L^\infty(\mathbb{R}^n)$.

We apply (4.4) to $Z = (X - Y)|X - Y|^{-n-1}$ and then put

$$\tilde{K}(X, Y) = \frac{X - Y}{|X - Y|^{n+1}} \cdot N(Y) \quad \text{and} \quad K_j(X, Y) = \frac{X - Y}{|X - Y|^{n+1}} \cdot T_j(Y),$$

for $1 \le j \le n$. This gives us the noteworthy decomposition

$$(4.5) \qquad \frac{X - Y}{|X - Y|^{n+1}} = \sum_1^n K_j(X, Y)T_j^\star(Y) + \tilde{K}(X, Y)N(Y).$$

The action of the operators of pointwise multiplication by $T^\star(Y)$ and $N(Y)$ on $L^2(\mathbb{R}^n)$ may be ignored, because the vectors involved belong to $L^\infty(\mathbb{R}^n)$.

So the problems of convergence that we need to resolve reduce to the corresponding questions about the kernels $K_j(X, Y)$ and $\tilde{K}(X, Y)$. But these kernels have a simpler structure than the kernel we started with. Indeed,

$$\omega_n K_j(X, Y) \, d\sigma(Y) = \frac{1}{n - 1}\frac{\partial}{\partial y_j}(|X - Y|^{-n+1}) \, dy$$

and

$$\omega_n \tilde{K}(X,Y)\, d\sigma(Y) = \frac{t - \phi(y) - (x-y)\cdot\nabla\phi(y)}{\left(|x-y|^2 + (t-\phi(y))^2\right)^{(n+1)/2}}\, dy\,.$$

The question whether $\int_V \tilde{K}(X,Y)f(Y)\, d\sigma(Y)$ converges non-tangentially is thus resolved by Theorem 1.

It remains to deal with each $\int_V K_j(X,Y)f(Y)\, d\sigma(Y)$. Since $X \notin V$, we may integrate by parts when f is in the Schwartz class $\mathcal{S}(\mathbb{R}^n)$. This gives

$$(4.6) \quad \omega_n \int_V K_j(X,Y)\, f(Y)\, d\sigma(Y) = -\frac{1}{n-1}\int_{\mathbb{R}^n} \frac{1}{|X-Y|^{n-1}}\frac{\partial f}{\partial y_j}\, dy\,,$$

where $Y = (y,\phi(y)) \in V$ and $X \notin V$.

As $X = (x,t)$ tends to $X_0 = (x_0, \phi(x_0))$, the limit of the right-hand side of (4.6) may be calculated by applying Lebesgue's dominated convergence theorem, after substituting $y - x = u - x_0$. The non-tangential convergence is irrelevant and the calculation is similar to that of the previous section. We get

$$(4.7) \quad -\frac{1}{n-1}\int_{\mathbb{R}^n} \left(|x_0 - y|^2 + (\phi(x_0) - \phi(y))^2\right)^{-(n-1)/2}\frac{\partial f}{\partial y_j}\, dy\,.$$

We then suppose that ϕ is differentiable at x_0 and proceed to an integration by parts in the opposite direction, but replacing $\int_{\mathbb{R}^n}$ in (4.7) by $\int_{|x_0 - y| \geq \varepsilon}$. This means that we can apply Green's theorem and, finally, after a calculation similar to that in the previous section, arrive at

$$(4.8) \quad \mathrm{PV}\int_{\mathbb{R}^n} \frac{x_{0,j} - y_j + (\phi(y) - \phi(x_0))\partial\phi/\partial y_j}{\left(|x_0 - y|^2 + (\phi(x_0) - \phi(y))^2\right)^{(n+1)/2}}\, f(y)\, dy\,.$$

This term is one of the principal-value distributions which appear in Theorem 5.

The jump term which appears in (4.3) comes from $\tilde{K}(X,Y)$, that is, from the double-layer potential which we have already analysed in Theorem 1.

We have proved Theorem 5. Here is a corollary.

Corollary. *With the hypotheses and notation of Theorem 5, for every function $f \in L^2(V, d\sigma)$ and almost all $X \in V$, we have*

$$(4.9) \quad \underset{Y \to X}{\mathrm{n.t.lim}}\, \nabla Sf(Y) \cdot N(X) = \pm\frac{1}{2}f(X) + K^*f(X)\,,$$

where $K^* : L^2(V, d\sigma) \to L^2(V, d\sigma)$ is the adjoint of the operator K, where the choice of sign $\pm$ is the opposite of the sign of $t - \phi(y)$, for $Y = (y,t)$, and where n.t.lim means non-tangential limit.

To prove this corollary, we let $T : (L^2(V, d\sigma))^{n+1} \to L^2(V, d\sigma)$ be the vector-valued Calderón-Zygmund operator whose distribution-kernel is $\mathrm{PV}((Y-X)/|Y-X|^{n+1})$. By Theorem 5, this operator exists. Let N :

$L^2(V, d\sigma) \to (L^2(V, d\sigma))^{n+1}$ be the operator of pointwise multiplication by the vector $N(Y)$. Then

$$Kf(X) = \frac{1}{\omega_n} \, \text{PV} \int_V \frac{Y - X}{|Y - X|^{n+1}} \cdot N(Y) f(Y) \, d\sigma(Y) = TNf(X).$$

Thus $K = TN$ and, passing to the adjoints, $K^* = N^* T^*$. But $T^* = -T$ and it follows that

$$(4.10) \qquad K^* f(X) = \frac{1}{\omega_n} N(X) \cdot \text{PV} \int_V \frac{X - Y}{|Y - X|^{n+1}} f(Y) \, d\sigma(Y),$$

which lets us identify the second term on the right-hand side of (4.9).

We have proved the corollary. By abuse of notation, we shall write $(\partial/\partial N)\nabla Sf(X)$ for the left-hand side of (4.9). Again, it will be necessary to state which side of V we are working from.

We can now draw some conclusions about the Dirichlet and Neumann problems.

For the Dirichlet problem, let us take an arbitrary function $f \in L^2(V, d\sigma)$ and suppose that there exists a function $g \in L^2(V, d\sigma)$ such that $(I/2 + K)g = f$. Then Theorem 1 tells us that $u = \mathcal{K}(g)$ is the solution of the Dirichlet problem.

Similarly, if $h \in L^2(V, d\sigma)$ is a function such that $(-I/2 + K^*)h = f$, then $v = S(h)$ gives the solution of the Neumann problem, as (4.9) shows, when we work from the open set Ω above V.

It remains to show that the operators $I/2 + K : L^2(V, d\sigma) \to L^2(V, d\sigma)$ and $-I/2 + K^* : L^2(V, d\sigma) \to L^2(V, d\sigma)$ are isomorphisms. At present, we do not have a symbolic calculus for generalized Calderón-Zygmund operators, moreover, K is not compact. The only resource we have is, following Verchota, to use the remarkable energy estimates of D. Jerison and C. Kenig.

5 The Jerison and Kenig identities

We continue with the same notation. We suppose that $f \in L^2(V, d\sigma)$, we write $u = S(f)$ for the single-layer potential generated by f, and we study the gradient ∇u of u. This gradient is defined in $\mathbb{R}^{n+1} \setminus V$ and, as it crosses V, suffers a discontinuity normal to V. Thus, for almost all $X \in V$, we can define the tangential gradient $\nabla_t u(X)$ by projecting the non-tangential limit of $\nabla u(Y)$, as $Y \to X$, onto the tangent plane at X. The tangential gradient defined in this way does not depend on the side from which the limit of V is taken.

As before, we let $N(X)$ denote the downwards-oriented unit vector normal to V at X.

With this notation, we get

Theorem 6. *If V is a Lipschitz graph, if $f \in L^2(V, d\sigma)$, and if $u = S(f)$, then*

$$(5.1) \qquad \int_V |\nabla_t u|^2 N \, d\sigma = \int_V \left(\frac{\partial u}{\partial N}\right)^2 N \, d\sigma + 2 \int_V \left(\frac{\partial u}{\partial N}\right) \nabla_t u \, d\sigma$$

and

$$(5.2) \qquad \int_V |\nabla_t u|^2 N \, d\sigma = \int_V \left(\frac{\partial u}{\partial N} + f\right)^2 N \, d\sigma + 2 \int_V \left(\frac{\partial u}{\partial N} + f\right) \nabla_t u \, d\sigma.$$

Let us begin by observing that (5.2) follows from the proof of (5.1) by applying that proof to the open set Ω_- below V, whereas, to prove (5.1) itself, we have to consider the open set Ω lying above V. When we replace Ω by Ω_-, N is replaced by $-N$. Then the normal derivative of the single-layer potential changes sign, but also involves a discontinuity. In keeping the same normal vector in (5.2) as in (5.1), the jump term in (4.9) is responsible for the change from $\partial u/\partial N$ to $\partial u/\partial N + f$.

So it is enough to prove (5.1). We first cheat a little, by pretending that Ω is a bounded, regular, open set and that $V = \partial\Omega$.

We now write $\nabla_t u = \nabla u - (\partial u/\partial N)N$ and (5.1) reduces to the vector equation

$$(5.3) \qquad \int_{\partial\Omega} |\nabla u|^2 N \, d\sigma = 2 \int_{\partial\Omega} \nabla u \frac{\partial u}{\partial N} \, d\sigma.$$

To establish (5.3), we first change it to a scalar identity by taking scalar products with an arbitrary constant vector $e \in \mathbb{R}^{n+1}$. It is enough to check that

$$(5.4) \qquad \int_{\partial\Omega} \left(|\nabla u|^2 N \cdot e - 2(\nabla u \cdot e)(\nabla u \cdot N)\right) d\sigma = 0.$$

We then apply the divergence theorem and see that it is enough to verify that $\operatorname{div}\left(|\nabla u|^2 e - 2(\nabla u \cdot e)\nabla u\right) = 0$ on Ω. But this divergence equals $-(\nabla u \cdot e)\Delta u$, which is zero, since u is harmonic.

To apply these considerations, we approximate to our original Ω by an increasing sequence Ω_m of bounded, regular, open sets. The boundary $\partial\Omega_m$ will be the union of two subsets V_m and W_m. In addition, we suppose that f is continuous, with compact support. We consider (5.3) for the open sets Ω_m, with $u = Sf$ being fixed and the operator S being taken with respect to V (not V_m.) Then we shall show that

$$(5.5) \qquad \int_{V_m} |\nabla u|^2 N_m \, d\sigma_m \to \int_V |\nabla u|^2 N \, d\sigma,$$

$$(5.6) \qquad \int_{V_m} \nabla u \frac{\partial u}{\partial N_m} \, d\sigma_m \to \int_V \nabla u \frac{\partial u}{\partial N} \, d\sigma,$$

and that the corresponding integrals over W_m tend to 0.

From this point, we get (5.3) by a simple passage to the limit from the corresponding identity for the harmonic function u, restricted to Ω_m.

We now describe the method of approximation and justify the limiting process.

Let ϕ_m be a sequence of real- or complex-valued Lipschitz functions defined on $\mathbb{R}^n$. We say that this sequence converges strictly to a Lipschitz function ϕ, if there exists a constant $C > 0$ such that, for every integer $m \in \mathbb{N}$, we have $\|\nabla\phi_m\|_\infty \leq C$ and $\nabla\phi_m(x) \to \nabla\phi(x)$, almost everywhere. Under these conditions, there exist normalization constants c_m such that the functions $\phi_m(x) + c_m$ converge to $\phi(x)$, uniformly on compact sets.

So let us start with a Lipschitz function ϕ and construct the ϕ_m by convolving with $m^n g(mx)$, where g is an infinitely differentiable function of compact support and mean 1. Then the functions ϕ_m converge strictly to ϕ. Changing the meaning of the subscript m, if necessary, and possibly adding a constant ε_m to each ϕ_m, we may suppose that $\phi(x) < \phi_m(x) < \phi(x) + 1/m$, if $|x| \leq m + 1$. We then join the graph V_m of ϕ_m above $|x| \leq m$ to the base of the cylinder defined by $|x| = m+1$, $\phi_m(x)+1/m \leq t \leq c'_m$ and then to the disk $|x| \leq m$, $t = c'_m$. All the joins are to be made so that the composite surface is the regular boundary $\partial\Omega_m$ of a regular bounded open set Ω_m, to which we can apply the divergence theorem. We also make sure that $\overline{\Omega}_m \subset \Omega$, so that the function u, which is harmonic in Ω, is harmonic in a neighbourhood of Ω_m.

To prove (5.5), we first observe that $\nabla u(x)$ is $O(|X|^{-n})$ at infinity, because f has compact support. As a consequence, the integral corresponding to $\int_{V_m} |\nabla u|^2 N_m \, d\sigma_m$, but where V_m is replaced by the graph of ϕ_m above $|x| > m$, has limit 0 when $m \to \infty$.

Having made this remark, (5.5) reduces to

$$(5.7) \quad \lim_{m \to \infty} \int_{\mathbb{R}^n} |\nabla u(x, \phi_m(x))|^2 \frac{\partial\phi_m}{\partial x_j} \, dx = \int_{\mathbb{R}^n} |\nabla u(x, \phi(x))|^2 \frac{\partial\phi}{\partial x_j} \, dx \,,$$

for $1 \leq j \leq n$, and a similar statement with $\partial\phi_m/\partial x_j$ and $\partial\phi/\partial x_j$ replaced by 1. To establish (5.7), we use the definition of strict convergence and Lebesgue's dominated convergence theorem. Indeed,

$$\omega(x) = \sup_{m \geq 1} |\nabla u(x, \phi_m(x))|^2 \in L^1(\mathbb{R}^n) \,,$$

as a consequence of Theorem 5.

The proof of (5.6) is identical.

The integrals over the outer boundaries of Ω_m tend to 0, because their surface area is $O(m^{n-1})$ and the integrands are $O(m^{-2n})$.

To conclude the proof of Theorem 6 and to pass to the general case, in which f is an arbitrary function of $L^2(V, d\sigma)$, we write $Q_1(f)$ and $Q_2(f)$ for the two sides of the identity (5.1). These quadratic forms are continuous on $L^2(V, d\sigma)$, by Theorem 5, and coincide on a dense linear subspace. They are thus equal throughout $L^2(V, d\sigma)$.

Corollary 1. *For $f \in L^2(V, d\sigma)$, the norms $\|(I/2 + K^\star)f\|_2$, $\|(I/2 - K^\star)f\|_2$, and $\|\nabla_t S(f)\|_2$ are equivalent.*

Indeed, let e denote the vector $(0, \ldots, 0, -1)$. The unit vector $N(Y)$, normal to V at Y and pointing downwards, has the property that the scalar product $N \cdot e$ lies in the interval $[(1 + M^2)^{-1/2}, 1]$, where M is such that $\|\nabla\phi\|_\infty \leq M$.

Taking the scalar products of e with (5.1) and (5.2) and finding upper bounds for the right-hand sides by the Cauchy-Schwarz inequality, we get

$$(5.8) \quad \int_V |\nabla_t u|^2 \, d\sigma \leq (1 + M^2)^{1/2} \left(\int_V \left| \frac{\partial u}{\partial N} \right|^2 d\sigma + 2 \left\| \frac{\partial u}{\partial N} \right\|_2 \|\nabla_t u\|_2 \right)$$

and

$$(5.9) \quad \int_V |\nabla_t u|^2 \, d\sigma$$
$$\leq (1 + M^2)^{1/2} \left(\int_V \left| \frac{\partial u}{\partial N} + f \right|^2 d\sigma + 2 \left\| \frac{\partial u}{\partial N} + f \right\|_2 \|\nabla_t u\|_2 \right).$$

But an inequality of the form $X^2 \leq C(Y^2 + 2XY)$ between two finite positive numbers X and Y necessarily implies that $X \leq C'Y$, where $C' = \sqrt{C + C^2} + C$. Thus, by Theorem 5, $\|\nabla_t u\|_2 \leq C'\|\partial u/\partial N\|_2$ and $\|\nabla_t u\|_2 \leq C'\|\partial u/\partial N + f\|_2$, where the norms are all that of $L^2(V, d\sigma)$.

Returning to (5.1), and taking the scalar product with e, once more, we get

$$(5.10) \quad (1 + M^2)^{1/2} \int_V |\nabla_t u|^2 \, d\sigma \geq \int_V \left| \frac{\partial u}{\partial N} \right|^2 d\sigma - 2 \left\| \frac{\partial u}{\partial N} \right\|_2 \|\nabla_t u\|_2,$$

which leads to $\|\partial u/\partial N\|_2 \leq C'\|\nabla_t u\|_2$. We similarly can use (5.2) to show that $\|\partial u/\partial N + f\|_2 \leq C'\|\nabla_t u\|_2$.

Corollary 2. *The norms $\|(I/2+K^\star)f\|_2$, $\|(I/2-K^\star)f\|_2$, and $\|f\|_2$ are equivalent on $L^2(V, d\sigma)$. The constants involved in the equivalences depend only on the constant M such that $\|\nabla\phi\|_\infty \leq M$ and not otherwise on the Lipschitz function ϕ itself.*

Indeed, the continuity of $K^\star$ gives $\|(I/2 \pm K^\star)f\|_2 \leq C\|f\|_2$. Conversely, $\|f\|_2 \leq \|(I/2 + K^\star)f\|_2 + \|(I/2 - K^\star)f\|_2$ and we can now use Corollary 1.

6 The rest of the proof of Theorems 2 and 3

Finishing the proof is a question of showing, in the unbounded case, that the two operators $I/2 \pm K^*$ are isomorphisms on $L^2(V, d\sigma)$. The same property will follow for their adjoints $I/2 \pm K$ and the modifications to deal with the bounded case can be found in Verchota's thesis ([232]). The following lemma is a rudimentary version of index theory.

Lemma 4. *Let H be a Hilbert space and let $\mathcal{L}(H)$ be the Banach algebra of continuous linear operators $T : H \to H$.*

Let $T(\lambda)$, $0 \leq \lambda \leq 1$, be a family of elements of $\mathcal{L}(H)$ with the following three properties:

(6.1) *there exists a constant $\delta > 0$ such that, for all $\lambda \in [0,1]$ and all $x \in H$, $\|T(\lambda)x\| \geq \delta \|x\|$;*

(6.2) *$T(0) : H \to H$ is surjective;*

(6.3) *there exists $\varepsilon > 0$ such that $\|T(\lambda') - T(\lambda)\| \leq \delta/2$ when $0 \leq \lambda \leq \lambda' \leq 1$ and $0 \leq \lambda' - \lambda \leq \varepsilon$.*

Then the operators $T(\lambda)$ are isomorphisms of H, for all $\lambda \in [0,1]$.

Let us start by remarking that, if $T_1 \in \mathcal{L}(H)$ is such that T_1 is surjective and there is a constant $\delta > 0$ such that $\|T_1(x)\| \geq \delta \|x\|$, for all $x \in H$, and if $T_2 \in \mathcal{L}(H)$ satisfies $\|T_2 - T_1\| < \delta$, then it follows that T_2 is also an isomorphism of H.

We then start from $T(0)$, which is an isomorphism, and, step by step, using (6.1) and (6.3), show that all the $T(\lambda)$, $0 \leq \lambda \leq 1$ are isomorphisms.

In the application we have in mind, (6.3) comes from a much stronger property, namely, the existence of a constant C_0 such that

$$\left\| \frac{d}{d\lambda} T(\lambda) \right\| \leq C_0.$$

To show that the operators $I/2 \pm K^* : L^2(V, d\sigma) \to L^2(V, d\sigma)$ are isomorphisms, we use Lemma 4. We take the graphs V_λ of the functions $\lambda\phi(x)$, $0 \leq \lambda \leq 1$, and we consider the corresponding operators $I/2 \pm K_\lambda^*$. These operators each act on different spaces $L^2(V_\lambda, d\sigma_\lambda)$. To apply Lemma 4, we identify each of these spaces with $L^2(\mathbb{R}^n, dx)$, by systematically using the variable $x \in \mathbb{R}^n$ to parametrize the surfaces V_λ by the equations $t = \lambda\phi(x)$.

After this transformation, the operators $I/2 \pm K^*$ become singular integral operators $T(\lambda)$ defined by the distribution-kernels $(\delta(x-y))/2 \pm$ PV $K_\lambda^*(x,y)$, where

$$K_\lambda^*(x,y) = \frac{\lambda}{\omega_n} \frac{\phi(y) - \phi(x) - (y - x) \cdot \nabla\phi(x)}{[|y - x|^2 + \lambda^2(\phi(x) - \phi(y))^2]^{(n+1)/2}}.$$

Clearly, $T(0) = I/2$ is an isomorphism and Corollary 2 of Theorem 6 gives (6.1).

To finish, we therefore need to show that

(6.4) $$\|T(\lambda') - T(\lambda)\| \leq C(M)(\lambda' - \lambda),$$

for $0 \leq \lambda < \lambda' \leq 1$. To do this, we consider the truncated kernels $K^\star_{\lambda,\varepsilon}$ corresponding to $K^\star_\lambda$ and the operators $T^{(\varepsilon)}(\lambda)$, corresponding to those truncated kernels. By Theorem 5, we know that

$$\|T(\lambda)(f) - f/2 - T^{(\varepsilon)}(\lambda)(f)\|_2 \to 0,$$

as $\varepsilon \downarrow 0$, for each $f \in L^2(\mathbb{R}^n)$. The estimate (6.3) will therefore follow from

(6.5) $$\|T^{(\varepsilon)}(\lambda') - T^{(\varepsilon)}(\lambda)\| \leq C(M)(\lambda' - \lambda).$$

To prove (6.5), we shall verify that $\|(d/d\lambda)T^{(\varepsilon)}(\lambda)\| \leq C(M)$. The computation of this derivative is immediate, because we have eliminated the kernel's singularity. The kernel of $(d/d\lambda)T^{(\varepsilon)}(\lambda)$ may be written as $A^{(\varepsilon)}(x,y) + B^{(\varepsilon)}(x,y)$, the truncated kernels of

$$A(x,y) = \frac{1}{\omega_n} \frac{\phi(y) - \phi(x) - (y-x)\cdot\nabla\phi(x)}{[|y-x|^2 + \lambda^2(\phi(x) - \phi(y))^2]^{(n+1)/2}}$$

and

$$B(x,y) = \frac{n+1}{\omega_n}\lambda^2 \frac{[\phi(x) - \phi(y)]^2[\phi(y) - \phi(x) - (y-x)\cdot\nabla\phi(x)]}{[|y-x|^2 + \lambda^2(\phi(x) - \phi(y))^2]^{(n+3)/2}}.$$

The L^2 continuity of the operators defined by these kernels is given by Theorem 11 of Chapter 9. The norms of the corresponding operators are bounded above by constants, depending only on M, where $\|\nabla\phi\|_\infty \leq M$. The uniform estimates on the truncated operators then follow from Chapter 7.

7 Appendix

For the convenience of the reader, we give a proof of Lemma 1. The proof follows the usual pattern. We show that

(7.1) $$\lim_{y \to 0} \frac{f(x+y) - f(x) - y\cdot\nabla f(x)}{|y|} = 0,$$

for a dense linear subspace E of an appropriate Banach space B, and we establish a maximal inequality corresponding to (7.1).

If we choose the Banach space B to be the space of Lipschitz functions $f : \mathbb{R}^n \to C$, we come up against the difficulty that the C^1 functions are not a dense linear subspace of B. This leads us to replace the Lipschitz space by the Sobolev space $L^{p,1}(\mathbb{R}^n)$ of $L^p(\mathbb{R}^n)$ functions whose gradients (in the sense of distributions) also lie in $L^p(\mathbb{R}^n)$.

So we intend to prove (7.1) for $f \in L^{p,1}(\mathbb{R}^n)$, $p > n$, for almost all $x \in \mathbb{R}^n$. Now, the Schwartz class $\mathcal{S}(\mathbb{R}^n)$ is dense in $L^{p,1}$, for $n < p < \infty$. It will therefore be enough to establish the maximal inequality

$$(7.2) \quad \left\| \sup_y \frac{|f(x+y) - f(x) - y \cdot \nabla f(x)|}{|y|} \right\|_{L^p(\mathbb{R}^n)} \leq C(p,n) \|\nabla f\|_{L^p(\mathbb{R}^n)},$$

for $n < p \leq \infty$.

The term $y \cdot \nabla f(x)$ is trivial and we may forget it, so that we can concentrate on $|y|^{-1}|f(x+y) - f(x)|$. Putting $r = 2|y|$ and letting B denote the ball, centre x, of radius r, (7.2) will follow from the inequality

$$(7.3) \quad \frac{|f(x+y) - f(x)|}{|y|} \leq C(p,n) \left(\frac{1}{|B|} \int_B |\nabla f|^p \, dy \right)^{1/p}.$$

Then the right-hand side of (7.3) is bounded above by $(|\nabla f|^p)^*$, which belongs to $L^r(dx)$, when $p < r < \infty$, by the Hardy-Littlewood theorem. As a consequence, (7.3), with $n < p < \infty$, gives (7.2), with p replaced by $p' > p$, which does not affect the result.

To establish (7.3), we may clearly suppose that $x = 0$. We have $|f(y) - f(0)| \leq |f(y) - f(r\nu)| + |f(r\nu) - f(0)|$, where $r = 2|y|$ and $\nu \in S^{n-1}$. Continuing, we take the mean over $\nu \in S^{n-1}$, having written

$$(7.4) \quad f(r\nu) - f(0) = \nu \cdot \int_0^r \nabla f(t\nu) \, dt$$

and

$$(7.5) \quad f(r\nu) - f(y) = (\nu - r^{-1}y) \cdot \int_0^r \nabla f(t\nu + (1 - t/r)y) \, dt.$$

This leads to two integrals of the form $\int_{S^{n-1}} \int_0^r \ldots dt \, d\sigma(\nu)$ followed by a change to Cartesian co-ordinates. The Jacobians involved are, respectively, $|x|^{-n+1}$ and $|x - y|^{-n+1}$, and we obtain

$$(7.6) \quad |f(y) - f(0)| \leq C \int_B |\nabla f(x)|(|x|^{-n+1} + |x - y|^{-n+1}) \, dx.$$

This is where the condition $n < p < \infty$ becomes necessary. We apply Hölder's inequality to the right-hand side of (7.6). The essential observation is that $|x|^{-n+1} \in L^q(B)$, if $q < n/(n-1)$, that is, if $p > n$.

The remaining details of the calculation are very easy and left to the reader.

In [40], Calderón uses an entirely different approach to prove Theorems 2 and 3.

16

Paradifferential operators

1 Introduction

Paradifferential operators were invented by J.M. Bony ([16]) to construct a calculus, similar to the pseudo-differential calculus, which would admit operators of multiplication by C^r functions ($r > 0$ is fixed from now on). However, he also wanted to include the constant coefficient differential operators. Naturally, these requirements were contradictory, because they immediately led to multiplying an arbitrary distribution by a C^r function. This was impossible, if the pointwise multiplication operator was to keep its usual meaning.

Bony's solution was to replace pointwise multiplication by paramultiplication $\pi(a, f)$ of a with f. The paraproduct is defined when a and f are tempered distributions. The algebra B_r ($r > 0$) of paradifferential operators includes all classical pseudo-differential operators $\sigma(x, D) \in Op\, S^0_{1,0}$ and all the operators given by taking a paraproduct with C^r functions.

This set-up gives results similar to those of the usual pseudo-differential calculus, the difference being that the "error terms" will be regularizing operators of order r, instead of smoothly regularizing operators.

The paradifferential calculus has a natural place in Calderón's programme, for several reasons.

Firstly, in 1965, that is, more than ten years before the birth of the paradifferential calculus, Calderón had specified a research programme which had as one of its objectives the investigation of the regularity of the solutions of non-linear partial differential equations using a pseudo-differential calculus which would include the operators of pointwise multiplication by C^r functions.

Secondly, if $a(x) \in L^\infty(\mathbb{R}^n)$, the paramultiplication operator $f \mapsto \pi(a, f)$, for $f \in L^2(\mathbb{R}^n)$, is a Calderón-Zygmund operator.

Lastly, Bony's linearization formula using the paraproduct coincides, in the holomorphic case, with a famous identity of Calderón's.

In this chapter we do not pretend to describe all the results arising from the paradifferential calculus. We refer the reader to the proceedings of the Seminar of the Ecole Polytechnique ([18]), where the work of Bony and his co-workers is to be found.

2 A first example of linearization of a non-linear problem

We shall use $\mathcal{F} : \mathcal{S}'(\mathbb{R}^n) \to \mathcal{S}'(\mathbb{R}^n)$ to denote the Fourier transform acting on tempered distributions. We shall also use $\hat{f}$ to mean $\mathcal{F}(f)$.

We fix a radial function ϕ in the Schwartz class $\mathcal{D}(\mathbb{R}^n)$ which satisfies $\phi(\xi) = 1$ for $|\xi| \le 1/2$ and $\phi(\xi) = 0$, when $|\xi| \ge 1$.

For each $j \in \mathbb{Z}$, the partial sum, or filtering, operator S_j is defined by $\mathcal{F}[S_j(f)](\xi) = \phi(2^{-j}\xi)\hat{f}(\xi)$. The "dyadic block" $\Delta_j(f)$ of f is defined by $\Delta_j(f) = S_{j+1}(f) - S_j(f)$ or, equivalently, by

$$(2.1) \quad \mathcal{F}[\Delta_j(f)](\xi) = \psi(2^{-j}\xi)\hat{f}(\xi) \qquad \text{where} \qquad \psi(\xi) = \phi(\xi/2) - \phi(\xi).$$

The support of the Fourier transform $\mathcal{F}(\Delta_j(f))$ is thus contained in the "dyadic shell"

$$(2.2) \qquad \Gamma_j = \{\xi \in \mathbb{R}^n : 2^{j-1} \le |\xi| \le 2^{j+1}\}.$$

With this notation, the "Littlewood-Paley decomposition" of f is the series

$$(2.3) \qquad f = S_0(f) + \Delta_0(f) + \cdots + \Delta_j(f) + \cdots,$$

which converges to f in $\mathcal{S}'(\mathbb{R}^n)$.

Of course, different choices of ϕ will lead to different decompositions.

A first example of linearization of a non-linear problem is given by the use of the Littlewood-Paley decomposition in the proof of the following classical theorem.

Theorem 1. *Let $F \in C^\infty(\mathbb{R})$ be a function vanishing at 0. If $1 < p < \infty$ and $s > n/p$, then, for all real-valued $f \in L^{p,s}(\mathbb{R}^n)$, the function $F(f)$ is also in $L^{p,s}(\mathbb{R}^n)$. If $s = n/p$ and the derivative F' of F is bounded, then we still have $F(f) \in L^{p,s}(\mathbb{R}^n)$ when $f \in L^{p,s}(\mathbb{R}^n)$.*

We know that $L^{p,s}$ is an algebra when $s > n/p$. Theorem 1 tells us that the infinitely differentiable functions operate on this algebra, in the sense of symbolic calculus. Since $L^{p,s}$ is not an algebra any more when $s = n/p$, the choice $F(t) = t^2$ must be excluded in this case, which is why we must suppose that F is sub-linear.

To prove the theorem for $s > n/p$, we put $f_j = S_j(f)$ and write

(2.4) $\quad F(f) = F(f_0) + (F(f_1) - F(f_0)) + \cdots + (F(f_{j+1}) - F(f_j)) + \cdots$.

The series in (2.4) is a telescopic series which converges uniformly to $F(f)$.

The term $F(f_0)$ presents no difficulties, because f_0 and all its derivatives belong to L^p. It follows that $F(f_0)$ and all its derivatives belong to L^p, as long as $F(0) = 0$.

We then write

(2.5) $$F(f_{j+1}) - F(f_j) = m_j \Delta_j(f) \,,$$

where

(2.6) $$m_j(x) = \int_0^1 F'(f_j(x) + t\Delta_j f(x)) \, dt \,.$$

The information we need about the functions f_j is contained in the following lemma.

Lemma 1. *If $f \in L^{p,s}(\mathbb{R}^n)$ and if $s > n/p$, or if $s = n/p$ and $F' \in L^\infty(\mathbb{R}^n)$, then there exist constants C_α, $\alpha \in \mathbb{N}^n$, such that, for all $j \in \mathbb{N}$, we have $\|\partial^\alpha m_j\|_\infty \le C_\alpha 2^{j|\alpha|}$.*

We assume this result for the moment and continue with the proof of the theorem.

Consider the linear operator $\mathcal{L}$ defined by

(2.7) $$\mathcal{L}u(x) = \sum_0^\infty m_j(x)\Delta_j u(x) \,.$$

The operator $\mathcal{L}$ belongs to $\mathcal{O}p\, S_{1,1}^0$. Indeed, $\sum_0^\infty m_j(x)\psi(2^{-j}\xi)$ is the symbol of $\mathcal{L}$ and the estimates defining an element of $S_{1,1}^0$ are precisely those given by Lemma 1. The operator $\mathcal{L}$ is thus bounded on all $L^{p,r}$, where $r > 0$, $1 < p < \infty$ (Chapter 10, section 6). In particular, $\mathcal{L}(f) \in L^{p,s}$ when $f \in L^{p,s}$.

We still have to prove Lemma 1. Let us start by looking at the case $s > n/p$. Then $\|f_j\|_\infty \le C$, since $f \in L^\infty(\mathbb{R}^n)$. Bernstein's theorem then gives $\|\partial^\alpha f_j\|_\infty \le C2^{(j+1)|\alpha|}$. To show that $F(f_j + t\Delta_j(f))$ satisfies inequalities of the same type, we apply the following lemma.

Lemma 2. *If the functions g_j satisfy $\|\partial^\alpha g_j\|_\infty \le C_\alpha 2^{j|\alpha|}$, where C_α does not depend on j, then the functions $F(g_j) = h_j$ satisfy $\|\partial^\alpha h_j\|_\infty \le C_\alpha' 2^{j|\alpha|}$.*

An amusing way to see this is to consider the auxiliary functions $\tilde{g}_j(x) = g_j(2^{-j}x)$, which satisfy $\|\partial^\alpha \tilde{g}_j\|_\infty \le C_\alpha$. Then the functions $\tilde{h}_j = F(\tilde{g}_j)$ satisfy $\|\partial^\alpha \tilde{h}_j\|_\infty \le C'_\alpha$, as can be seen by applying the Faa di Bruno formula for the derivative of the composition of functions. We now recall that formula.

To compute $\partial^\alpha F(g)$, where $\alpha = (\alpha_1, \ldots, \alpha_n)$, we let q denote an integer with $1 \le q \le |\alpha|$ and split the vector $\alpha \in \mathbb{N}^n$ into all the possible vector sums $\beta_1 + \cdots + \beta_q$, where $\beta_1, \ldots, \beta_q \in \mathbb{N}^n$. We then form all the corresponding products $F^{(q)}(g)\partial^{\beta_1}g \cdots \partial^{\beta_q}g$. We then take the sum over all the possible decompositions of α, keeping q fixed, and then take the sum over all values of $q \in [1, |\alpha|]$. This gives $\partial^\alpha F(g)$.

If $s = n/p$ and $F' \in L^\infty(\mathbb{R}^n)$, then, clearly, $m_j \in L^\infty(\mathbb{R}^n)$. We obtain the bounds on the derivatives of m_j by observing that $\|\partial^\alpha f_j\|_\infty \le C_\alpha 2^{j|\alpha|}$, when $|\alpha| \ge 1$, even though it is not the case that $\|f_j\|_\infty \le C_0$.

3 A second linearization of the non-linear problem

The paraproduct $\pi(a, f)$ of the tempered distributions a and f is defined by

$$(3.1) \qquad \pi(a, f) = \sum_{2}^{\infty} S_{j-2}(a)\Delta_j(f).$$

It is not at all obvious that the series converges in the topology of $\mathcal{S}'(\mathbb{R}^n)$, but we can verify that it converges, by applying the following simple lemma.

Lemma 3. *Let u_j, $j \in \mathbb{N}$, be $C^\infty(\mathbb{R}^n)$ functions of polynomial growth satisfying*

(3.2) *there exist constants $\beta > \alpha > 0$ such that the support of the Fourier transform of the distribution u_j lies in $\alpha 2^j \le |\xi| \le \beta 2^j$;*

(3.3) *there exist a constant C and an exponent m such that $|u_j(x)| \le C2^{mj}(1 + |x|)^m$, for all $j \in \mathbb{N}$.*

Then $\sum_0^\infty u_j(x)$ converges to a tempered distribution in $\mathcal{S}'(\mathbb{R}^n)$.

Conversely, if S is a tempered distribution, $u_j = \Delta_j(S)$ satisfies (3.3).

Under the given conditions, $|S_{j-2}(a)(x)| \le C2^{pj}(1 + |x|)^p$, for some integer p, and the support of the Fourier transform of $S_{j-2}(a)$ is contained in the ball B_j, centre 0 and radius 2^{j-2}. Similarly, $|\Delta_j(f)(x)| \le C2^{qj}(1 + |x|)^q$ and the support of $\mathcal{F}(\Delta_j(f))$ is contained in the dyadic shell Γ_j defined by $2^{j-1} \le |\xi| \le 2^{j+1}$. Thus, the support of the Fourier transform of the product $S_{j-2}(a)\Delta_j(f)$ is contained in $B_j + \Gamma_j$, that is, in the region $\alpha 2^j \le |\xi| \le \beta 2^j$, where $\alpha = 1/4$ and $\beta = 9/4$. The conditions of the lemma are satisfied.

In definition (3.1), the gap between the two subscripts plays an essential rôle: we may not replace $S_{j-2}(a)\Delta_j(f)$ by $S_j(a)\Delta_j(f)$. The support of the Fourier transform of the latter product is contained in the ball $|\xi| \le 3 \cdot 2^j$ and the resulting series would diverge. To understand the difference between $\sum_2^\infty S_{j-2}(a)\Delta_j(f)$ and $\sum_0^\infty S_j(a)\Delta_j(f)$ better, we subtract one from the other. This gives $\sum(\Delta_{j-2}(a) + \Delta_{j-1}(a))\Delta_j(f)$. Now, using the Littlewood-Paley decomposition to calculate the product af of our two distributions, we get, essentially,

$$\sum_{j \ge 0}\sum_{j' \ge 0} \Delta_j(a)\Delta_{j'}(f)$$
$$= \sum_{j \le j'-3}\sum \Delta_j(a)\Delta_{j'}(f) + \sum_{j' \le j-3}\sum \Delta_j(a)\Delta_{j'}(f) + \sum_{|j-j'| \le 2}\sum \Delta_j(a)\Delta_{j'}(f)$$
$$= \pi(a,f) + \pi(f,a) + \sum_{|\varepsilon| \le 2}\sum \Delta_{j-\varepsilon}(a)\Delta_j(f).$$

The two paraproducts are well-defined: the obstructions to taking the product of two distributions are the defective series $\sum \Delta_{j-\varepsilon}(a)\Delta_j(f)$, $|\varepsilon| \le 2$, to which Lemma 3 cannot be applied.

On the other hand, we may replace $S_{j-2}(a)$ in (3.1) by $S_{j-20}(a)$. Such a change would not affect the results to come (Theorems 2 and 3).

Let us return to the linearization of the non-linear operation which changes $f \in L^{p,s}$ to $F(f)$. We then have

Theorem 2. *Let $1 < p < \infty$ and let $r = s - n/p > 0$. Suppose that $f : \mathbb{R}^n \to \mathbb{R}$ belongs to $L^{p,s}(\mathbb{R}^n)$ and that $F \in C^\infty(\mathbb{R})$ satisfies $F(0) = 0$. Then*

(3.4) $$F(f) = \pi(F'(f), f) + g$$

where $g \in L^{p,s+r}(\mathbb{R}^n)$.

The purpose of the paradifferential calculus is to localize, as precisely as possible, the singularities of solutions of non-linear partial differential equations. But these singularities will be relative to a threshold of regularity: everything which is more regular than the threshold is not analysed and is considered as an error term. In (3.4), g is such an error term.

It is worth describing a holomorphic variant of (3.4), because it involves the bilinear operation which foreshadowed the paraproduct in Calderón's work.

If f and g are holomorphic functions in the (open) upper half-plane P and if f and g vanish at infinity, in a certain sense, Calderón defines a function h, which is holomorphic in P and satisfies $h'(z) = f(z)g'(z)$ and $h(i\infty) = 0$. We put $\Pi(f,g) = h$.

Now, let f be a function which is holomorphic in P and which, together with its derivative f', belongs to the holomorphic Hardy space $\mathbb{H}^2(P)$. We let K denote the closure of $f(P)$ in $\mathbb{C}$. Let F be a function which vanishes at 0 and is holomorphic in a neighbourhood of K. Then

$$(3.5) \qquad\qquad F(f) = \Pi(F'(f), f).$$

Indeed,

$$F(f)(i\infty) = F(0) = 0 \qquad \text{and} \qquad \frac{d}{dz}F(f) = F'(f(z))f'(z).$$

Theorem 2 generalizes, and slightly improves, Bony's theorem ([16]), where $p = 2$ and $s + r$ was replaced by $s + r - \varepsilon$, for arbitrary $\varepsilon > 0$.

The proof of (3.4) consists of comparing the operator $\mathcal{L}$ of (2.7) (which arose in the first linearization) with the operator T_a defined by $T_a(f) = \pi(a,f)$.

We need the followimg lemma.

Lemma 4. *Let f belong to $L^{p,s}(\mathbb{R}^n)$, where $r = s - n/p > 0$ and let $\mathcal{L}$ be defined by (2.7). If $a = F'(f)$, then $\mathcal{L} - T_a$ belongs to the Hörmander class $Op\,S^{-r}_{1,1}$.*

Assuming this, the proof of (3.4) is immediate. The first linearization gave $F(f) = F(S_0(f)) + \mathcal{L}(f)$, where $F(S_0(f))$ may already be considered as an error term, since it belongs to all the $L^{p,m}$ spaces, for $m \geq 0$. So, if we put $R(f) = \mathcal{L}(f) - T_a(f)$, by the lemma and section 6 of Chapter 10, $R(f) \in L^{p,s+r}$.

To prove, as asserted by Lemma 4, that R belongs to $Op\,S^{-r}_{1,1}$, we compute its symbol $\rho(x,\xi)$. It is given by

$$\rho(x,\xi) = \sum_{2}^{\infty} q_j(x)\psi(2^{-j}\xi) + m_0(x)\psi(\xi) + m_1(x)\psi(\xi/2),$$

where

$$(3.6) \qquad q_j(x) = \int_0^1 F'(S_j(f) + t\Delta_j(f))\,dt - S_{j-2}(F'(f)).$$

We want to show that there exist constants C_α, $\alpha \in \mathbb{N}^n$, such that, for all $j \in \mathbb{N}$,

$$(3.7) \qquad\qquad \|\partial^\alpha q_j(x)\|_\infty \leq C_\alpha 2^{-jr}2^{j|\alpha|}.$$

The terms $m_0(x)\psi(\xi)$ and $m_1(x)\psi(\xi/2)$ of $\rho(x,\xi)$ lead to functions which belong to all the $L^{p,m}$ spaces, for $m \geq 0$, so we need take no further notice of them.

We concentrate on (3.7). It will be enough to establish these inequalities for $|\alpha| > r$ and for $\alpha = 0$. The classical logarithmic convexity inequalities will take care of the cases in between, that is, when $1 \leq |\alpha| \leq r$.

If $\alpha = 0$, we use the Sobolev embedding $L^{p,s} \subset C^r$, where $r = s - n/p$ and where C^r denotes the inhomogenous Hölder space which we have defined several times in the course of this work.

Thus $\|f - S_j(f)\|_\infty \leq C2^{-rj}$ and it follows immediately that

$$\|F'(f) - F'(S_j(f) + t\Delta_j(f))\|_\infty \leq C2^{-rj} .$$

But $F'(f)$ also belongs to C^r and, thus,

$$\|F'(f) - S_{j-2}(F'(f))\|_\infty \leq C2^{-rj} .$$

Finally, (3.7) follows, for $\alpha = 0$, by applying the triangle inequality.

We still have to prove (3.7) when $|\alpha| > r$. We no longer use the fact that $q_j(x)$ is the difference of two terms, but show separately that

$$(3.8) \qquad \|\partial^\alpha S_{j-2}(F'(f))\|_\infty \leq C_\alpha 2^{-jr}2^{j|\alpha|}$$

and

$$(3.9) \qquad \|\partial^\alpha F'(S_j(f) + t\Delta_j(f))\|_\infty \leq C_\alpha 2^{-jr}2^{j|\alpha|} .$$

The proof of (3.8) depends on the following lemma.

Lemma 5. *If $g \in C^r(\mathbb{R}^n)$, for $r > 0$, there exist constants C_α such that, for all $j \in \mathbb{N}$ and $|\alpha| > r$,*

$$(3.10) \qquad \|\partial^\alpha S_j(g)\|_\infty \leq C_\alpha 2^{j(|\alpha|-r)} .$$

We start by writing $S_j(g) = S_0(g) + \Delta_0(g) + \cdots + \Delta_j(g)$. The term $S_0(g)$ is unproblematic, because $\|\partial^\alpha S_0(g)\|_\infty \leq C_\alpha$, for all $\alpha \in \mathbb{N}^n$. Next, $\|\Delta_k(g)\|_\infty \leq C2^{-kr}$, so $\|\partial^\alpha \Delta_k(g)\|_\infty \leq C2^{|\alpha|(k+1)}2^{-kr}$, by Bernstein's inequality. Since $|\alpha| > r$, we have $1 + \cdots + 2^{j(|\alpha|-r)} \leq C2^{j(|\alpha|-r)}$, which concludes the proof of Lemma 5.

We move to the proof of (3.9). We shall change the notation a little by putting $f_j = S_j(f) + t\Delta_j(f)$. We shall keep only the following properties of the f_j:

$$(3.11) \qquad \|f_{j+1} - f_j\|_\infty \leq C2^{-jr} \qquad \text{for } j \in \mathbb{N}$$

and

$$(3.12) \qquad \|\partial^\alpha f_j\|_\infty \leq C_\alpha 2^{j(|\alpha|-r)} \qquad \text{for } |\alpha| > r \text{ and } j \in \mathbb{N}.$$

We then apply

Lemma 6. *Let f_j, $j \in \mathbb{N}$, be a sequence of infinitely differentiable functions satisfying the estimates (3.11) and (3.12). Then, for any function $F \in C^\infty(\mathbb{R})$, $F(f_j)$ still satisfies (3.11) and (3.12), the constants C and C_α being replaced by new constants C' and C'_α.*

The statement is obvious for (3.11) and we shall concentrate on (3.12).

The only difficulty is that the derivatives $\partial^\alpha F(f_j)$, $|\alpha| > r$, involve derivatives of the form $\partial^\beta f_j$, with $|\beta| \le r$, to which (3.12) does not apply. The remarks which follow are designed to get round this difficulty.

First of all, we observe that (3.11) and (3.12) give

$$(3.13) \qquad \|\partial^\alpha(f_{j+1} - f_j)\|_\infty \le C_\alpha 2^{j|\alpha|} 2^{-rj} \qquad \text{for all } \alpha \in \mathbb{N}^n.$$

In our particular situation this gives no new information. However, with the hypotheses of Lemma 6, (3.13) follows by convexity from the cases $\alpha = 0$ and $|\alpha| > r$.

Thus (and from now on we may ignore (3.11))

$$(3.14) \qquad \|\partial^\beta f_j\|_\infty \le C \qquad \text{if } |\beta| < r$$

and

$$(3.15) \qquad \|\partial^\beta f_j\|_\infty \le C(1+j) \qquad \text{if } |\beta| = r.$$

We then compute $\partial^\alpha F(f_j)$ using the Faa di Bruno identity. We must find an upper bound for $\|\partial^{\beta_1} f_j\|_\infty \cdots \|\partial^{\beta_q} f_j\|_\infty$ when $\beta_1 + \cdots + \beta_q = \alpha$ and $|\alpha| > r$. To do this, we summarize (3.14), (3.15) and (3.12) as

$$(3.16) \qquad \|\partial^\beta f_j\|_\infty \le C 2^{j|\beta|(1-r/|\alpha|)} \qquad \text{for } 0 \le |\beta| \le |\alpha|, \ |\alpha| > r.$$

The identities (3.16) are less precise than (3.14), (3.15) and (3.12), except when $|\beta| = |\alpha|$, when (3.16) is the same as (3.12).

Since $|\alpha| = |\beta_1| + \cdots + |\beta_q|$, we get, as promised,

$$\|\partial^{\beta_1} f_j\|_\infty \cdots \|\partial^{\beta_q} f_j\|_\infty \le C' 2^{j|\alpha|(1-r/|\alpha|)} = C' 2^{-jr} 2^{j|\alpha|}.$$

We have now proved Theorem 2. In our intended applications, it will be the multidimensional version of Theorem 2, rather than the theorem itself, which will allow us to localize the zones of irregularity of solutions of non-linear partial differential equations.

Let $F : \mathbb{R}^N \to \mathbb{C}$ be an infinitely differentiable function of N real variables $u_1, \ldots, u_N$. We suppose that $F(0) = 0$ and let F_j denote the N partial derivatives $\partial F/\partial u_j$, $1 \le j \le N$.

Let $r = s - n/p > 0$, for some p with $1 < p < \infty$, and let $f = (f_1, \ldots, f_N)$ be an $\mathbb{R}^N$-valued function, belonging to $L^{p,s}(\mathbb{R}^n)$. Then $F(f) = F(f_1, \ldots, f_N)$ has the following expansion as a sum of para-products.

Theorem 3. *Under the above hypotheses,*

$$(3.17) \qquad F(f) = \sum_1^N \pi(F_j(f), f_j) + g,$$

where $g \in L^{p,s+r}$.

The proof of Theorem 3 is the same as that of Theorem 2 and thus left to the reader.

Before finishing this section, we give a lemma which will be useful for studying paradifferential operators.

Lemma 7. Let f_j, $j \in \mathbb{N}$, be a sequence of functions in $C^\infty(\mathbb{R}^n)$ whose behaviour as $j \to \infty$ is described by the following conditions:

(3.18) there exist $r > 0$ and a constant C_0 such that, for all $j \in \mathbb{N}$, we have $\|f_j\|_{C^r} \leq C_0$;

(3.19) there exists a family of constants C_β such that, if $|\beta| > r$, then, for all $j \in \mathbb{N}$, we have $\|\partial^\beta f_j\|_\infty \leq C_\beta 2^{j(|\beta|-r)}$.

Then we can decompose f_j as

(3.20) $$f_j = g_j + h_j,$$

where

(3.21) the Fourier transform of g_j has support contained in the ball $|\xi| \leq 2^{j-10}$,

(3.22) $$\|g_j\|_{C^r} \leq C_0', \qquad \text{for all } j \in \mathbb{N},$$

and

(3.23) $$\|\partial^\beta h_j\|_\infty \leq C_\beta' 2^{j(|\beta|-r)}, \qquad \text{for all } \beta \in \mathbb{N}^n.$$

To see this, we perform a Littlewood-Paley decomposition on f_j, writing

$$f_j = S_{j-10}(f_j) + \Delta_{j-10}(f_j) + \cdots$$

and putting

$$g_j = S_{j-10}(f_j) \qquad \text{and} \qquad h_j = \sum_{l \geq j-10} \Delta_l(f_j).$$

The estimates (3.21) and (3.22) are now obvious. For (3.23), we use (3.19); further, $\|\Delta_l(f_j)\|_\infty \leq C_m 2^{(j-l)m} 2^{-jr}$ for every integer $m \in \mathbb{N}$, which, combined with Bernstein's ineqality, gives (3.23).

We may remark that, conversely, (3.22) and (3.23) imply (3.19).

4 Paradifferential operators

Let r be a strictly positive real number which we shall keep fixed. Following Bony, we shall construct an algebra of operators, contained in G. Bourdaud's algebra, and a symbolic calculus which will enable us to invert the operators in our algebra, modulo operators which are regularizing of order r, that is, operators whose symbols lie in the class $S_{1,1}^{-r}(\mathbb{R}^n \times \mathbb{R}^n)$, which we used in the proof of Theorem 2. An operator which is regularizing of order r is a continuous mapping from $L^{p,s}(\mathbb{R}^n)$

to $L^{p,s+r}(\mathbb{R}^n)$, as long as $s + r > 0$, for $1 < p < \infty$ (Chapter 10, Section 5), and there is a similar statement for Besov spaces.

Bourdaud's algebra consists of operators T which, together with their adjoints T^*, belong to $\mathcal{O}p\,S_{1,1}^0$. This is described in Chapter 9, but has the disadvantage of not having a reasonable symbolic calculus. In a way, Bourdaud's algebra is too big and we are going to improve it by incorporating additional regularity with respect to x into the symbols we use. That regularity is measured by the fixed real number $r > 0$.

It is no extra effort to define symbols of order m, although the calculations we shall need to carry out involve only operators of order 0.

Lastly, in the definition which follows, C^r will denote the inhomogeneous Hölder space: the norm of f in C^r includes the L^∞ norm. We have written enough about these spaces in the preceding chapters for the reader to need no further details.

Definition 1. *Let r be a positive real number. A function $\sigma(x,\xi)$ belongs to the class A_r^m of symbols if $\sigma(x,\xi) \in C^\infty(\mathbb{R}^n \times \mathbb{R}^n)$ and the two following conditions hold:*

(4.1) *there exist constants $C(\alpha)$, $\alpha \in \mathbb{N}^n$, such that the functions of the variable x defined by $\partial_\xi^\alpha \sigma(x,\xi)$ all belong to $C^r(\mathbb{R}^n)$ and satisfy*
$$\|\partial_\xi^\alpha \sigma(x,\xi)\|_{C^r(\mathbf{R}^n)} \leq C(\alpha)(1+|\xi|)^{m-|\alpha|} \,;$$
(4.2) *there exist constants $C(\alpha,\beta)$, $\alpha,\beta \in \mathbb{N}^n$, such that, if $|\beta| > r$,*
$$|\partial_\xi^\alpha \partial_x^\beta \sigma(x,\xi)| \leq C(\alpha,\beta)(1+|\xi|)^{m-|\alpha|+|\beta|-r} \,.$$

In other words, we get regularity with respect to x "for free", as long as that regularity is not greater than r. After that, we have to pay.

If $r = \infty$, we are back to the usual class $S_{1,0}^0$. Using a symbol of this type, we define the operator $\sigma(x, D)$ by the usual formalism

(4.3) $\sigma(x,D)[e^{ix\cdot\xi}] = \sigma(x,\xi)e^{ix\cdot\xi} \,.$

The next definition is closer to Bony's approach ([16]).

Definition 2. *The set B_r^m consists of the symbols $\sigma(x,\xi) \in A_r^m$ which satisfy the following condition: for each fixed ξ, the support of the (partial) Fourier transform of $\sigma(x,\xi)$, regarded as a function of x, is contained in the ball $|\eta| \leq |\xi|/10$.*

We could just as well define B_r^m without specifying that $\sigma(x,\xi) \in A_r^m$, but requiring (4.1) to hold, as well as the last condition of Definition 2. Condition (4.2) then follows automatically, as can be seen by rereading the proof of Lemma 7.

The reader is invited to check that, for all real m,
$$S_{1,0}^m \subset A_r^m \subset S_{1,1}^m \subset A_r^{m+r}.$$
Clearly $A_s^m \subset A_r^m$, if $s \geq r$, and B_r^m is contained in A_r^m, as we have already remarked.

If $m = 0$, we write A_r and B_r, instead of A_r^0 and B_r^0.

As in the usual case, $\sigma(x, \xi)$ belongs to A_r^m or B_r^m, if and only if $(1 + |\xi|^2)^{-m/2}\sigma(x, \xi)$ belongs to A_r or B_r, which enables us to reduce many problems to the case $m = 0$.

The following lemma clarifies the relationship of A_r to B_r.

Lemma 8. *For a function $\sigma(x, \xi) \in C^\infty(\mathbb{R}^n \times \mathbb{R}^n)$, the two following properties are equivalent:*

(4.4) $$\sigma(x, \xi) \in A_r\,;$$

(4.5) $$\sigma(x, \xi) = \tau(x, \xi) + \rho(x, \xi)\,,$$

where $\tau(x, \xi) \in B_r$ and $\rho(x, \xi) \in S_{1,1}^{-r}$.

As we have already remarked, $S_{1,1}^{-r}$ is contained in A_r. It is then clear that (4.5) implies (4.4). Conversely, we use the notation of the Littlewood-Paley decomposition to write $1 = \phi(\xi) + \sum_0^\infty \psi(2^{-j}\xi)$ and then

(4.6) $$\sigma(x, \xi) = \sigma(x, \xi)\phi(\xi) + \sum_0^\infty \sigma_j(x, \xi)\,,$$

where $\sigma_j(x, \xi) = \sigma(x, \xi)\psi(2^{-j}\xi)$.

We then use Lemma 7 to decompose each of the functions $\sigma_j(x, \xi)$ with respect to x.

The importance of A_r lies in its being an algebra under ordinary multiplication. More precisely, we have

Lemma 9. *Let $\sigma_1(x, \xi), \ldots, \sigma_N(x, \xi)$ be N real-valued symbols belonging to A_r. For every function $F \in C^\infty(\mathbb{R}^N)$, $F(\sigma_1(x, \xi), \ldots, \sigma_N(x, \xi))$ lies in A_r.*

There is no difficulty about the proof of (4.1), while the verification of (4.2) is identical to the proof of Lemma 6, with 2^j replaced by $1 + |\xi|$.

Corollary 1. *Let $\sigma_1(x, \xi)$ and $\sigma_2(x, \xi)$ be two real- or complex-valued symbols in A_r. Suppose that there is a a constant $\delta > 0$ such that $|\sigma_2(x, \xi)| \geq \delta$ whenever $\sigma_1(x, \xi) \neq 0$. Let us define $\sigma_3(x, \xi)$ by putting $\sigma_3(x, \xi) = 0$, when $\sigma_1(x, \xi) = 0$, and $\sigma_3(x, \xi) = \sigma_1(x, \xi)/\sigma_2(x, \xi)$ otherwise. Then $\sigma_3(x, \xi)$ also belongs to A_r.*

Indeed, let $F(u, v)$ be an infinitely differentiable function of two real variables u and v, which coincides with $(u + iv)^{-1}$ when $(u^2 + v^2)^{1/2} \geq \delta$.

We let $\alpha_2(x, \xi)$ denote the real part of $\sigma_2(x, \xi)$ and let $\beta_2(x, \xi)$ denote its imaginary part. We then form $F(\alpha_2(x, \xi), \beta_2(x, \xi))$. The new symbol belongs to A_r, by Lemma 9, and the same is true for the product $\sigma_1(x, \xi)F(\alpha_2(x, \xi), \beta_2(x, \xi)) = \sigma_3(x, \xi)$.

Corollary 2. *Let $\sigma(x, \xi)$ be a symbol in A_r. Suppose that, for some $x_0 \in \mathbb{R}^n$ and $\xi_0 \in \mathbb{R}^n \setminus \{0\}$,*

$$(4.7) \qquad \liminf_{\lambda \to \infty} |\sigma(x_0, \lambda\xi_0)| > 0.$$

Then there exist an infinitely differentiable function $u(x)$, which is 1 at x_0, an infinitely differentiable function $v(\xi)$, satisfying $v(\lambda\xi) = v(\xi)$, for $|\xi| \geq 1$, $\lambda > 1$, and $v(\lambda\xi_0) = 1$, if $\lambda \geq \lambda_0 > 0$, and, lastly, a symbol $\tau(x, \xi) \in A_r$ such that

$$(4.8) \qquad \sigma(x, \xi)\tau(x, \xi) = u(x)v(\xi).$$

Indeed, let 2δ be the lim inf of (4.7). The regularity of $\sigma(x, \xi)$, given by (4.1), implies that $|\sigma(x', \xi') - \sigma(x, \xi)| \leq \varepsilon$, if $|x' - x| \leq \eta(\varepsilon)$ and $|\xi' - \xi|/|\xi| \leq \eta(\varepsilon)$. In particular, $|\sigma(x, \lambda\xi)| \geq \delta > 0$ if $|x - x_0| \leq \eta$, $|\xi - \xi_0| \leq \eta$, and $\lambda \geq R$.

We choose $u \in \mathcal{D}(\mathbb{R}^n)$ equal to 1 at x_0 and zero outside $|x - x_0| \leq \eta$. Similarly, the function $v(\xi)$ will be zero in a neighbourhood of 0 and outside a cone of revolution with vertex 0, generated by the ball $|\xi - \xi_0| \leq \eta$ with, further, $v(\lambda\xi_0) = 1$, if $\lambda \geq \lambda_0 > 0$.

It is then enough to apply Corollary 1 to the quotient $u(x)v(\xi)/\sigma(x, \xi)$.

5 The symbolic calculus for paradifferential operators

The symbolic calculus for paradifferential operators follows from the next result.

Theorem 4. *Let $\tau(x, \xi)$ be a symbol in the Hörmander class $S_{1,1}^0$ and let $\sigma(x, \xi)$ be a symbol in B_r, for some $r > 0$. Then*

$$(5.1) \qquad \tau(x, D) \circ \sigma(x, D) = \gamma(x, D) + \rho(x, D),$$

where

$$(5.2) \qquad \gamma(x, \xi) = \sum_{|\alpha| \leq r} \frac{(-i)^{|\alpha|}}{\alpha!} \partial_\xi^\alpha \tau(x, \xi) \partial_x^\alpha \sigma(x, \xi),$$

and

$$\rho(x, \xi) \in S_{1,1}^{-r}.$$

Naturally, the formula for computing $\gamma(x, \xi)$ is the same as that encountered in the usual symbolic calculus. The only difference is that we stop at order r. The classical pseudo-differential calculus can, therefore, be regarded as the limiting case of (5.1), as r tends to infinity.

To apply the theorem, we need to arrange the terms of (5.2) in a hierarchy. We have $\partial_\xi^\alpha \tau(x,\xi) \in S_{1,1}^{-|\alpha|}$ and, similarly, if τ belonged to B^r, would have $\partial_\xi^\alpha \tau(x,\xi) \in B_r^{-|\alpha|}$. But this advantage is lost immediately by the weakness of $\partial_x^\alpha \sigma(x,\xi)$, where the most we can say is that $\partial_x^\alpha \sigma(x,\xi) \in B_r^{|\alpha|}$, or $B_{r-|\alpha|}$. That is to say, in the best case (in which both σ and τ belong to B_r) all the terms in the right-hand side of (5.2) have the same order in A_r and we cannot establish a hierarchy.

On the other hand, as soon as we put these terms into the classes $S_{1,1}^m$, order reigns. If r is not an integer, $\partial_\xi^\alpha \tau(x,\xi)\partial_x^\alpha \sigma(x,\xi) \in S_{1,1}^{-|\alpha|}$, as long as $0 \le |\alpha| < r$, and the error term cannot be confused with any of the other terms. If r is an integer and if $|\alpha| = r$, we still have $\partial_x^\alpha \sigma(x,\xi) \in S_{1,1}^\varepsilon$ for every $\varepsilon > 0$, but not for $\varepsilon = 0$ (because the Hölder space C^r has to be defined in terms of the Zygmund class when $r \in \mathbb{N}$).

Let us pass to the proof of Theorem 4. It is convenient to use several well-known properties of the Wiener algebra $A(\mathbb{R}^n)$. It consists of continuous functions vanishing at infinity which are the Fourier transforms $\hat{f}$ of functions $f \in L^1(\mathbb{R}^n)$. By definition, the norm of $\hat{f}$ in $A(\mathbb{R}^n)$ is that of f in $L^1(\mathbb{R}^n)$.

Lemma 10. *If $u(x) \in A(\mathbb{R}^n)$, then, for all $t > 0$, $u(tx)$ also belongs to $A(\mathbb{R}^n)$ and has the same norm as u.*

Lemma 11. *The Sobolev space $H^s(\mathbb{R}^n)$ is contained in $A(\mathbb{R}^n)$ for $s > n/2$.*

To prove Theorem 4, we make the same calculations as in the case of the usual pseudo-differential operators. Only the estimates differ.

In particular, $\rho(x,\xi)$ is given by a classical formula which we shall recall. We let $S(\eta,\xi)$ denote the distribution which, for each fixed ξ, is the Fourier transform of $\sigma(x,\xi)$ regarded as a function of the variable x. This distribution has support in the ball $|\eta| \le |\xi|/10$, since $\sigma \in B_r$. By abuse of language, we shall write $\int \phi(\eta)S(\eta,\xi)\,d\eta$ instead of $\langle S(\cdot,\xi), \phi(\cdot)\rangle$. Under these conditions,

$$(5.3) \quad \rho(x,\xi) = \frac{1}{(2\pi)^n} \int_{\mathbb{R}^n} \Big\{\tau(x,\xi+\eta) - \sum_{|\alpha| \le r} \frac{\eta^\alpha}{\alpha!}\partial_\xi^\alpha \tau(x,\xi)\Big\} e^{i\eta\cdot x} S(\eta,\xi)\,d\eta .$$

We shall just establish that $|\rho(x,\xi)| \le C(1+|\xi|)^{-r}$, asking the reader to check that the proof of the inequalities

$$|\partial_\xi^\alpha \partial_x^\beta \rho(x,\xi)| \le C(\alpha,\beta)(1+|\xi|)^{-r+|\beta|-|\alpha|}$$

is the same.

To analyse $\rho(x, \xi)$, we once again use a Littlewood-Paley decomposition, in the form $1 = \phi(\eta) + \sum_0^\infty \psi(2^{-j}\eta)$. Exactly as in the usual calculation, ϕ is supported by $|\eta| \le 1$ and ψ has support in $1/3 \le |\eta| \le 1$. The products $S(\eta, \xi)\psi(2^{-j}\xi)$ are denoted by $S_j(\eta, \xi)$ and we put

$$T_j(x, \eta, \xi) = \{\tau(x, \xi + \eta) - \sum_{|\alpha| \le r} \frac{\eta^\alpha}{\alpha!} \partial_\xi^\alpha \tau(x, \xi)\}\psi(2^{-j}\eta).$$

Having done this, we write

(5.4) $$\rho_j(x, \xi) = \frac{1}{(2\pi)^n} \int_{\mathbb{R}^n} T_j(x, \eta, \xi) e^{i\eta \cdot x} S_j(\eta, \xi)\, d\eta.$$

Finally, we set

(5.5) $$\rho(x, \xi) = \tilde{\rho}(x, \xi) + \sum_0^J \rho_j(x, \xi),$$

where J is the smallest integer j such that $2^{j+1} > 3|\xi|/10$ and where $\tilde{\rho}(x, \xi)$ is the analogue of $\rho_0(x, \xi)$, with $\psi(\xi)$ replaced by $\phi(\xi)$.

Let $m \in \mathbb{N}$ be the integer part of r, that is, $m \le r < m+1$. We intend to verify the estimate

(5.6) $$|\rho_j(x, \xi)| \le C2^{j(m+1-r)}(1 + |\xi|)^{-m+1}.$$

The same proof will work for $\tilde{\rho}(x, \xi)$ and summing the inequalities for $0 \le j \le J$ will give $|\rho(x, \xi)| \le C(1 + |\xi|)^{-r}$.

In proving (5.6), we may forget about the variables x and ξ, which just play the rôles of parameters. We concentrate on the variable η and define $A_j \in L^1(\mathbb{R}^n)$ and $B_j \in L^\infty(\mathbb{R}^n)$ by

$$\hat{B}_j(\eta) = e^{i\eta \cdot x} S_j(\eta, \xi)$$

and

$$\hat{A}_j(\eta) = \{\tau(x, \xi + \eta) - \sum_{|\alpha| \le r} \frac{\eta^\alpha}{\alpha!} \partial_\xi^\alpha \tau(x, \xi)\}\psi(2^{-j}\eta).$$

Since $\sigma(x, \xi) \in C^r$, the characterization of the Hölder spaces by the Littlewood-Paley decomposition immediately gives $\|B_j\|_\infty \le C2^{-jr}$.

Now

$$\rho_j(x, \xi) = \frac{1}{(2\pi)^n} \int \hat{A}_j(\eta)\hat{B}_j(\eta)\, d\eta = \int A_j(u)B_j(-u)\, du,$$

so $|\rho_j(x, \xi)| \le \|A_j\|_1 \|B_j\|_\infty$.

It remains to evaluate $\|A_j\|_1$. To do this, we first use Lemma 10 and try to estimate the A norm of

$$g_j(y) = \{\tau(x, \xi + 2^j y) - \sum_{|\alpha| \le m} \frac{(2^j y)^\alpha}{\alpha!} \partial_\xi^\alpha \tau(x, \xi)\}\psi(y).$$

Recall that $j \in \mathbb{N}$ has to satisfy the condition $2^j \le 3|\xi|/10$ and therefore, we have $|\xi + 2^j y| \ge |\xi| - 2^j \ge 7|\xi|/10$ when $\psi(y) \ne 0$. We then use

the regularity of the symbol $\tau(x,\xi)$ with respect to ξ and get
$$|g_j(y)| \le C2^{j(m+1)}(1+|\xi|)^{-m-1}$$
and, similarly,
$$|\partial_y^\alpha g_j(y)| \le C_\alpha \left(2^j(1+|\xi|)^{-1}\right)^{|\alpha|+m+1}.$$
The norm of $g_j(y)$ in $H^s(\mathbb{R}^n)$ thus does not exceed
$$\left(2^j(1+|\xi|)^{-1}\right)^{m+1}$$
which gives (5.6).

As we have indicated, the estimates on the derivatives of $\rho(x,\xi)$ with respect to x and ξ are obtained similarly, giving $\rho(x,\xi) \in S_{1,1}^{-r}$.

Corollary. Let $\sigma(x,\xi)$ be a symbol in B_r. Suppose that, for some $x_0 \in \mathbb{R}^n$ and some $\xi_0 \in \mathbb{R}^n \setminus \{0\}$, we have $\liminf_{\lambda\to\infty} |\sigma(x_0,\lambda\xi_0)| > 0$. Then there exist a symbol $\tau(x,\xi)$ in the algebra A_r, and two functions $u(x)$ and $v(\xi)$ satisfying the conditions of Corollary 2 of Lemma 9, and such that
$$(5.7) \qquad \tau(x,D) \circ \sigma(x,D) = u(x)v(D) + \rho(x,D),$$
where ρ belongs to the Hörmander class $S_{1,1}^{-r}$.

To prove (5.7), we follow the usual method and look for $\tau(x,\xi)$ of the form $\tau = \tau_0 + \cdots + \tau_m$, where $\tau_q \in S_{1,1}^{-q} \cap A_r$, for $0 \le q \le m$, where m is again defined by $m \le r < m+1$. If $r = m$, the condition $\tau_m \in S_{1,1}^{-m} \cap A_r$ is replaced by $\tau_m \in S_{1,1}^{-m+\varepsilon} \cap A_r$, for all $\varepsilon > 0$.

We use Theorem 4 to compute $\tau_q(x,D) \circ \sigma(x,D)$ and get $\gamma_q(x,D) + \rho_q(x,D)$, where
$$(5.8) \quad \gamma_q(x,\xi) = \tau_q(x,\xi)\sigma(x,\xi) + \sum_{1\le|\alpha|\le r-q} \frac{(i)^{|\alpha|}}{\alpha!} \partial_\xi^\alpha \tau_q(x,\xi)\partial_x^\alpha \sigma(x,\xi)$$
and $\rho_q \in S_{1,1}^{-r}$.

The terms on the right-hand side of (5.8) belong, respectively, to $S_{1,1}^{-q}$, $S_{1,1}^{-q-1}$, and so on. When $q = 0$ and $|\alpha| = m = r$, the last terms belong to $S_{1,1}^{-m+\varepsilon}$, for $\varepsilon > 0$. Finally, when $q = m$, the right-hand side of (5.8) involves only one term, namely $\tau_m(x,\xi)\sigma(x,\xi)$.

We apply Corollary 2 of Lemma 9 to define $\tau_0(x,\xi)$, $u(x)$ and $v(\xi)$ satisfying (4.8). After this, our strategy is to amend all the error terms on the right-hand side of (5.8) by an appropriate choice of τ_{q+1}. We proceed in the simplest possible way by requiring
$$\tau_{q+1}(x,\xi)\sigma(x,\xi) + \sum_{1\le|\alpha|\le r-q} \frac{(i)^{|\alpha|}}{\alpha!} \partial_\xi^\alpha \tau_q(x,\xi)\partial_x^\alpha \sigma(x,\xi) = 0.$$
To make sure this is possible, we observe that the supports of the

symbols $\tau_q(x,\xi)$ are contained in that of the product $u(x)v(\xi)$. Since $|\sigma(x,\xi)| \geq \delta > 0$ on the support of $u(x)v(\xi)$, we may apply Corollary 1 of Lemma 9.

6 Application to non-linear partial differential equations

We shall state and prove a theorem about the regularity of solutions of non-linear partial differential equations. Since this result was obtained, Bony and his co-workers have established many others: we refer the reader to [18].

We start by relating the paraproduct operator $\pi(a,f)$, with $a(x) \in C^r(\mathbb{R}^n)$, to the class A_r of Definition 2.

The symbol of $\pi(a,f)$ is $\sigma(x,\xi) = \sum_2^\infty S_{j-2}(a)\psi(2^{-j}\xi)$. Conditions (4.1) and (4.2) are obviously satisfied.

As far as the class B_r is concerned, the support of the partial Fourier transform of $\sigma(x,\xi)$ is contained in the ball $|\eta| \leq |\xi|/2$, rather than $|\eta| \leq |\xi|/10$. We could obtain the coefficient $1/10$ by replacing $S_{j-2}(a)$ by $S_{j-5}(a)$. What would be the effect of such a modification? Nothing changes in Theorems 2 and 3. Indeed, our modification leads to error terms $\sum_5^\infty \Delta_{j-3}(a)\Delta_j(f), \ldots, \sum_5^\infty \Delta_{j-5}(a)\Delta_j(f)$, and these error terms obviously belong to $L^{p,s+r}$ if $f \in L^{p,s}$ and $a \in C^r$.

Throughout this section, we shall therefore suppose that the paraproduct is defined by $\sum_5^\infty S_{j-5}(a)\Delta_j(f)$.

We now consider a non-linear partial differential equation

$$(6.1) \qquad F(x, f(x), \ldots, \partial^\alpha f(x), \ldots) = 0\,,$$

where $|\alpha| \leq m$, $F \in C^\infty(\mathbb{R}^N)$, and $f : \mathbb{R}^n \to \mathbb{R}$.

The function F is given, m is the order of the equation, $N-1$ is the number of the $\alpha = (\alpha_1, \ldots, \alpha_n) \in \mathbb{N}^n$ of height $|\alpha| \leq m$, and f is the unknown function.

The non-linear character of this equation leads us to make a global regularity hypothesis, which will let us avoid products of distributions in (6.1). We use either $f(x) \in C^{s+m}(\mathbb{R}^n)$, with $s > 0$, or $f(x) \in L^{p,s+m}$, where $s - n/p = r > 0$. Then all the derivatives $\partial^\alpha f(x)$, $|\alpha| \leq m$, appearing in (6.1), belong to C^r and the meaning of (6.1) is as simple as possible.

Our intention is to show that $f(x)$ is much more regular, on a micro-local scale. The gain in regularity is equal to r.

The singular region is a closed set Γ in $\mathbb{R}^n \times \mathbb{R}^n \setminus \{0\}$ and the conclusion of Theorem 5 below is that $f(x)$ is micro-locally C^{r+s+m} (or $L^{p,s+m+r}$) outside the singular region.

Let us be more precise. We begin by defining Γ. In fact, $\Gamma = \Gamma(f)$ depends on the solution f of (6.1), whose regularity we are investigating, and is the set of $(x, \xi) \in \mathbb{R}^n \times \mathbb{R}^n \setminus \{0\}$ such that

$$(6.2) \qquad \sum_{|\alpha|=m} (i\xi)^\alpha c_\alpha(x) = 0 \,,$$

where

$$(6.3) \qquad c_\alpha(x) = \frac{\partial F}{\partial u_\alpha}(x, f(x), \dots, \partial^\beta f(x), \dots) \,.$$

We have used $u_\alpha \in \mathbb{R}$ to denote the variables on which F depends: u_α is replaced by $\partial^\alpha f$ in (6.1).

Theorem 5. *Every solution f of (6.1) which is globally in the class $L^{p,s}$, where $s = m + n/p + r$, $r > 0$, $1 < p < \infty$, is micro-locally in the class $L^{p,s+r}$ outside Γ.*

We observe that Γ is a conical subset of $\mathbb{R}^n \times \mathbb{R}^n \setminus \{0\}$: if $(x, \xi) \in \Gamma$, the same holds for $(x, \lambda\xi)$, for all $\lambda > 0$ (indeed, for all $\lambda \in \mathbb{R}$). We then consider a point $(x_0, \xi_0) \notin \Gamma$, $\xi_0 \neq 0$, and take functions $u(x)$, $v(\xi)$ similar to those of Corollary 2 of Lemma 9, such that the support of $u(x)v(\xi)$ is disjoint from Γ. The conclusion of Theorem 5 means that the operator $u(x)v(D)$ transforms f into a function belonging to $L^{p,s+r}$.

To prove Theorem 5, we first use Theorem 3. For all $f \in L^{p,s}$, we get the identity

$$(6.4) \qquad F(x, f(x), \dots, \partial^\alpha f(x), \dots) = \sum_{|\alpha|\leq m} L_\alpha(\partial^\alpha f)(x) + g(x) \,,$$

where $g \in L^{p,s+r}$ and $L_\alpha(f) = \pi(c_\alpha(x), f)$.

If f is the solution of (6.1), then

$$(6.5) \qquad \sum_{|\alpha|\leq m} L_\alpha(\partial^\alpha f) = -g \,.$$

We shall use Theorem 4. Put $L = \sum_{|\alpha|\leq m} \partial^\alpha (I - \Delta)^{-m/2}$. This operator is in $\mathcal{O}p\, B_r$, because of the change we made to the definition of the paraproduct. The symbol of L is

$$(6.6) \qquad \lambda(x, \xi) = \sum_{|\alpha|\leq m}\sum_{j\geq 5} S_{j-5}(c_\alpha)(x)\psi(2^{-j}\xi)(i\xi)^\alpha(1 + |\xi|^2)^{-m/2} \,.$$

We suppose that $(x_0, \xi_0) \notin \Gamma$, $\xi_0 \neq 0$, and show that

$$(6.7) \qquad \liminf_{t\to\infty} |\lambda(x_0, t\xi_0)| > 0 \,.$$

Observe that

$$\sum_N^\infty \psi(2^{-j}\xi) = 1 - \phi(2^{-N}\xi) = 1 \qquad \text{if } |\xi| \geq 2^N .$$

Furthermore $|S_j(c_\alpha(x)) - c_\alpha(x)| \leq C2^{-jr}$, since $c_\alpha(x) \in C^r$. These properties imply that, if $\xi \neq 0$,

(6.8) $$\lim_{t \to \infty} \sum_{j \geq 5} S_{j-5}(c_\alpha(x))\psi(2^{-j}t\xi) = c_\alpha(x).$$

Finally, if $\xi \neq 0$,

$$\lim_{t \to \infty} \lambda(x, t\xi) = \sum_{|\alpha|=m} c_\alpha(x)(i\xi)^\alpha |\xi|^{-m}$$

and (6.7) is a consequence of the definition of Γ given by (6.2).

Having made these remarks, we return to (6.5), which we write as $L(h) = -g$, where $h = (I - \Delta)^{m/2}f$. The corollary of Theorem 4 gives $u(x)v(D)h \in L^{p,s+r-m}$ and we get $u(x)v(D)f \in L^{p,s+r}$, as claimed.

The same arguments give the analogue of Theorem 5, when the regularity is measured in terms of Hölder spaces.

7 Paraproducts and wavelets

Consider the linear operator T_a which takes a function or distribution f to the paraproduct $\pi(a, f)$ and suppose that $a \in C^r(\mathbb{R}^n)$, where $0 < r < 1$. We end this volume by rediscovering wavelets as eigenfunctions of a realisation $\tilde{T}_a$ of T_a, with $\tilde{T}_a - T_a$ belonging to $Op\,S_{1,1}^{-r}$.

More precisely, let ψ_λ, $\lambda \in \Lambda$, be the wavelets arising from a multiresolution approximation as in Chapter 3 of *Wavelets and Operators*. We have $\Lambda = \bigcup_{-\infty}^{\infty} \Lambda_j$, where $\Lambda_j = 2^{-j-1}\mathbb{Z}^n \setminus 2^{-j}\mathbb{Z}^n$. In fact, we shall only use $j \geq 0$, which means that we need to complete the orthonormal system ψ_λ, $\lambda \in \Lambda_j$, $j \geq 0$, by adding the sequence ϕ_k, $k \in \mathbb{Z}^n$, defined by $\phi_k(x) = \phi(x - k)$, $k \in \mathbb{Z}^n$.

Let $a(x)$ be a bounded, continuous function on $\mathbb{R}^n$. We define the operator $\tilde{T}_a : L^2(\mathbb{R}^n) \to L^2(\mathbb{R}^n)$ by

(7.1) $$\tilde{T}_a(\psi_\lambda) = a(\lambda)\psi_\lambda, \qquad \lambda \in \Lambda_j, \quad j \in \mathbb{N},$$

and

(7.2) $$\tilde{T}_a(\phi_k) = a(k)\phi_k.$$

The collection of the operators $\tilde{T}_a$ is a commutative, self-adjoint subalgebra of $\mathcal{L}(L^2(\mathbb{R}^n), L^2(\mathbb{R}^n))$. The operator norm $\|\tilde{T}_a\|$ is $\sup_{\lambda \in \Lambda} |a(\lambda)| = \sup_{x \in \mathbb{R}^n} |a(x)| = \|a\|_\infty$ and it follows that the algebra consisting of all the operators $\tilde{T}_a$ is a Banach algebra.

Theorem 6. *If $a(x) \in C^r(\mathbb{R}^n)$ and if $0 < r < 1$, then $\tilde{T}_a - T_a \in Op\,S_{1,1}^{-r}$.*

The proof of Theorem 6 consists, to begin with, of comparing T_a with the operator $\mathcal{T}_a$ defined by $\mathcal{T}_a(f) = \sum_0^\infty E_j(a)D_j(f)$, where, following the notation of Chapter 2 of *Wavelets and Operators*, the operators E_j and D_j come from the Littlewood-Paley multiresolution approximation. But, to do this, we must use yet another variant of the operator T_a.

We observe that T_a would be a martingale transform operator if, instead of the Littlewood-Paley multiresolution approximation, we used the multiresolution approximation for which the E_j are the conditional expectation operators associated with a dyadic martingale.

This modification of the paraproduct operator does not require the condition $r < 1$. This will be needed when we try to replace the function $E_j(a)$ by its sampling on $2^{-j}\mathbb{Z}^n$.

Even before we compare T_a with $\mathcal{T}_a$, we must make a preliminary alteration. We let u and v be functions in $\mathcal{S}(\mathbb{R}^n)$ whose Fourier transforms $\hat{u}(\xi)$ and $\hat{v}(\xi)$ satisfy $\hat{u}(\xi) = \hat{v}(\xi) = 1$ in a neighbourhood of 0, and $\hat{u}(\xi) = \hat{v}(\xi) = 0$, for sufficiently large $|\xi|$. We then put $u_j(x) = 2^{nj}u(2^j x)$ and let U_j denote the operator of convolution with u_j. We define v_j and V_j similarly.

The first modification to the definition of the paraproduct consists of defining

$$(7.3) \qquad \tilde{\pi}(a, f) = \sum_0^\infty U_j(a)(V_{j+1}(f) - V_j(f)) \,.$$

To compare $\tilde{\pi}(a, f)$ to the "genuine" paraproduct, we look at the operator $R_a(f) = \tilde{\pi}(a, f) - \pi(a, f)$. We shall show that

$$(7.4) \qquad R_a \in \mathcal{O}p\, S_{1,1}^{-r} \qquad \text{if } a \in C^r.$$

Let ρ denote the difference $v - \phi$. The Fourier transform of ρ has compact support and vanishes in a neighbourhood of 0. Denoting the operator of convolution with $2^{nj}\rho(2^j x)$ by R_j, we see that $\sum_3^\infty \Delta_{j-3}(a)R_j(f)$ is an error term. Indeed, its symbol is $\sum_3^\infty \Delta_{j-3}(a)\hat{\rho}(2^{-j}\xi)$ and this belongs to $S_{1,1}^{-r}$, because $\|\Delta_j(a)\|_\infty \leq C2^{-jr}$.

Now put $\pi_1(a, f) = \sum_2^\infty S_{j-2}(a)(V_{j+1}(f) - V_j(f))$ and let us verify that, up to an operator whose symbol lies in $S_{1,1}^{-r}$, π_1 and π coincide. Their difference is exactly $\sum_2^\infty S_{j-2}(a)(R_{j+1}(f) - R_j(f))$, a series to which we apply the Abel transform. Modulo a trivial operator we get the error term $-\sum_3^\infty \Delta_{j-3}(a)R_j(f)$.

After this, the comparison of $\pi_1(a, f)$ with $\tilde{\pi}(a, f)$ is easier. Their difference belongs to $\mathcal{O}p\, S_{1,1}^{-r}$.

We return to the wavelets and to Proposition 9 of *Wavelets and Operators*, Chapter 2. To simplify the notation a little, we restrict our attention to dimension 1. With the notation of that chapter, we have

$$(7.5) \qquad E_j = S_j + R_j^+ + R_j^-$$

where S_j is the operator of convolution with $2^j\theta(2^jx)$, and $\hat{\theta}(\xi) = (\hat{\phi}(\xi))^2 = 1$, on $[-2\pi/3, 2\pi/3]$ and $= 0$, if $|\xi| \geq 4\pi/3$.

The operators R_j^+ and R_j^- are of the form M_jN_j, where N_j is an operator of convolution with $2^j\eta(2^jx)$ or $2^j\zeta(2^jx)$. Here η and ζ belong to $\mathcal{S}(\mathbb{R})$ and their Fourier transforms vanish in a neighbourhood of 0 and have compact supports. On the other hand, M_j is the operator of pointwise multiplication by $e^{\pm2\pi i2^jx}$. With this notation, we have

Lemma 12. *If $a(x)$ is in $C^r(\mathbb{R})$, the paraproduct $\pi(a, f)$ may be defined by $\sum_0^\infty E_j(a)D_j(f)$, modulo an operator belonging to $Op\,S_{1,1}^{-r}$.*

Here, again, we replace $E_j(a)$ by the principal term of (7.5), namely $S_j(a)$. As for the error terms, we have $\|R_j^\pm(a)\|_\infty \leq C2^{-jr}$, because $a(x) \in C^r$. We get bounds for the derivatives of $R_j^\pm$ using Bernstein's inequalities.

This leaves $\sum_0^\infty S_j(a)D_j(f)$. We have

$$D_j = E_{j+1} - E_j = S_{j+1} - S_j + R_{j+1}^+ - R_j^+ + R_{j+1}^- - R_j^- .$$

The principal term is $\sum_0^\infty S_j(a)(S_{j+1} - S_j)(f)$, that is, one of the forms of $\tilde{\pi}(a, f)$. The two error terms are dealt with by Abel's transform.

We let $\Pi(a, f)$ denote the form of the paraproduct given by Lemma 12. For this, we have

(7.6) $\Pi(a, \psi_\lambda) = E_j(a)\psi_\lambda$,   if $\lambda \in \Lambda_j$, $j \in \mathbb{N}$,

and

(7.7) $\Pi(a, \phi_k) = 0$,   for $k \in \mathbb{Z}$.

We write $a_j(x) = E_j(a)(x)$ and suppose, as before, that $0 < r < 1$. We then let $\tilde{\Pi}_a$ denote the diagonal operator with respect to the wavelet basis, defined by

$$\tilde{\Pi}_a(\psi_\lambda) = a_j(\lambda)\psi_\lambda\,, \text{ if } \lambda \in \Lambda_j, \quad \text{and} \quad \tilde{\Pi}_a(\phi_k) = 0\,, \text{ if } k \in \mathbb{Z}.$$

Then

$$\Pi(a, \psi_\lambda) - \tilde{\Pi}_a(\psi_\lambda) = (a_j(x) - a_j(\lambda))\psi_\lambda(x) = q_{j,k}(2^jx)\psi_\lambda(x) .$$

Since $a(x) \in C^r$, we have $|q_{j,k}(u)| \leq C2^{-jr}|u - (k + 1/2)|^r$ and, since $0 < r < 1$, we get, for every integer $l \geq 1$,

(7.8) $$\left|\left(\frac{d}{du}\right)^l q_{j,k}(u)\right| \leq C_l 2^{-rj} .$$

The symbol of the difference $\Pi(a, \cdot) - \tilde{\Pi}_a$ is $\sum_0^\infty \gamma_j(2^jx, 2^{-j}\xi)\bar{\hat{\psi}}(2^{-j}\xi)$, where

(7.9) $$\gamma_j(u, v) = \sum_{k=-\infty}^\infty q_{j,k}(u)\psi(u - k)e^{-i(u-k)v} .$$

It is easy to see that

$$\left| \left(\frac{\partial}{\partial u} \right)^l \left(\frac{\partial}{\partial v} \right)^m \gamma_j(u,v) \right| \le C(l,m) 2^{-rj}$$

and it follows that $\Pi(a,\cdot) - \tilde{\Pi}_a \in \mathcal{O}p\, S_{1,1}^{-r}$.

The final modification is to replace $a_j(\lambda)$ by $a(\lambda)$, which is immediate, because $\|a - a_j\|_\infty \le C2^{-rj}$. We have proved Theorem 6.

If $m \le r < m+1$, we would have to replace the operator $\tilde{T}_a$ of Theorem 6 by

$$\tilde{T}_a^{(m)}(\psi_\lambda)(x) = \left(a(\lambda) + \cdots + \frac{(x-\lambda)^m}{m!} a^{(m)}(\lambda) \right) \psi_\lambda(x)$$

and a similar proof would give $T_a - \tilde{T}_a^{(m)} \in \mathcal{O}p\, S_{1,1}^{-r}$.

References and Bibliography

[1] AHLFORS, L. Zur Theorie der Überlagerungsflächen. *Acta Math.*, **65**, 1935, 157-194.

[2] ARSAC, J. *Transformation de Fourier et théorie des distributions.* Dunod, Paris, 1961.

[3] AUSCHER, P. Wavelets on chord-arc curves. *Wavelets. Time-frequency methods and phase-space. Proc. Inter. Conf. Marseilles, 1987.* Combes, J.M., Grossman, A., and Tchamitchian, P., eds., Springer-Verlag, 1990, 253-258.

[4] AUBIN, J.P. *Approximation of elliptic boundary-value problems.* Pure and Applied Mathematics, Krieger Publishing Co., Huntington, N.Y., 1980.

[5] BAERNSTEIN II, A. and SAWYER, E.T. Embedding and multiplier theorems for $H^p(\mathbb{R}^n)$. *Memoirs of the Amer. Math. Soc.*, **53**, 1985.

[6] BALIAN, R. Un principe d'incertitude fort en théorie du signal ou en mécanique. *C.R. Acad. Sci. Paris*, **292**, *Série II*, 1981, 1357-1361.

[7] BALSLEV, E., GROSSMANN, A. and PAUL, T. A characterization of dilation analytic operators. *Ann. I.H.P., Physique Théorique*, **45**, 1986.

[8] BATTLE, G. A block spin construction of ondelettes, Part I: Lemarié functions. *Comm. Math. Phys.* **110**, 1987, 601-615.

[9] BATTLE, G. A block spin construction of ondelettes, Part II: the QFT connection. *Comm. Math. Phys.* **114**, 1988, 93-102.

[10] BATTLE, G. and FEDERBUSH, P. Ondelettes and phase cluster expansion, a vindication. *Comm. Math. Phys.* **109**, 1987, 417-419.

[11] BEALS, R. Characterization of pseudo-differential operators and applications. *Duke Math. J.* **44**, 1977, 45-57.

[12] BENEDEK, A., CALDERÓN, A. and PANZONE, R. Convolution operators on Banach space valued functions. *Proc. Nat. Acad. Sci. U.S.A.*, **48**, 1962, 356-365.

[13] BERKSON, E. On the structure of the graph of the Franklin analyzing wavelet. *Analysis at Urbana, I.* Berkson, E., Peck, T. and Uhl, J., eds., London Math. Soc. Lecture Note Series, **137**, Cambridge University Press, Cambridge, 1989, 366-394.

[14] BESOV, O.V. Théorèmes de plongement des espaces fonctionnels. *Actes du Congrès Int. Math., Nice, 1970.* Gauthier-Villars, Paris, 1971, II, 467-463.

[15] BEURLING, A. Construction and analysis of some convolution algebras. *Ann. Inst. Fourier (Grenoble)*, **14**, 1962, 1-32.

[16] BONY, J.M. Calcul symbolique et propagation des singularités pour les équations aux dérivés partielles non-linéaires. *Ann. Scient. E.N.S.*, **14**, 1981, 209-246.

[17] BONY, J.M. Propagation et interaction des singularités pour les solutions des équations aux dérivés partielles non-linéaires. *Proc. Int. Congress Math., Warszawa, 1983.* North-Holland, Amsterdam, 1984, 1133-1147.

[18] BONY, J.M. Interaction des singularités pour les équations aux dérivés partielles non-linéaires. *Séminaire E.D.P.*, Centre de Mathématique, École Polytechnique, 91128 Palaiseau, 1979-80, no. 22, 1981-82, no. 2 and 1983-84, no. 10.

[19] BOOLE, G. On the comparison of transcendents with certain applications to the theory of definite integrals. *Phil. Trans. Roy. Soc.*, **47**, 1857, 745-803.

[20] BOCHKARIEV, S.V. Existence of bases in the space of analytic functions and some properties of the Franklin system. *Mat. Sbornik*, **98**, 1974, 3-18.

[21] BOURDAUD, G. *Sur les opérateurs pseudo-différentiels à coéfficients peu réguliers.* Université de Paris VII, Mathématique, tour 45-55, 5ième étage, 2 Place Jussieu, 75251 Paris Cedex 05.

[22] BOURDAUD, G. Réalisations des espaces de Besov homogènes. *Arkiv för Mat.*, **26**, 1988, 41-54.

[23] BOURDAUD, G. Localisation et multiplicateurs des espaces de Sobolev homogènes. *Manuscripta. Math.*, **60**, 1988, 93-103.

[24] BOURGAIN, J. On the L^p-bounds for maximal functions associated to convex bodies. *Israel J. Math.*, **54**, 1986, 307-316.

[25] BOURGAIN, J. Geometry of Banach spaces and harmonic analysis. *Proc. Int. Congress Math., Berkeley, Calif., 1986.* American Mathematical Society, Providence, R.I., 1987, 871-878.

[26] BOURGAIN, J. Extension of a result of Benedek, Calderon and Panzone. *Arkiv för Math.*, **22**, 1984, 91-95.

[27] BOURGAIN, J. Some remarks on Banach spaces in which martingale difference sequences are unconditional. *Arkiv för Math.*, **21**, 1983, 163-168.

[28] BOURGAIN, J. On square functions on the trigonometric system. *Bull. Soc. Math. Belg., Sér. B*, **37**, 1, 1985, 20-26.

[29] BOURGAIN, J. Vector-valued singular integrals and the H^1-BMO duality. *Probability theory and harmonic analysis.* Chao, J.-A. and Wojczinski, W.A., eds., Marcel Dekker, New York, 1986, 1-19.

[30] BURKHOLDER, D.L. A geometric condition that implies the existence of certain singular integrals of Banach space valued functions. *Conference in harmonic analysis in honor of Antoni Zygmund.* Beckner, W., Calderón, A.P., Fefferman, R. and Jones, P.W., eds., Wadsworth, Belmont, Calif., 1983, I, 270-286.

[31] BURKHOLDER, D.L. Martingale theory and harmonic analysis in Euclidean spaces. *Harmonic analysis in Euclidean spaces. Proc. Symp. Pure Math.*, **35**, 1979, 2, 283-301.

[32] BURKHOLDER, D.L. Martingales and Fourier analysis in Banach spaces. *Probability and analysis, Varenna (Como), 1985.* Lecture Notes in Mathematics, **1206**, Springer-Verlag, Berlin, 1986, 61-108.

[33] BURKHOLDER, D. A geometric characterization of Banach spaces in which martingale difference sequences are unconditional. *Ann. Prob., Vol.* **9**, **6**, 1981, 997-1011.

[34] BURKHOLDER, D., GUNDY, R. and SILVERSTEIN, M. A maximal function characterization of the class H^p. *Trans. Amer. Math. Soc.*, **157**, 1971, 137-153.

[35] CALDERÓN, A.P. Intermediate spaces and interpolation, the complex method. *Studia Math.*, **24**, 1964, 113-190.

[36] CALDERÓN, A.P. Uniqueness in the Cauchy problem for partial differential equations. *Amer. J. Math.*, **80**, 1958, 15-36.

[37] CALDERÓN, A.P. Algebra of singular integral operators. *Singular Integrals. Proc. Symp. Pure Math.*, **10**, 1967, 18-55.

[38] CALDERÓN, A.P. Cauchy integrals on Lipschitz curves and related operators. *Proc. Nat. Acad. Sci. U.S.A.*, **74**, 1977, 1324-1327.

[39] CALDERÓN, A.P. Commutators, singular integrals on Lipschitz curves and applications. *Proc. Int. Congress Math., Helsinki, 1978.* Academia Scientiarum Fennica, Helsinki, 1980, 85-96.

[40] CALDERÓN, A.P. Boundary value problems for the Laplace equation in Lipschitzian domains. *Recent progress in Fourier analysis.* North-Holland Math. Studies, **111**, Peral, I. and Rubio de Francia, J.L., eds., North-Holland, Amsterdam 1983, 33-49.

[41] CALDERÓN, A.P. and TORCHINSKY, A. Parabolic maximal functions associated with a distribution. *Advances in Math.*, **16**, 1975, 1-63 and **24**, 1977, 101-171.

[42] CALDERÓN, A.P. and ZYGMUND, A. On the existence of certain singular integrals. *Acta Math.*, **88**, 85-139.

[43] CALDERÓN, A.P. and ZYGMUND, A. Singular integrals and periodic functions. *Studia Math.*, **14**, 1954, 249-271.

[44] CALDERÓN, A.P. and ZYGMUND, A. On singular integrals. *Amer. J. Math.*, **78**, 1956, 289-309 and 310-320.

[45] CALDERÓN, A.P. and ZYGMUND, A. Singular integral operators and differential equations. *Amer. J. Math.*, **79**, 1957, 901-921.

[46] CARLESON, L. An explicit unconditional basis in H^1. *Bull. des Sciences Math.*, **104**, 1980, 405-416.

[47] CARLESON, L. Interpolation of bounded analytic functions and the corona problem. *Ann. of Math.*, **76**, 1962, 547-559.

[48] CARLESON, L. Two remarks on H^1 and BMO. *Analyse harmonique*. Publications mathématiques d'Orsay, **164**, Université de Paris-Sud, Département de mathématiques, 91405 Orsay, 1975.

[49] CHANG, A. Two remarks about H^1 and BMO on the bidisc. *Conference in harmonic analysis in honor of Antoni Zygmund*. Beckner, W., Calderón, A.P., Fefferman, R. and Jones, P.W., eds., Wadsworth, Belmont, Calif., 1983, II, 373-393.

[50] CHANG, A. and CIESIELSKI, Z. Spline characterizations of H^1. *Studia Math.*, **75**, 1983, 183-192.

[51] CHANG, S.Y. and FEFFERMAN, R. A continuous version of the duality of H^1 with BMO. *Ann. of Math.*, **112**, 1980, 179-201.

[52] CHRIST, M. Weighted norm inequalities and Schur's lemma. *Studia Math.*, **78**, 1984, 309-319.

[53] CHRIST, M. and JOURNÉ, J.L. Polynomial growth estimates for multilinear singular integral operators. *Acta Math.*, **159**, 1987, 51-80.

[54] CHRIST, M. and RUBIO DE FRANCIA, J.L. Weak type (1,1)-bounds for rough operators. *Inv. Math.*, **93**, 1988, 225-237.

[55] CIESIELSKI, Z. Properties of the orthonormal Franklin system. *Studia Math.*, **23**, 1963, 141-157 and **27**, 1966, 289-323.

[56] CIESIELSKI, Z. Bases and approximations by splines. *Proc. Int. Congress Math., Vancouver, 1974*, II, 47-51.

[57] CIESIELSKI, Z. Haar orthogonal functions in analysis and probability. *Alfred Haar memorial conference*. Colloq. Math. Soc. János Bolyai, Budapest, 1985, 25-26.

[58] CIESIELSKI, Z. and FIGIEL,T. Spline approximations and Besov spaces on compact manifolds. *Studia Math.*, **75**, 1982, 13-16.

[59] CIESIELSKI, Z. and FIGIEL,T. Spline bases in classical function spaces on compact manifolds. *Studia Math.*, **76**, 1983, 95-136.

[60] COIFMAN, R.R. A real variable characerization of H^p. *Studia Math.*, **51**, 1974, 269-274

[61] COIFMAN, R.R., DAVID, G. and MEYER, Y. La solution des conjectures de Calderón. *Advances in Math.*, **48**, 1983, 144-148.

[62] COIFMAN, R.R., DENG, D.G. and MEYER, Y. Domaine de la racine carrée de certains opérateurs différentiels accrétifs. *Ann. Inst. Fourier (Grenoble)*, **33**, 2, 1983, 123-134.

[63] COIFMAN, R.R. and FEFFERMAN, C. Weighted norm inequalities for maximal functions and singular integrals. *Studia Math.*, **54**, 1974, 241-250.

[64] COIFMAN, R.R., JONES, P.W. and SEMMES, S. Two elementary proofs of the L^2-boundedness of the Cauchy integral on Lipschitz curves. *J. Amer. Math. Soc.*, **2**, 3, 1989, 553-564.

[65] COIFMAN, R.R., McINTOSH, A. and MEYER, Y. L'intégrale de Cauchy définit un opérateur borné sur les courbes Lipschitziennes. *Ann. of Math.*, **116**, 1982, 361-387.

[66] COIFMAN, R.R., McINTOSH, A. and MEYER, Y. The Hilbert transform on Lipschitz curves. *Miniconference on partial differential equations, Canberra, July 9-10, 1981*. Price, P.F., Simon, L.M. and Trudinger, N.S., eds.

[67] COIFMAN, R.R. and MEYER, Y. Lavrentiev curves and conformal mappings. *Institut Mittag-Leffler, Report no.* **5**, 1983.

[68] COIFMAN, R.R. and MEYER, Y. Au delà des opérateurs pseudo-différentiels. *Astérisque,* **57**, 1978.

[69] COIFMAN, R.R. and MEYER, Y. Fourier analysis of multilinear convolutions., Calderón's theorem and analysis on Lipschitz curves. *Euclidean harmonic analysis, proceedings.* Lecture Notes in Math., **779**, Springer-Verlag, Berlin, 1979, 109-122.

[70] COIFMAN, R.R. and MEYER, Y. Non-linear harmonic analysis, operator theory and P.D.E., *Beijing Lectures in Harmonic Analysis.* Stein, E.M., ed., *Ann. of Math. Studies,* **112**, Princeton University Press, Princeton, N.J., 1986, 3-45.

[71] COIFMAN, R.R., MEYER, Y. and STEIN, E.M. Some new function spaces and their applications to harmonic analysis. *J. Funct. Anal.,* **62**, 1985, 304-335.

[72] COIFMAN, R.R., MEYER, Y. and STEIN, E.M. Un nouvel espace fonctionnel adapté à l'étude des opérateurs définis par des intégrales singulières. *Harmonic Analysis, Cortona 1982.* Lecture Notes in Math., **992**, Springer-Verlag, Berlin, 1983, 1-15.

[73] COIFMAN, R.R., ROCHBERG, R. Representation theorems for holomorphic and harmonic functions in L^p. TAIBLESON, M.H. and WEISS, G. The molecular characterization of certain Hardy spaces. *Astérisque* **77**, 1980.

[74] COIFMAN, R.R., ROCHBERG, R. and WEISS, G. Factorization theorems for Hardy spaces in several complex variables. *Ann. of Math.,* **103**, 1976, 611-635.

[75] COIFMAN, R.R. and WEISS, G. Extensions of Hardy spaces and their use in analysis. *Bull. Amer. Math. Soc.,* **83**, 1977, 569-645.

[76] COIFMAN, R.R. and WEISS, G. *Analyse harmonique non commutative sur certains espaces homogènes.* Lecture Notes in Math., **242**, Springer-Verlag, Berlin, 1971.

[77] COIFMAN, R.R. and WEISS, G. Transference methods in analysis. *Regional Conference Series in Mathematics, no.* **31**, American Mathematical Society, Providence, R.I.

[78] CORDOBA, A. Maximal functions, covering lemmas and Fourier multipliers. *Harmonic analysis in Euclidean spaces. Proc. Symp. Pure Math.,* **35**, 1979, I, 29-50.

[79] CORDOBA, A. and FEFFERMAN, R. A geometric proof of the strong maximal theorem. *Ann. of Math.,* **102**, 1975, 95-100.

[80] COWLING, M. Harmonic analysis on semigroups. *Ann. of Math.,* **117**, 1983, 267-283.

[81] DAHLBERG, B. Estimates of harmonic measures. *Arch. for Rat. Mech. and Anal.,* **65**, 1977, 275-288.

[82] DAHLBERG, B. Real analysis and potential theory. *Proc. Int. Congress Math., Warszawa, 1983.* North-Holland, Amsterdam, 1984, 2, 953-959.

[83] DAHLBERG, B. Poisson semigroups and singular integrals. *Proc. Amer. Math. Soc.,* **97**, 1986, 41-48.

[84] DAHLBERG, B., JERISON, D. and KENIG, C. Area integral estimates for elliptic differential operators with non-smooth coefficients. *Arkiv för Mat.,* **22**, 1984, 97-108.

[85] DAHLBERG, B. and KENIG, C. The L^p Neumann problem for Laplace's equation on Lipschitz domains. *Ann. of Math.,* **125**, 1987, 435-465.

[86] DAHLBERG, B., KENIG, C. and VERCHOTA, G. The Dirichlet problem for the bi-Laplacian on Lipschitz domains. *Ann. Inst. Fourier (Grenoble),* **36**, 1986, 3, 109-135.

[87] DAUBECHIES, I. The wavelet transform, time-frequency localization and signal analysis. *IEEE Trans. Inf. Theory,* **36**, 1990, 961-105.

[88] DAUBECHIES, I. Orthonormal basis of compactly supported wavelets. *Comm. Pure Appl. Math.,* **46**, 1988, 909-996.

[89] DAUBECHIES, I., GROSSMANN, A. and MEYER, Y. Painless nonorthonormal expansions. *J. Math. Phys.,* **27**, 5, May 1986, 1271-1283.

[90] DAUBECHIES, I. and PAUL, T. Wavelets and applications. *Proc. VIIIth Int. Congress on Math. Phys.* Mebkhout, M. and Seneor, R., eds., World Scientific Publishers, Teaneck, N.J., 1987.

[91] DAUBECHIES, I., KLAUDER, J.R. and PAUL, T. Wiener measures for path integrals with affine kinematic variables. *J. Math. Phys.,* **28**, 1987, 85-102.

[92] DAVID, G. Opérateurs de Calderon-Zygmund. *Proc. Int. Congress Math., Berkeley, Calif., 1986.* American Mathematical Society, Providence, R.I., 1987, 890-899.

[93] DAVID, G. Opérateurs intégraux singuliers sur certaines courbes du plan complexe. *Ann. Sci. E.N.S.,* **17**, 1984, 157-189.

[94] DAVID, G. *A lower estimate for the norm of the Cauchy integral operator on Lipschitz curves.* Preprint.

[95] DAVID, G. Opérateurs d'intégrale singulière sur les surfaces régulières. *Ann. Sci. E.N.S.,* **21**, 1988, 225-258.

[96] DAVID, G. and JOURNÉ, J.L. A boundedness criterion for generalized Calderón-Zygmund operators. *Ann. of Math.,* **120**, 1984, 371-397.

[97] DAVID, G., JOURNÉ, J.L. and SEMMES, S. Opérateurs de Calderón-Zygmund, fonctions para-accrétives et interpolation. *Revista Mat. Ibero-Americana,* **1**, 1985, 4, 1-56.

[98] DAVID, G., JOURNÉ, J.L. and SEMMES, S. *Calderón-Zygmund operators, para-accretive functions, and interpolation.* Preprint, Thèse, Centre de Mathématique, Centre de Mathématique, École Polytechnique, 91128 Palaiseau and Dept. Math., Yale University, Newhaven, Conn.

[99] DAVID, G. and SEMMES, S. L'opérateur défini par PV $\int (|A(x) - A(y)|/ |x - y|)(x - y)^{-1} f(y)\, dy$ est borné sur $L^2(\mathbb{R})$ lorsque A est Lipschitzienne. *C.R. Acad. Sci., Paris,* **303**, Série I, 1986, 499-502.

[100] DESLAURIERS, G. and DUBUC, S. Dyadic interpolation. *Fractals, non-integral dimensions, and applications.* Cherbit, G., ed., Masson, Paris, 1987 and Wiley, Chichester, 1991, 27-35.

[101] DUBUC, S. Interpolation through an iterative scheme. *J. Math. Anal. and Appl.,* **114**, 1, 1986, 185-204.

[102] DUOANDIKOETXEA, J. and RUBIO DE FRANCIA, J.L. Maximal and singular integral operators via Fourier transform estimates. *Inv. Math.*, **84**, 1986, 541-561.

[103] DUREN, P. *Theory of H^p spaces.* Academic Press, New York, 1970.

[104] FABES, E., JODEIT, M. and RIVIÈRE, N. Potential techniques for boundary value problems on C^1 domains. *Acta Math.*, **141**, 1978, 165-186.

[105] FABES, E., JERISON, D. and KENIG, C. Multilinear Littlewood-Paley estimates with applications to partial differential equations. *Proc. Nat. Acad. Sci., U.S.A.*, **79**, 1982, 5746-5750.

[106] FABES, E., JERISON, D. and KENIG, C. Multilinear square functions and partial differential equations. *Amer. J. Math.*, **107**, 1985, 1325-1367.

[107] FEDERBUSH, P. Quantum theory in ninety minutes. *Bull. Amer. Math. Soc.*, **17**, 1987, 93-103.

[108] FEFFERMAN, C. Recent progress in classical Fourier analysis. *Proc. Int. Congress Math., Vancouver, 1974.* I, 95-118.

[109] FEFFERMAN, C. and STEIN, E.M. H^p spaces of several variables. *Acta Math.*, **129**, 1972, 137-193.

[110] FEFFERMAN, R. Multiparameter Fourier analysis. *Beijing lectures in harmonic analysis.* Stein, E.M., ed., *Ann. of Math. Studies*, **112**, Princeton University Press, Princeton, N.J., 1986, 47-130.

[111] FEFFERMAN, R. Bounded mean oscillation on the polydisk. *Ann. of Math.*, **110**, 1979, 395-406.

[112] FEFFERMAN, R. Calderón-Zygmund theory for product domains H^p spaces. *Proc. Nat. Acad. Sci., U.S.A.* **83**, 1986, 840-843.

[113] FRAZIER, M. and JAWERTH, B. Decomposition of Besov spaces. *Indiana University Math. J.*, **34**, 1985, 777-799.

[114] FRAZIER, M. and JAWERTH, B. The ϕ-transform and applications to distribution spaces. *Function spaces and applications, Lund, 1986.* Lecture Notes in Math., **1302** Springer-Verlag, Berlin, 1988, 223-246.

[115] GARCIA-CUERVA, J. and RUBIO DE FRANCIA, J.L. *Weighted norm inequalities and related topics.* North-Holland Math. Studies, North-Holland, Amsterdam, 1985.

[116] GARNETT, J. *Bounded analytic functions.* Academic Press, New York, 1981.

[117] GARNETT, J. Corona problems, interpolation problems and inhomogeneous Cauchy-Riemann equations. *Proc. Int. Congress Math., Berkeley, Calif., 1986.* American Mathematical Society, Providence, R.I., 1987, 917-923.

[118] GILBERT, J.E. Nikishin-Stein theory and factorization with applications. *Harmonic analysis in Euclidean spaces, Proc. Symp. Pure Math.*, **35**, 2, 1979, 233-267.

[119] GIRAUD, G. Equations à intégrales principales. *Ann. Sci. E.N.S.*, 1934, 251-372.

[120] GLIMM, J. and JAFFE, A. *Quantum physics, a functional integral point of view.* Springer-Verlag, New York, 1981.

[121] GOUPILLAUD, P., GROSSMANN, A. and MORLET, J. Cycle-octave and related transforms in seismic signal analysis. *Geoexploration*, **23**, Elsevier, Amsterdam, 1984-85, 85-102.

[122] GRÖCHENIG, K. Analyse multiéchelle et bases d'ondelettes. *C.R. Acad. Sci. Paris*, **305**, Série I, 1987, 13-17.

[123] GROSSMAN, A., HOLSCHNEIDER, M., KRONLAND-MARTINET, R. and MORLET, J. *Detection of abrupt changes in sound signals with the help of the wavelet transform.* Preprint, Centre de Physique Théorique, C.N.R.S. Luminy Case 907, 13288 Marseilles Cedex 9.

[124] GROSSMAN, A. and MORLET, J. Decomposition of Hardy functions into square integrable wavelets of constant shape. *SIAM J. Math. Anal.*, **15**, 1984, 723-736.

[125] GROSSMAN, A., MORLET, J. and PAUL, T. Integral transforms associated to square integrable representations I, *J. Math. Phys*, **26**, 1985, 2473-2479, and II, *Ann. Inst. Henri Poincaré, Phys. Théorique*, **45**, 1986, 293-309.

[126] DE GUZMAN, M. *Differentiation of integrals in* $\mathbb{R}^n$. Lecture Notes in Mathematics, **481**, Springer-Verlag, Berlin, 1975.

[127] DE GUZMAN, M. *Real variable methods in Fourier analysis*. Notas de Matematica, North-Holland Math. Studies, **46**, North-Holland, Amsterdam.

[128] HAAR, A. Zur Theorie der orthogonalen Funktionensysteme. *Math. Ann.*, **69**, 1910, 331-371.

[129] HARDY, G.H. and LITTLEWOOD, J.E. A maximal theorem with function theoretic applications. *Acta Math.*, **54**, 1930, 81-116.

[130] HELSON, H. *Harmonic analysis*. Addison-Wesley, Reading, Mass., 1983.

[131] HOLSCHNEIDER, M. On the wavelet transformation of fractal objects. *J. Stat. Phys.*, **50**, 1988, 963-993.

[132] HOLSCHNEIDER, M., KRONLAND-MARTINET, R., MORLET, J. and TCHAMITCHIAN, P. A real-time algorithm for signal analysis with the help of the wavelet transform. *Wavelets. Time-frequency methods and phasespace. Proc. Inter. Conf. Marseilles, 1987*. Combes, J.M., Grossman, A., and Tchamitchian, P., eds., Springer-Verlag, 1990, 286-297.

[133] HÖRMANDER, L. *The analysis of linear partial differential equations*, I, II, and III. Springer-Verlag, New York, 1983-85.

[134] JAFFARD, S. and MEYER, Y. Bases d'ondelettes dans des ouverts de $\mathbb{R}^n$. *J. Math. Pures et Appliqués*, **68**, 1989, 95-108.

[135] JANSON, S. On functions with derivatives in H^1. *Harmonic analysis and partial differential equations, El Escorial, 1987*. Garcia-Cuerva, J., ed., Lecture Notes in Math., **1384**, Springer-Verlag, Berlin, 1989, 193-201.

[136] JERISON, D. and KENIG, C. An identity with applications to harmonic measures. *Bull. Amer. Math. Soc.*, **2**, 1980, 447-451.

[137] JERISON, D. and KENIG, C. The Dirichlet problem in non-smooth domains. *Ann. of Math.*, **113**, 1981, 367-382.

[138] JERISON, D. and KENIG, C. Boundary value problems on Lipschitz domains. *Studies in partial differential equations*. Littman, W., ed., *MAA Studies in Math.*, **23**, 1982, 1-68.

[139] JERISON, D. and KENIG, C. The Neumann problem on Lipschitz domains. *Bull. Amer. Math. Soc.*, **4**, 1981, 203-207.

[140] JOHN, F. and NIRENBERG, L. On functions of bounded mean oscillation. *Comm. Pure Appl. Math.*, **18**, 1965, 415-426.

[141] JOHNSON, R. Application of Carleson measures to partial differential equations and Fourier multiplier problems. *Harmonic analysis, Cortona 1982*. Lecture Notes in Math., **992**, Springer-Verlag, Berlin, 1983, 16-72.

[142] JONES, P.W. Some topics in the theory of Hardy spaces. *Topics in modern harmonic analysis, Proc. Sem. Torino-Milano, May-June 1982.* Istituto Nazionale di alta matematica Francesco Severi, **2**, 551-569.

[143] JONES, P.W. Recent advances in the theory of Hardy spaces. *Proc. Int. Congress Math., Warszawa, 1983.* North-Holland, Amsterdam, 1984, 829-838.

[144] JONES, P.W. Square functions, Cauchy integrals, analytic capacity and harmonic measure. *Harmonic analysis and partial differential equations, El Escorial, 1987.* Garcia-Cuerva, J., ed., Lecture Notes in Math., **1384**, Springer-Verlag, Berlin, 1989, 24-68.

[145] JONES, P.W. and ZINSMEISTER, W. Sur la transformation conforme des domaines de Lavrentiev. *C.R. Acad. Sci. Paris*, **295**, *Série I*, 1982, 563-566.

[146] JOURNÉ, J.L. *Calderón-Zygmund operators, pseudo-differential operators and the Cauchy integral of Calderón.* Lecture Notes in Mathematics, **994**, Springer Verlag, Berlin, 1983.

[147] JOURNÉ, J.L. Calderón-Zygmund operators on product spaces. *Revista Mat. Ibero-Americana*, **1**, 1985, 55-91.

[148] KAHANE, J.P. *Séries de Fourier absolument convergentes.* Ergebnisse der Math., **50**, Springer-Verlag, Berlin, 1970.

[149] KAHANE, J.P. *Some random series of functions.* Cambridge studies in advanced mathematics, **5**, Cambridge University Press, Cambridge, 1968.

[150] KAHANE, J.P., KATZNELSON, Y. and DE LEEUW, K. Sur les coefficients de Fourier des fonctions continues. *C.R. Acad. Sci. Paris*, **285**, *Série I*, 1977, 1001-1003.

[151] KATO, T. *Perturbation theory for linear operators.* Springer-Verlag, New York, 1966.

[152] KATO, T. Scattering theory. *MAA studies in Math.*, **7**, Taub, A.H., ed., 1971, 90-115.

[153] KATO, T. and PONCE, G. *On the Euler and Navier-Stokes equations in Lebesgue spaces $L^{p,s}(\mathbb{R}^n)$.* Preprint, Dept. of Math., University of California, Berkeley, Calif., 94720.

[154] KATZNELSON, Y. *An introduction to harmonic analysis.* John Wiley, New York, 1968.

[155] KELDYSH, M.V. and LAVRENTIEV, M.A. Sur la représentation conforme des domaines limités par des courbes rectifiables. *Ann. Sci. E.N.S.*, **54**, 1937, 1-38.

[156] KENIG, C. Weighted Hardy spaces on Lipschitz domains. *Harmonic analysis in Euclidean spaces. Proc. Symp. Pure Math.*, **35**, 1979, I, 263-274.

[157] KENIG, C. Weighted H^p spaces on Lipschitz domains. *Amer. J. Math.*, **102**, 1980, 129-163.

[158] KENIG, C. Recent progress on boundary value problems on Lipschitz domains. *Pseudodifferential operators and applications. Proc. Symp. Pure Math.*, **43**, 1985, 175-205.

[159] KENIG, C. Elliptic boundary value problems on Lipschitz domains. *Beijing Lectures in Harmonic Analysis.* Stein, E.M., ed., *Ann. of Math. Studies*, **112**, Princeton University Press, Princeton, N.J., 1986, 131-183.

[160] KENIG, C. and MEYER, Y. The Cauchy integral on Lipschitz curves and the square root of second order accretive operators are the same. *Recent progress in Fourier analysis.* North-Holland Math. Studies, **111**, North-Holland, Amsterdam, 1985, 123-145.

[161] KENIG, C. and TOMAS, P. Maximal operators defined by Fourier multipliers. *Studia Math.*, **68**, 1980, 79-83.

[162] KOOSIS, P. *Introduction to H^p-spaces.* London Math. Soc. Lecture Notes, **40**, Cambridge University Press, Cambridge, 1980.

[163] LATTER, R.H. A characterization of $H^p(\mathbb{R}^n)$ in terms of atoms. *Studia Math.*, **62**, 1978, 93-101.

[164] LEMARIÉ, P.G. Continuité sur les espaces de Besov des opérateurs définis par des intégrales singulières. *Ann. Inst. Fourier (Grenoble)*, **35**, 1985, 4, 175-187.

[165] LEMARIÉ, P.G. Bases d'ondelettes sur les groupes de Lie stratifiés. *Bull. Soc. Math. France*, **117**, 1989, 211-232.

[166] LEMARIÉ, P.G. and MEYER, Y. Ondelettes et bases hilbertiennes. *Revista Mat. Ibero-Americana*, **2**, 1986, 1-18.

[167] LIÉNARD, J.S. *Speech analysis and reconstruction using short-time, elementary wave-forms.* LIMSI-CNRS, Orsay.

[168] LINDENSTRAUSS, J. and PELCZYNSKI, A. Contributions to the theory of classical Banach spaces. *J. Funct. Anal.*, **8**, 1971, 225-241.

[169] LINDENSTRAUSS, J. and PELCZYNSKI, A. *Classical Banach spaces, I.* Springer-Verlag, New York, 1977.

[170] LITTLEWOOD, J.E. and PALEY, R. Theorems on Fourier series and power series. *J. London Math. Soc.*, **6**, 1931, 230-233.

[171] LITTLEWOOD, J.E. and PALEY, R. Theorems on Fourier series and power series. *Proc. London Math. Soc.*, **42**, 1937, 52-89.

[172] McINTOSH, A. *Square roots of elliptic operators.* Centre for Math. Analysis, Australian National University, GPO Box 4, Canberra, ACT 2601, Australia.

[173] McINTOSH, A. Square roots of elliptic operators. *J. Funct. Anal.*, **61**, 1985, 3, 307-327.

[174] McINTOSH, A. Functions and derivations of C^*-algebras. *J. Funct. Anal.*, **30**, 1978, 2, 264-275.

[175] McINTOSH, A. Counter-example to a question on commutators. *Proc. Amer. Math. Soc.*, **29**, 1971, 430-434.

[176] McINTOSH, A. On representing closed accretive sesquilinear forms as $(A^{1/2}u, A^{*1/2}v)$. *Macquarie Math. Reports*, August 1981.

[177] McINTOSH, A. Clifford algebras and the higher dimensional Cauchy integral. *Approximation theory and function spaces, 1986.* Banach Centre, Warsaw, Poland.

[178] MALLAT, S. Multiresolution approximation and wavelet orthonormal bases of $L^2(\mathbb{R})$. *Trans. Amer. Math. Soc.*, **315**, 1989, 69-87.

[179] MALLAT, S. A theory for multiresolution signal decomposition: the wavelet representation. *IEEE Trans. on Pattern Analysis and Machine Intelligence, Technical Report MS-CIS-87-22*, University of Pennsylvania, 1989.

[180] MALLAT, S. Dyadic wavelets energy zero-crossings. *IEEE Trans. Inf. Theory, Technical Report MS-CIS-88-30*, 1989, University of Pennsylvania, 1989.

[181] MALLAT, S. *Multiresolution representations and wavelets.* Ph.D. thesis in Electrical Engineering, University of Pennsylvania, Philadelphia, Pennsylvania, 19104, USA.

[182] MANDELBROT, B. *The fractal geometry of nature.* W.H. Freeman, New York, 1983.

[183] MARCINKIEWICZ, J. Sur les multiplicateurs des séries de Fourier. *Studia Math.*, **8**, 1939, 78-91.

[184] MAUREY, B. Isomorphismes entre espaces H^1. *Acta Math.*, **145**, 1980, 79-120.

[185] MAUREY, B. Le système de Haar. *Séminaire Maurey-Schwartz, 1974-75.* Centre de Mathématique, École Polytechnique, 91128 Palaiseau.

[186] MEYER, Y. Wavelets and operators. *Analysis at Urbana I.* London Math. Soc. Lecture Notes, **137**, Cambridge University Press, Cambridge, 1989, 256-365.

[187] MEYER, Y. Real analysis and operator theory. *Pseudo-differential operators and applications. Proc. Symp. Pure Math.*, **43**, 1985, 219-235.

[188] MEYER, Y. Ondelettes, fonctions splines et analyses graduées. *Cahiers Math. de la Décision*, **8703**. Ceremade, Université de Paris-Dauphine.

[189] MEYER, Y. Intégrales singulières, opérateurs multilinéaires, analyse complexe et équations aux dérivées partielles. *Proc. Int. Congress Math., Warszawa, 1983.* North-Holland, Amsterdam, 1984, 1001-1010.

[190] MEYER, Y. Intégrales singulières, opérateurs multilinéaires et équations aux dérivées partielles. *Séminaire Goulaouic-Schwartz.* Centre de Mathématique, École Polytechnique, 91128 Palaiseau.

[191] MEYER, M. Une classe d'espaces de type BMO. Applications aux intégrales singulières. *Arkiv för Mat.*, **27**, 1989, 305-318.

[192] MUCKENHOUPT, B. Weighted norm inequalities for classical operators. *Harmonic analysis in Euclidean spaces. Proc. Symp. Pure Math.*, **35**, I, 69-83.

[193] MURAI, T. *Boundedness of singular integral operators of Calderón type (V and VI).* Nagoya University preprint series, 1984

[194] MURAI, T. *A real variable method for the Cauchy transform and analytic capacity.* Lecture Notes in Math., **1307**, Springer-Verlag, Berlin, 1988.

[195] NAGEL, A., RIVIÈRE, N. and WAINGER, S. On Hilbert transformations along curves. *Bull. Amer. Math. Soc.*, **80**, 1974, 106-108.

[196] NAGEL, A., RIVIÈRE, N. and WAINGER, S. On Hilbert transforms along curves. *Amer. J. Math.*, **98**, 1976, 395-403.

[197] NECAS, J. *Les méthodes directes en théorie des équations elliptiques.* Academia, Prague, and Masson, Paris, 1967.

[198] O'NEILL, R., SAMPSON, G. and SOARES DE SOUZA, G. Several characterizations of the special atom space with applications. *Revista Mat. Ibero-Americana*, **2**, 1986, 3, 333-355.

[199] PAUL, T. Functions analytic on the half plane as quantum mechanic states. *J. Math. Phys*, **25**, 1984, 3252-3256.

[200] PAUL, T. Wavelets and path integrals. *Wavelets. Time-frequency methods and phase-space. Proc. Inter. Conf. Marseilles, 1987.* Combes, J.M., Grossman, A., and Tchamitchian, P., eds., Springer-Verlag, 1990, 204-208.

[201] PEETRE, J. On convolution operators leaving $L^{p,\lambda}$ spaces invariant. *Ann. Math. Pura Appl.*, **72**, 1966, 295-304.

[202] PEETRE, J. *New thoughts on Besov spaces.* Duke University Math. Dept., Durham, N. Carolina, 1976.

[203] PHONG, D.H. and STEIN, E.M. Hilbert integrals, singular integrals and Radon transforms. *Acta Math.,* **157**, 1985, 99-157.

[204] PONCE, G. *Propagation of $L^{q,k}$ smoothness for solutions of the Euler equation.* Preprint, Dept. of Math., University of Chicago, Chicago, Ill., 60637.

[205] RICCI, F. and WEISS, G. A characterization of $H^1(\Sigma_{n-1})$. *Harmonic analysis in Euclidean spaces. Proc. Symp. Pure Math.,* **35**, 1979, I, 289-294.

[206] RODET, X. Time-domain formant-wave-function synthesis. *Computer Music J.,* **8**, 1985, Part 3.

[207] RUBIO DE FRANCIA, J.L. A new technique in the theory of A_p weights. *Topics in modern harmonic analysis, Proc. Sem. Torino-Milano, May-June 1982.* Istituto Nazionale di alta matematica Francesco Severi, **2**, 571-580.

[208] RUBIO DE FRANCIA, J.L. A Littlewood-Paley inequality for arbitrary intervals. *Revista Mat. Ibero-Americana,* **1**, 1985, 2, 1-13.

[209] RUDIN, W. *Fourier analysis on groups.* Interscience, John Wiley, New York, 1962.

[210] SADOSKY, C. *Interpolation of operators and singular integrals.* Marcel Dekker, New York, 1979.

[211] SAKS, C. and ZYGMUND, A. *Analytic functions.* Monografie Matematyczne, Warszawa-Wroclaw, 1952.

[212] SEMMES, S.W. *A criterion for the boundedness of singular integrals on hypersurfaces.* Preprint, Dept. of Math., Yale University, New Haven, Conn., 06520.

[213] SJÖLIN, P. and STRÖMBERG, J.O. Spline systems as bases in Hardy spaces. *Israel J. Math.,* **45**, 1983, 2-3, 147-156.

[214] SJÖLIN, P. and STRÖMBERG, J.O. Basis properties of Hardy spaces. *Arkiv för Mat.,* **21**, 1983, 111-125.

[215] SPANNE, S. Sur l'interpolation entre les espaces $\mathcal{L}_k^{p\Phi}$. *Ann. Scuola Norm. Sup. Pisa,* **20**, 1966, 625-648.

[216] STEGENGA, D.A. Multipliers of the Dirichlet space. *Illinois J. Math.,* **24**, 1980, 113-139.

[217] STEIN, E.M. *Singular integrals and differentiability properties of functions.* Princeton University Press, Princeton N.J., 1970.

[218] STEIN, E.M. *Topics in harmonic analysis.* Ann. of Math. Studies, **63**, Princeton University Press, Princeton, N.J., 1970.

[219] STEIN, E.M. On limits of sequences of operators. *Ann. of Math.,* **74**, 1961, 140-170.

[220] STEIN, E.M. and WAINGER, S. Problems in harmonic analysis related to curvature. *Bull. Amer. Math. Soc.,* **84**, 1978, 1239-1295.

[221] STEIN, E.M. and WEISS, G. *Introduction to Fourier analysis on Euclidean spaces.* Princeton University Press, Princeton, N.J., 1971.

[222] STEIN, E.M. and WEISS, G. On the theory of H^p spaces. *Acta Math.,* **103**, 1960, 25-62.

[223] STRÖMBERG, J.O. A modified Franklin system and higher-order spline systems on $\mathbb{R}^n$ as unconditional bases for Hardy spaces. *Conference in harmonic analysis in honor of Antoni Zygmund.* Wadsworth, Belmont, Calif., 1983, II, 475-493.

[224] STRÖMBERG, J.O. Bounded mean oscillation with Orlicz norms and duality of Hardy spaces. *Indiana University Math. J.,* **28**, 1979, 511-544.

[225] STRÖMBERG, J.O. and TORCHINSKY, J. Weights, sharp maximal functions and Hardy spaces. *Bull. Amer. Math. Soc.,* **3**, 1980, 1053-1056.

[226] TAKAGI, T. A simple example of a continuous function without derivative. *The collected papers of Teiji Takagi,* 5-6 (III.78), Iwanami Shoten, Tokyo, 1973.

[227] TCHAMITCHIAN, P. Calcul symbolique sur les opérateurs de Calderón-Zygmund et bases inconditionelles de $L^2(\mathbb{R})$. *C.R. Acad. Sci. Paris,* **303**, *Série I,* 1986, 215-218.

[228] TCHAMITCHIAN, P. *Ondelettes adaptées à l'analyse complexe.* Preprint.

[229] TITCHMARSH, E.C. *Introduction to the theory of Fourier integrals.* Oxford at the Clarendon Press, 1937.

[230] TRIEBEL, H. *Theory of function spaces.* Monographs in Mathematics, **78**, Birkhäuser, Basel, 1983.

[231] UCHIYAMA, A. A constructive proof of the Fefferman-Stein decomposition for BMO($\mathbb{R}^n$). *Acta Math.,* **148**, 1982, 215-241.

[232] VERCHOTA, G. Layer potentials and regularity for the Dirichlet problem for Laplace's equation in Lipschitz domains. *J. Funct. Anal.,* **59**, 572-611.

[233] WAINGER, S. On some aspects of differentiation theory. *Topics in modern harmonic analysis, Proc. Sem. Torino-Milano, May-June 1982.* Istituto Nazionale di alta matematica Francesco Severi, **2**, 667-700.

[234] WAINGER, S. Averages and singular integrals over lower dimensional sets. *Beijing lectures in harmonic analysis.* Stein, E.M., ed., *Ann. of Math. Studies,* **112**, Princeton University Press, Princeton, N.J., 1986, 357-421.

[235] WILSON, J.M. A simple proof of the atomic decomposition for $H^p(\mathbb{R}^n)$. *Studia Math.,* **74**, 1982, 25-33.

[236] WILSON, J.M. *On the atomic decomposition for Hardy spaces.* Preprint.

[237] WOJTASZCZYK, P. The Franklin system is an unconditional basis for H^1. *Arkiv för Mat.,* **20**, 1982, 293-300.

[238] ZINSMEISTER, M. *Domaines de Lavrentiev.* Publications mathématiques d'Orsay, **204**, Université de Paris-Sud, Département de mathématiques, 91405 Orsay, 1986.

[239] ZYGMUND, A. *Trigonometric Series.* Second edition, Cambridge University Press, Cambridge, 1968.

References and Bibliography
for the English Edition

[E1] ALINAC, S. and GÉRARD, P. *Opérateurs pseudo-defférentiels et théorème de Nash-Moser.* Savoirs Actuels, Inter-Editions, CNRS, 1991.

[E2] AUSCHER, P. and Tchamitchian, P. Calcul fonctionnel précisé pour des opérateurs elliptiques complexes. *Annales de l'Institut Fourier,* **45**, 1995, 721-778.

[E3] BEYLKIN, G., COIFMAN, R., and ROKHLIN, V. Fast wavelet transforms and numerical algorithms 1. *Comm. Pure Appl. Math.,* **43**, 1991, 141-183.

[E4] BEYLKIN, G. et al., eds. *Wavelets.* Jones and Bartlett, Boston, Mass., 1992.

[E5] CANNONE, M. *Ondelettes, paraproduits et Navier-Stokes.* Diderot Editeur, Art et Sciences, 1995.

[E6] CHEMIN, J.Y. Fluides parfaits incompressibles. *Astérisque,* **230**, 1995.

[E7] COIFMAN, R.R. Adapted multiresolution analysis, computation, signal processing and operator theory. *ICM 90, Kyoto.* Springer-Verlag, Berlin, 1992.

[E8] COIFMAN, R., LIONS, P.L., and MEYER, Y. Compensated compactness and Hardy spaces. *J. Math. Pures et Appliqués,* **72**, 1993, 247-286.

[E9] DAVID, G. *Wavelets and singular integrals on curves and surfaces.* Lecture Notes in Mathematics, **1465**, Springer-Verlag, Berlin, 1991.

[E10] DONOHO, D. Non-linear solution of linear inverse problems by wavelet-vaguelette decomposition. *Applied and Computational Harmonic Analysis,* **2 (2)**, 1995, 101-126.

[E11] EVANS, L.C. and MÜLLER, S. Hardy spaces and the two-dimensional Euler equations with non-negative vorticity. J. Amer. Math. Soc., **7**, 1994, 199-219.

[E12] FARGE, M., KEVLAHAN, N., PERRIER, V., and GOIRAND, E. Wavelets and turbulence. *Proc. IEEE,* **84 (4)**, 1996, 639-669.

[E13] FEDERBUSH, P. Navier and Stokes meet the wavelet. *Comm. Math. Phys.,* **155**, 1993, 219-248.

[E14] GASQUET, C. and WITOMSKI, P. *Analyse de Fourier et applications. Filtrage, calcul numérique, ondelettes.* Masson, Paris, 1990.

[E15] KENIG, C. Harmonic analysis techniques for second order elliptic boundary value problems. *CBMS Regional Conference Series in Mathematics,* **83.** American Mathematical Society, Providence, R.I., 1994.

[E16] LEMARIÉ, P.G. *Les ondelettes en 1989.* Lecture Notes in Mathematics, **1438**, Springer-Verlag, Berlin, 1990.

[E17] LEMARIÉ-RIEUSSET, P.G. Analyses multirésolutions non orthogonales, commutateur entre projecteur et derivation et ondelettes à divergence nulle. *Revista Mat. Ibero-Americana,* **8**, 1992, 221-236.

[E18] MELNIKOV, M.S. and VERDERA, J. A geometric proof of the L^2 boundedness of the Cauchy integral on Lipschitz graphs. *International Math. Research Notices,* **7**, 1995, 325-331.

[E19] MEYER, Y., ed. *Wavelets and applications. Proc. Inter. Conf., Marseille, France, May 1989.* Masson, Paris & Springer-Verlag, Berlin, 1992.

[E20] MEYER, Y. Wavelets and applications. *ICM 90, Kyoto.* Springer-Verlag, Berlin, 1992.

[E21] TAYLOR, M.E. *Pseudodifferential operators and non-linear PDE.* Birkhauser, Boston, Mass., 1991.

Index

Cet ouvrage a été achevé d'imprimer en février 1997
dans les ateliers de Normandie Roto Impression s.a.
61250 Lonrai
N° d'imprimeur : 970320
Dépôt légal : février 1997

Imprimé en France

Printed in the United States
By Bookmasters